Chemical Engineering Design and Analysis

Students taking their first chemical engineering course plunge into the "nuts and bolts" of mass and energy balances and often miss the broad view of what chemical engineers do. This innovative text offers a well-paced introduction to chemical engineering. Through a series of real-world examples and extensive exercises, students learn the basic engineering concepts of design and analysis.

The text has two main objectives:

- To have students practice engineering. Students are introduced to the fundamental steps in design and three methods of analysis: mathematical modeling, graphical methods, and dimensional analysis. In addition, students apply engineering skills, such as how to simplify calculations through assumptions and approximations, how to verify calculations, determine significant figures, use spreadsheets, prepare graphs (standard, semilog, and log–log), and use data maps.
- To introduce the chemical engineering profession. Students learn about chemical engineering by designing and analyzing chemical processes and process units to assess product quality, economics, safety, and environmental impact.

This text will help undergraduate chemical engineering students develop engineering skills early in their studies and encourage an informed decision about whether to pursue this profession. Students in related fields such as chemistry, biology, materials science, and mechanical engineering can use this book to learn the underlying principles of chemical processes and their far-reaching applications.

T. Michael Duncan is an associate professor of chemical engineering at Cornell University and Associate Director of the undergraduate program in the School of Chemical Engineering.

Jeffrey A. Reimer is a professor of chemical engineering at the University of California at Berkeley and faculty scientist at the E. O. Lawrence Berkeley National Laboratory.

CAMBRIDGE SERIES IN CHEMICAL ENGINEERING

Chemical Engineering Design and Analysis

An Introduction

T. Michael Duncan and Jeffrey A. Reimer

CAMBRIDGE
UNIVERSITY PRESS

CAMBRIDGE UNIVERSITY PRESS
Cambridge, New York, Melbourne, Madrid, Cape Town,
Singapore, São Paulo, Delhi, Tokyo, Mexico City

Cambridge University Press
32 Avenue of the Americas, New York, NY 10013-2473, USA

www.cambridge.org
Information on this title: www.cambridge.org/9780521639569

First published 1998
10th printing 2011

A catalog record for this publication is available from the British Library.

Library of Congress Cataloging in Publication Data

Duncan, T. Michael.
Chemical engineering design and analysis : an introduction / T.
Michael Duncan, Jeffrey A. Reimer.
   p.   cm. – (Cambridge series in chemical engineering)
Includes bibliographical references and index.
ISBN 0-521-63041-X (hb). – ISBN 0-521-63956-5 (pb)
1. Chemical engineering.   I. Reimer, Jeffrey A. (Jeffrey Allen).
II. Title.   III. Series.
TP155.D74   1998   98-16452
660′.2 – dc21   CIP

ISBN 978-0-521-63956-9 Paperback

To my son, Maxwell

    T. Michael Duncan

To Karen, Jennifer, Jonathan, Charlotte, and Martin

    Jeffrey A. Reimer

Contents

Contents

Contents

Preface

Traditional chemical engineering curricula present the first formal course for the major in the sophomore year; it is customarily a course in mass and energy balances. Courses taught earlier in the student calendar are usually either a survey course, in which chemical process industries (or the research interests of the faculty) are summarized, or a course in stoichiometry, emphasizing mass balances in steady-state systems. We concluded that a different freshman course was needed. We wanted a course to strengthen traditional curricula and to encourage students with diverse backgrounds to join the chemical engineering profession.

Professor Duncan had previously assumed responsibility for Engineering 112 at Cornell University, a survey course intended to introduce chemical engineering, but one not required for the major. This course was one of several introductory courses created by Cornell's College of Engineering in the early 1980s. A course to introduce design and analysis was developed and, although well received, suffered from the lack of a suitable textbook. Professor Reimer, who was responsible for the mass and energy balance course at Berkeley, was discouraged by the disparate student motivation and performance in the first required course for the major. Furthermore, the introductory mass and energy balance course was becoming overburdened with multiple (and sometimes conflicting) goals, including application of conservation principles, mathematical modeling, process spreadsheeting, computer methods, problem solving, and reviews of chemical technology. Finally, it was apparent to both of us that some groups were underrepresented in the chemical engineering profession.

In the fall of 1993 we set out to produce a text that dealt with sophisticated issues of engineering design yet assumed only the precollege mathematics, chemistry, and physics typical of a secondary school education. We wanted an inclusive text that described contemporary problems in chemical engineering design and practice, demonstrated various learning and teaching styles, and could be used in all postsecondary school educational formats, including two-year colleges and continuing education programs. This text was intended to allow students to decide early in their undergraduate education whether or not to become a chemical engineer. It was also designed so students could take full advantage of the remainder of their degree program by providing an appropriate context for the ensuing coursework in chemical engineering.

The book is organized so that each concept is introduced within the two most important paradigms of engineering practice: design and analysis. We believe that chemical engineering education should start with the same emphasis with which it ends: design. Therefore we emphasize that students should devise specific plans for chemical and physical processes that are based upon sound economic strategies and thoughtful analysis of key physical and chemical phenomena. We introduce three methods of analysis modeling based upon: (i) fundamental physical laws and constitutive equations, (ii) empirical (and usually graphical) correlations, and (iii) dimensional analysis.

Our text has a number of unusual features vis-à-vis other introductory textbooks. First, we adopt the "just in time" philosophy for introducing chemical engineering concepts, a philosophy that we discovered in Richard Feynman's *Lectures on Physics*. Attempts to comprehensively cover concepts such as energy balances would hopelessly swamp an introductory text with information. Instead we introduce only what students need to know to deal with the problem at hand. Thus we cover the enormous scope of chemical engineering concepts but treat each concept with only cursory depth. Second, we discard the usual assumptions that freshman students cannot comprehend complex phenomena such as combined reaction and diffusion. On the contrary, we believe that many chemical engineering curricula fragment chemical engineering concepts so much that students have difficulty integrating these concepts to solve complex problems after graduation.

Acknowledgments

I am grateful to Bill Olbricht for his ideas when I was developing this syllabus and for his continued encouragement as this textbook was written. Discussions with Thatcher Root refined and expanded my ideas and directed me to seminal resources. Finally, I acknowledge the School of Chemical Engineering at Cornell University for the opportunity to develop this text.

Mike Duncan
Ithaca, New York

I thank Cornell's School of Chemical Engineering, and Bill Olbricht in particular, for the generous support and encouragement given to me in 1993. This support resulted in the start of our manuscript. I also thank my research group during the period 1993–98; their generosity with my time gave me the opportunity to work on this text. I am especially grateful to Jacqueline Mintz of UC Berkeley's GSI Teaching and Resource Center for her kind mentorship on the scholarship of teaching. Finally, I am greatly indebted to Morton M. Denn for his patience and guidance during all my Berkeley years; he is truly one of Berkeley's greatest teachers.

Jeff Reimer
Richmond, California

Several textbooks influenced this work. Two fine textbooks on design – *Process Synthesis* by Dale Rudd, Gary Powers, and Jeffrey Siirola and *Process Modeling* by Morton Denn – inspired us and spawned the material on process design (Chapter 2), mathematical modeling (Chapter 3), and transient processes (Chapter 6).

The topic of mass and energy balances, introduced in Chapter 3 as examples of mathematical modeling, is a mature one in chemical engineering. We are grateful to the authors of two excellent textbooks – Richard Felder and Ronald Rousseau, *Elementary Principles of Chemical Processes*, and William Luyben and Leonard

Wenzel, *Chemical Process Analysis: Mass and Energy Balances* – for permission to adapt and reprint examples and exercises.

We are grateful to the many colleagues, students, and friends that unselfishly gave their time to listen to our ideas, read chapters, work exercises, and make suggestions. We are particularly grateful to Thatcher Root, Alan Foss, Mort Denn, and Claude Cohen.

An Overview of
Chemical Engineering

THIS TEXTBOOK has two goals. The first is to describe the chemical engineering profession. We will use contemporary applications of chemical engineering to introduce fundamental concepts. The applications include case studies and chemical processes from the technical literature and the popular press. The second goal is to introduce and develop basic engineering skills. Chief among these skills are design – the ability to conceive and develop plans – and analysis – the methodology to model and evaluate chemical and physical processes.

Some of the concepts introduced in this text are complex and usually require an entire course and its prerequisites to appreciate fully. You must be willing, therefore, to set aside questions about the basis for certain material or the origins of certain equations or relationships. We will, however, attempt to provide at least a heuristic description of the material's origin and point to where in the chemical engineering curriculum the material is discussed in more detail.

Finally, in this text we attempt to appeal to a variety of learning and thinking styles. We appreciate that not all students prefer to think globally, reason deductively, or perceive visually. In each of the exercise sets we have attempted to invoke different styles of learning to make learning chemical engineering as inclusive as possible.

1.1   Chemical Engineering

Chemical engineers create processes based upon physical and chemical change. The processes may yield marketable items, such as gasoline or penicillin, or noncommercial items, such as clean air or clean water. The processes are created by integrating principles from basic sciences – traditionally chemistry, physics, and mathematics – with consideration of economics, environmental impact, and employee safety. Several textbooks in chemical engineering have introductory chapters on the chemical engineering profession. At the end of this chapter we list some of the most frequently used texts. We encourage you to browse through the introductory chapters of these books. Their subsequent chapters will give you a glimpse of topics in the chemical engineering curriculum.

The chemical engineering profession, barely 100 years old, began as an interface between chemistry and mechanical engineering. The principal goal in the early days of chemical engineering was to commercialize chemical reactions developed at a chemist's bench. In 1983 a list of the top ten achievements of chemical engineering was compiled on the occasion of the seventy-fifth anniversary of the American Institute of Chemical Engineers (AIChE), a national organization with approximately 60,000 members. The AIChE used two criteria to form this list: first, the degree to which the achievement was an innovative and creative response to a societal need, and second, the historical impact of the process. These achievements are summarized as follows:

> *Synthetic rubber.* Elastic materials, such as automobile tires and drive belts, are an integral part of everyday life. The annual production of rubber in 1983 was twenty-two billion pounds. Remarkably, this industry was developed in only two years, just in time to replace shortages of natural rubber during World War II.
>
> *Antibiotics.* In 1918 an influenza epidemic killed twenty million people worldwide, one-half million in the United States alone. Venereal diseases were incurable. Until the 1950s polio crippled millions. Discovering medicines was only part of the solution. After it was observed that a mold inhibited bacterial growth in a Petri dish, chemical engineering developed the technology to ultimately produce millions of pounds per year of penicillin. Chemical engineering made possible the mass production of medicines and the subsequent availability to people worldwide.
>
> *Polymers.* Plastics – such as PVC, nylon, polystyrene, and polyethylene – are the predominant materials for consumer products. Plastics have replaced wood, metal, and glass in many applications because of their superior strength/weight ratio, chemical resistance, and mechanical properties.
>
> *Synthetic fibers.* Methods to produce fine threads of polymers allow us to rely less on exploiting plants and animals for clothing, carpets, and fabrics.
>
> *Cryogenic separation of air into O_2 and N_2.* The present production is about 10^{12} cubic feet per year. N_2 is a key reagent for fertilizer and is used as a cryogen. O_2 is used in medicine and metals processing.
>
> *Separation of nuclear isotopes:* $^{235}U/^{238}U$; $^{12}C/^{14}C$; $^{16}O/^{18}O$. Isotopically enriched uranium changed the world for better and for worse in 1945. Nuclear energy continues to be a viable supplement to fossil fuels. Medical research, diagnostics, and treatments require isotopically enriched elements.
>
> *Catalytic cracking of crude oil.* Crude oil was once distilled into light and heavy fractions (kerosene, gasoline, lubricating oil); the range of oil products was limited by the physical mixture of the raw material. Catalytic cracking systematically decomposes oil molecules into molecular building blocks that may be

used to construct complex chemicals. The ability to make high octane fuel was a crucial factor in the Battle of Britain and World War II.

Pollution control. Chemical engineers can work to design processes with minimal offending by-products and devise strategies to restore polluted sites.

Fertilizers, especially ammonia. New fertilizers have improved agricultural productivity and helped to feed the world.

Biomedical engineering. Chemical engineering principles have been used to model the processes of the human body as well as to develop artificial organs, such as the kidney, heart, and lungs.

The contributions of chemical engineers influenced the evolution of modern society. Most of the top ten achievements listed above came during the heyday of engineering – when it seemed that society's needs could be met by technology, with engineers being the purveyors of technology. Around the mid-1950s, however, technology came to be perceived as dangerous. People began to feel that society and the environment were dominated by technology, even victimized by technology. This perception remains today. The chemical engineering curricula attempts to sensitize students to these issues by encouraging studies in humanities, social sciences, and ethics.

Contemporary chemical engineers are increasingly involved in services, compared to the historical emphasis on manufacturing. This trend will probably continue as chemical engineers are enlisted to remedy environmental contamination and modify existing processes to meet modern business and manufacturing agendas. Some frontier areas of chemical engineering include:

Production of novel materials. Chemical engineers will design processes to produce ceramic parts for engines, high-temperature superconductors, polymer-composites for structural components, and specialty chemicals produced in small amounts to exacting specifications. Chemical processes will shift from the traditional area of petrochemicals to inorganic compounds, from liquids to solids, and from large scale to small scale.

Biotechnology. Chemical engineers will improve methods of isolating bioproducts, design processes for chemical production from biomass, and capitalize on advances in genetic engineering to produce drugs, foods, and materials. Whereas chemical engineering has traditionally sought new reaction paths to produce established chemical commodities, biotechnology will seek ways to produce new chemicals, such as secondary metabolites and so-called fancy proteins. Whereas chemical processes are typically continuous – reactants constantly enter and products constantly leave – bioprocesses tend to be batch – add reactants, wait, then remove products. Finally, whereas traditional chemical processes, such as petrochemicals, tolerate rough separations ($\sim$99.44% pure), bioproducts will require more rigorous isolation.

Solid wastes. Chemical engineers will invent methods to treat landfills as well as to remedy contaminated sites. Chemical engineers will also design alternatives for waste, such as incineration, biological decomposition, and recycling. Whereas the reactants entering traditional chemical processes are well characterized and invariable from day to day, processing wastes requires designs that accept reactants with ill-defined compositions that may change daily.

Pollution control. Chemical engineers will continue to reduce pollution at its sources, for example, by recycling intermediate outputs, redesigning chemical reactors, and reengineering entire processes. Gone are the days when a public waterway was designated "for industrial use." Waste water today sometimes exceeds the purity of the public waterway it enters. Chemical engineers will design processes that not only meet current regulations but anticipate regulations. Chemical engineers will aspire to the ultimate goal of "zero emissions."

Energy. Chemical engineers will continue to improve the efficiency of present energy sources as well as develop new sources.

Process control. Chemical engineers will develop and implement better sensors for temperature, pressure, and chemical composition. Processes will be designed to integrate artificial intelligence for process control, monitoring, and safety.

Recent demographic data for new chemical engineers provide another perspective on the profession. The number of bachelor's degrees conferred in chemical engineering is cyclic, with a period of about 10–15 years, as illustrated in Figure 1.1. The historical average was around 3,000 degrees per year. The strength of the demand for chemical engineers is reflected by the median weekly salary for chemical engineers, shown in Figure 1.2. Of the forty-nine occupations and professions profiled, chemical engineers had the second-highest median weekly salaries – below lawyers but above physicians and airline pilots.

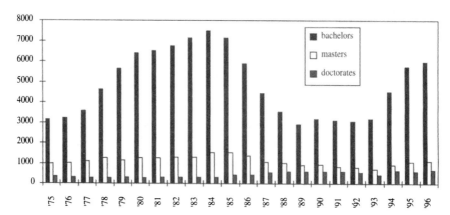

Figure 1.1. Chemical engineering degrees awarded during the period 1975–1996. (Data from *Chemical and Engineering News*, Annual Reports on Professional Training, 1975 through 1997.)

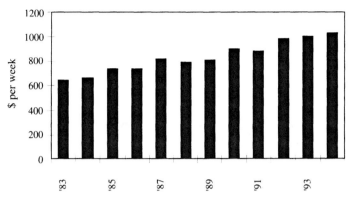

Figure 1.2. Median weekly salary for chemical engineers as compiled by the Department of Labor's Bureau of Labor Statistics. (Data from *New York Times*, May 14, 1995.)

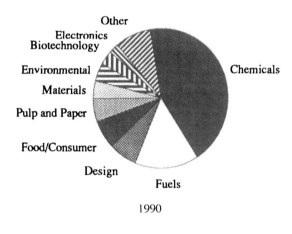

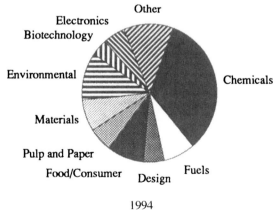

Figure 1.3. Chemical engineering specialities for 1990 and 1994 graduates. (Data from *Chemical Engineering Progress*, July 1991, p. 19 and *AIChExtra*, September 1995.)

Figure 1.3 shows the deployment of degree recipients for 1990 and 1994. These data are consistent with the long-term trend toward fewer new jobs in chemicals and fuels and increasing opportunities in materials, biotechnology, electronics, and environmental applications.

The chemical industry is one of the few U.S. industries with a favorable balance of trade. In 1991, when the U.S. trade deficit was $65 billion, the chemical industry had a trade surplus of $19 billion. In 1991 the balance of trade for the automobile industry was a deficit of $35 billion. By 1997 U.S. chemical exports reached $71 billion and the chemical trade surplus was a record $20.5 billion. This favorable balance of trade is due, in part, to the ingenious chemistry derived by chemists and materials scientists developed into commercial processes by chemical engineers.

The chemical engineering profession continues to be a lively and rewarding field. This is due, in part, to the continuing demand for the design and operation of processes governed by molecular and atomic phenomena. The chemical engineering curriculum is designed to meet the needs of future engineers by teaching basic concepts and skills rather than specific products or processes. Perhaps most important, the chemical engineering curriculum strives to instill a desire and capacity for lifelong learning.

1.2   Summary

To practice the profession of chemical engineering one must be

- knowledgeable of the fundamentals of chemistry, physics, and mathematics;
- able to analyze by applying tools of engineering science;
- skilled in chemical process design (innovative, creative, and adaptive); and
- ethical.

REFERENCES

Felder, R. M., and Rousseau, R. W. 1986. *Elementary Principles of Chemical Processes*, Wiley, New York.

Fogler, H. S. 1986. *Elements of Chemical Reaction Engineering*, Prentice Hall, NJ.

Kirk–Othmer Concise Encyclopedia of Chemical Technology, 4th ed., Wiley-Interscience, New York, 1994.

Luyben, W. L., and Wenzel, L. A. 1988. *Chemical Process Analysis: Mass and Energy Balances*, Prentice Hall, NJ.

McGraw-Hill Encyclopedia of Science and Technology, 8th ed., McGraw-Hill, New York, 1997.

Peters, M. 1984. *Elementary Chemical Engineering*, McGraw-Hill, New York.

Reklaitis, G. V. 1983. *Introduction to Material and Energy Balances*, Wiley, New York.

Shreve, R. N. 1985. *Shreve's Chemical Process Industries*, 5th ed., McGraw-Hill, New York.

2

Process Design

I N THE FIRST CHAPTER we stated that chemical engineers create processes based on physical and chemical changes. In this chapter we develop designs for five chemical processes. The step-by-step examples will also introduce strategies for design and conventions for depicting a chemical process.

2.1   The Synthesis of Ammonia

Let's design a chemical process based on the chemical change of N_2 and H_2 into ammonia, NH_3, used primarily as a fertilizer. The U.S. chemical industry produced 35 *billion* pounds of ammonia in 1995, the sixth largest production of all chemicals. The chemical reaction is

$$N_2 + 3H_2 \rightarrow 2NH_3. \tag{2.1}$$

Indeed, reaction (2.1) is the key to the process. Before a catalyst was developed to promote this reaction, large-scale production of nitrogenous fertilizer was not feasible. The device that conducts this reaction, the *reactor*, is the core of the process. The details of this reactor are key to the viability of the process, but these are not important now. We will assume a viable reactor is available – that's our incentive for starting the design. Our task is to design a process around this reaction.

We need a means of describing our design. We could describe the process with words, such as "a mixture of N_2 and H_2 is piped into the reactor." This would be cumbersome. A chemical process is a situation where a picture is worth a thousand words. A chemical process is represented by a diagram known as a process *flowsheet*, a key tool in chemical engineering design.

A process flowsheet comprises *units*, represented by simple shapes, such as rectangles or circles:

The pipes that conduct material between units are called *streams*. The streams are represented by arrows:

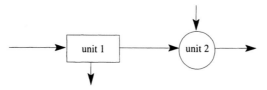

The specifics of a stream, such as its composition, may be written above the stream. For example, the ammonia reactor can be represented as

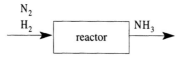

The ammonia reactor is a *continuous* process; material constantly moves through the unit. A *batch* process, in contrast, is characterized by chemical or physical change without material moving in or out, except at the beginning or end of a cycle. A chemical reaction you conduct in a flask in your introductory chemistry laboratory is a batch process. Continuous processes are rarely encountered in undergraduate chemical laboratories, yet they dominate in the chemical industry. We shall explore the reasons for the preeminence of continuous processes later.

The ammonia reactor above is idealized, however. It is rare that the reactants are entirely converted into products. There is almost always some residual reactant in the effluent stream. A realistic description includes N_2 and H_2 in the reactor effluent, as shown below:

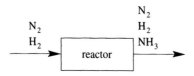

The customer for our NH_3 would probably not be satisfied with N_2 and H_2 in the product. And we would be wasting raw materials. We need to purify the NH_3. How might this be done? To separate substances, we need a physical or chemical basis. The most common basis for separations is *states of matter*, or *phase*. A gas phase is easily separated from a liquid phase. Immiscible liquids (such as oil and water) are easily separated as well. N_2, H_2, and NH_3 are gases at room temperature and pressure. Let's explore the possibility of condensing one or more of the compounds without condensing the others. A handbook of chemical data gives the information listed in

Table 2.1. Boiling points for the ammonia process

Compound	Boiling point at 1 atm, in °C
NH_3	−33
H_2	−253
N_2	−196

Table 2.1. These data show that if we cool the gaseous mixture to below −33°C, NH_3 will condense. Thus we add to our flowsheet a unit to condense ammonia, aptly called a *condenser*:

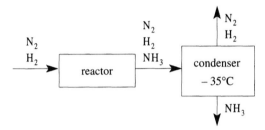

We have assumed that the separation of NH_3 from N_2 and H_2 is perfect, which of course is idealistic. In Chapter 4 we will consider practical separators and methods to design for a desired product purity. Obviously we need something to remove the heat from the condenser, but we will ignore that issue for now. Note that it is conventional to show the liquid effluent leaving the bottom of a unit and the gaseous effluent leaving the top.

What shall we do with the N_2 and H_2 exiting the condenser? Both are innocuous and could be discharged to the air (note, however, that there is an explosion hazard associated with H_2). But this is wasteful. It is more efficient to *recycle* the reactants back into the reactor, as shown below:

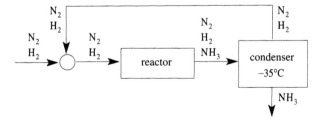

This flowsheet contains a new feature, a *combiner*, which is represented by a circle with two or more streams entering and one stream leaving. A combiner merely combines streams; no mixing is implied. But in this case no mixing is needed because the two streams combined have the same composition.

A process based on this rudimentary design will produce ammonia. One would purchase N_2 and H_2 and sell NH_3. But gases such as N_2 and H_2 are expensive. Furthermore, gases are bulky and must be compressed for efficient transport. H_2 has the added problem of explosion risk. Let's modify the process to produce the reactants N_2 and H_2. If our potential supplier can produce N_2 and H_2, perhaps we can, too. Consider first a source for N_2. Air is an obvious choice. Air is about 79% nitrogen and 21% oxygen and air is inexpensive. A subtle point is that air contains nitrogen in the chemical form we desire. Thus we need only a separator and not a reactor. (As a rule, reactors are more expensive to build and operate.) It seems reasonable to consider separating nitrogen from air. (In 1996 the U.S. chemical industry produced 80 *billion* pounds of nitrogen gas!) Again we first consider separation based on phase – liquid versus gas. We add the boiling point of oxygen to our table (Table 2.2).

Table 2.2. More boiling points for the ammonia process

Compound	Boiling point at 1 atm, in °C
NH_3	−33
H_2	−253
N_2	−196
O_2	−183

We thus add a condenser to provide N_2 and our process becomes:

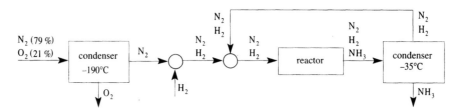

What is a good source for hydrogen? Unlike nitrogen, hydrogen has extremely low abundance in its desired chemical form, H_2 (hydrogen is less than 0.00005% of air). We will need to produce hydrogen by chemical reaction. What is a practical source of hydrogen? Water is 67% hydrogen and water, like air, is inexpensive. One possibility is to decompose, or "crack," water:

$$H_2O_{(gas)} + 242\,kJ \rightarrow H_2 + \frac{1}{2}O_2. \tag{2.2}$$

Decomposing water requires 242 kilojoules (kJ) per mole, a great deal of energy. Indeed, the reverse reaction is an H_2/O_2 torch, which releases a great deal of energy. We could drive the reaction with electricity. You have probably seen this demonstrated on a small scale in secondary school. We thus add an electrolytic reactor to produce H_2:

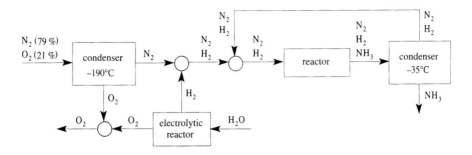

A process based on this design takes in air and water (and energy) and produces ammonia as well as oxygen. Again, this process is workable, but it too can be improved. Let's focus on the source of hydrogen. Electricity is an expensive source of energy. So we look to another means to provide the energy to decompose water into hydrogen and oxygen. Consider the *chemical* energy stored in fuels such as CH_4, the principal component of natural gas:

$$CH_4 + 2O_2 \rightarrow CO_2 + 2H_2O_{(gas)} + 803 \text{ kJ}. \tag{2.3}$$

Each mole of CH_4 provides enough energy to decompose 3.3 moles of H_2O. (803 kJ/mole CH_4 divided by 242 kJ/mole H_2O equals 3.3 moles of H_2O.) So if one combines 3.3 times chemical reaction (2.2) with reaction (2.3) the result is

$$3.3H_2O + 3.3 \times 242 \text{ kJ} \rightarrow 3.3H_2 + \frac{3.3}{2}O_2$$

$$CH_4 + 2O_2 \rightarrow CO_2 + 2H_2O + 803 \text{ kJ}$$

$$\overline{CH_4 + 0.35O_2 + 1.3H_2O \rightarrow CO_2 + 3.3H_2 + {\sim}0 \text{ kJ}} \tag{2.4}$$

Reaction (2.4) is convenient because it supplies the energy needed and it produces H_2, a reactant in our process. Our methane reactor is thus

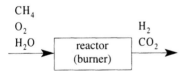

As usual, reaction (2.4) does not convert all the reactants to products. We must include reactants in the effluent:

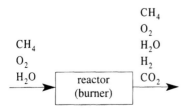

The hydrogen must be separated from the methane, oxygen, water, and carbon dioxide. As before we first consider separation by condensing some of the gases to liquids. We augment our table with more boiling points.

Table 2.3. Even more boiling points for the ammonia process

Compound	Boiling point at 1 atm, in °C
NH_3	−33
H_2	−253
N_2	−196
O_2	−183
CH_4	−164
H_2O	100
CO_2	−79

Oxygen is probably the most crucial chemical to be removed because it will ultimately react with hydrogen (and perhaps nitrogen) in the ammonia synthesis reactor. As luck would have it, the boiling points reveal that oxygen is the most difficult to separate from hydrogen. Therefore, we could design a condenser similar to the N_2/O_2 separator that operates below −183°C. Again, such a design would be workable, but there is a better way. A better way to eliminate oxygen from the effluent is to consume all the oxygen in the reactor. Instead of using stoichiometric proportions indicated in reaction (2.4), use excess methane. Our reactor (burner) thus becomes

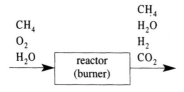

With oxygen eliminated from the effluent, the requirement on the condenser is set by methane; we must operate the condenser below −164°C to condense methane. This is not much of an improvement over the requirement dictated by oxygen. Must methane be separated from hydrogen? If we allow the methane (which will be only a small amount) to continue with the hydrogen, we can focus on removing the major by-product, carbon dioxide. We consult with our reactor experts who conduct some experiments and report that a small amount of methane in the ammonia reactor is not a problem. Fine. Our hydrogen production system, then, looks like this:

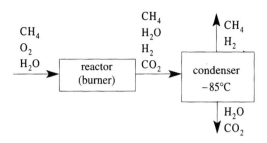

It is not obvious that producing H_2 by burning CH_4 is better than electrolysis of water. There are at least three reasons why the contrary seems true: The methane process requires an additional unit, methane impurity enters the ammonia synthesis reactor, and we no longer produce oxygen as a by-product – we instead consume some of the oxygen produced by the air separator. And it remains to be seen if the air separator produces sufficient oxygen. Also, the original motivation for using methane (cheaper energy) could be undone by the energy requirements of operating the condenser. As we shall see when we develop the tools to compare these alternatives, the methane burner is indeed better.

We replace the electrolytic reactor with a methane burner and our ammonia synthesis process becomes:

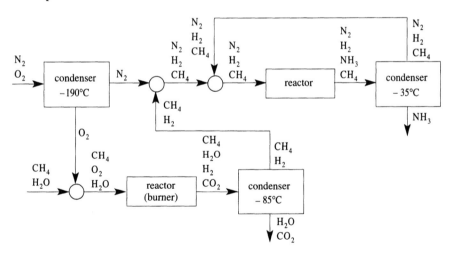

What is the fate of the methane that tags along with the hydrogen? Our reactor experts assure us that the methane passes through the N_2–H_2 reactor unchanged. Because the boiling point of methane is lower than $-35°C$, the ammonia condenser does not condense methane. That's good – we do not need to separate methane from our ammonia product. The methane leaves the ammonia condenser via the N_2–H_2 recycle stream. The methane recycles to the reactor and accumulates indefinitely in our process. That's bad – there is no outlet for unreacted methane. Similarly, other impurities in air, notably argon, will also accumulate indefinitely in the recycle loop!

This is a drawback of any recycling scheme: the accumulation of undesired substances. We could add a unit to separate nitrogen and hydrogen from methane (and argon) but this is expensive. It is simpler (and cheaper) to *purge* a small amount of the recycle stream, perhaps 10%. The revised flowsheet is then:

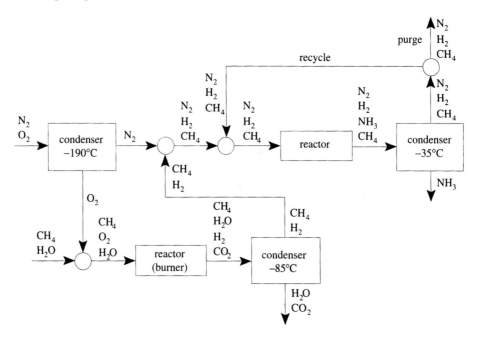

The purge stream is produced by a new type of unit, a *splitter*. As shown in the ammonia process and in the figure below, a splitter is represented by a circle:

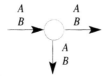

A splitter divides a stream into two or more streams of *identical* composition. A splitter only splits streams without separating mixtures, just as a combiner only combines streams without mixing.

We need to check another consequence of adding the methane burner: Does the air condenser yield enough oxygen to supply the methane burner? Let's estimate the flow rates based on a stoichiometric reaction of one mole of N_2 and three moles of H_2. One mole of N_2 requires $1/0.79$ moles of air into the separator, which yields $0.21/0.79 = 0.27$ moles of oxygen. From reaction (2.4), 3.3 moles of hydrogen require 0.35 moles of oxygen, so 3 moles of hydrogen require $(3/3.3)0.35 = 0.32$ moles of oxygen. We have a slight oxygen deficit of $0.32 - 0.27 = 0.05$ moles. Thus we must either buy extra oxygen or run the air condenser at a rate $0.32/0.27 = 1.2$ times faster than needed for the stoichiometric amount of nitrogen. We will assume that it is cheaper to produce oxygen than to buy it and we will sell the excess nitrogen.

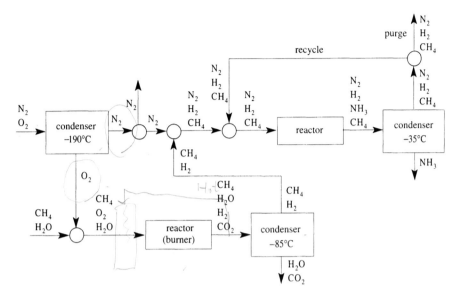

We now have a qualitative process to make ammonia. The mass flows are arranged logically; we have carefully reused by-products such as oxygen, nitrogen, and hydrogen. But what of the energy flow carried inherently by the mass? Are we discarding energy? Let's examine the water/methane reactor. We need to heat the reactants and we need to cool the products (to $-85°C$ in the condenser). We need a *heat exchanger*.

A heat exchanger transfers heat from one stream to another stream without mixing the streams. An example of a heat exchanger is shown below:

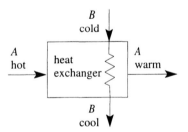

The zigzag line drawn inside the heat exchanger represents tubing that contains B and reminds us that the two streams are in thermal contact, but not in physical contact. The radiator on an automobile is a heat exchanger: Liquid coolant (usually a mixture of ethylene glycol and water) enters the radiator hot from the engine and flows through tubes. Air also flows through the radiator but on the outside of the tubes. The liquid coolant is cooled and the air is heated. If the radiator is operating properly, the liquid coolant does not mix with the air.

A heat exchanger can be inserted before the reactor in our ammonia process.

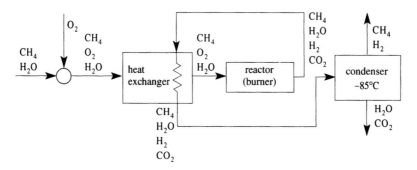

The air condenser also wastes energy: The air must be cooled to −190°C but products of the condenser – oxygen and nitrogen – must be heated before they enter their respective reactors. Two heat exchangers can satisfy both needs.

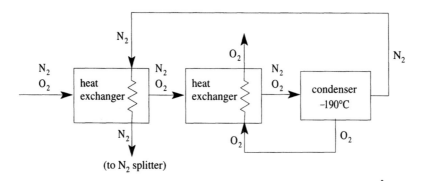

(to N₂ splitter)

Similarly, the ammonia synthesis (presumably) occurs at high temperature. Thus the reactor effluent is hot and the reactants must be heated. Again, a heat exchanger could be added.

In summary, we have divided a process to synthesize ammonia into discrete units, such that each unit has a specific function. We designed the process to minimize waste of both materials and energy.

2.2 Unit Operations and Flowsheets

Every chemical process is a collection of units interconnected by streams. This is the concept of *unit operations*, the first, and most venerable, paradigm adopted by the fledgling discipline of chemical engineering near the beginning of the twentieth century.

The five most common units are reactors, heat exchangers, pumps, mixers, and separators. The design and operation of these five units, which exist in hundreds of incarnations, are the focus of the traditional chemical engineering curriculum. Furthermore, most chemical engineering curricula have a senior-level laboratory course entitled *unit operations*.

As we saw in the ammonia process, the reactor is usually the key unit. The reactor will often dictate whether a chemical process is possible. Separators are secondary only to the reactor in most processes; the performance of the separator(s) will often determine if a process is profitable. Examples of separators include dryers, filters, absorbers, adsorbers, and centrifuges. The method used to separate a mixture – distillation, condensation, or filtering – is usually indicated by a specific symbol on the flowsheet. We will introduce these specific units later, as needed.

Additional units could be added to the ammonia process. Reactants should be mixed before entering a reactor, not just combined. Simply *combining* flour, eggs, water, and baking soda is not enough. The reactants must be *mixed* to bake a cake. The simplest *mixer* combines two or more streams into one stream:

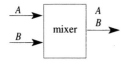

A propeller at the end of a shaft may be added to indicate mixing:

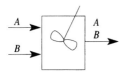

Mixers are important for solids and liquids. Gases, such as N_2 and H_2, tend to mix on their own.

We neglected to add pumps to the ammonia process. Fluids are usually moved about a process by pumps. A pump increases a fluid's pressure and the fluid flows from high pressure to lower pressure. Or, fluid pressure might be increased to condense the fluid (from a gas to a liquid) or speed a chemical reaction. A generic pump may be represented as

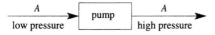

Because liquids and gases respond differently to an increase in pressure (a gas is compressed whereas a liquid is not) different units are used to distinguish this difference. A centrifugal pump for liquids may be represented as

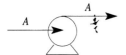

whereas a centrifugal gas compressor may be represented as

or simply

A gas expander, or turbine, is similar to the gas compressor above, but with the flows reversed.

We need some valves to control the flows. The simplest representation of a valve is

Finally, we offer two stylistic conventions for flowsheets. Avoid crossing streams. If streams must cross, indicate explicitly whether the streams mix or don't mix. A flowsheet with two streams as shown below on the left is ambiguous. Do the contents of the streams mix? If the contents of the streams combine (and split), indicate this explicitly with a splitter or combiner, as shown in the middle diagram below. If the streams cross without mixing, this should be indicated as shown on the right below. The primary stream should be unbroken when streams cross.

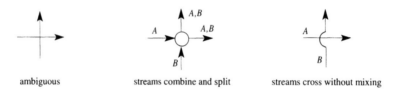

Only horizontal and vertical lines should be used for streams on a flowsheet. Arrowheads should appear only when a stream terminates at a unit, not at a bend in a stream.

2.3   Purifying Heptane – Creative Problem Solving

Creating a process flowsheet for a chemical or physical process is at the heart of engineering design. These flowsheets, however, are often difficult to conceive and an engineer must use many skills to develop the optimal design. A chief stumbling block for many designs is the failure to *properly define the problem to be solved*. We redefined the problem during our design of the ammonia process. For example, the problem with the effluent from the methane burner was not "separate hydrogen from oxygen" but rather "don't have oxygen in the effluent." Another example was the methane in the hydrogen stream – the goal was not to purify hydrogen but rather to have a hydrogen source suitable for reacting with nitrogen to produce ammonia. Chemically inert substances mixed with the hydrogen are acceptable. Let's consider some other examples of properly defining the problem to be solved.

You design a chemical process that uses *n*-heptane, $CH_3(CH_2)_5CH_3$, as a solvent.

The *n*-heptane must be 99.999% pure when it enters your process. During the process the *n*-heptane acquires an impurity: 10% *i*-propanol, or isopropyl alcohol, $CH_3CH(OH)CH_3$.

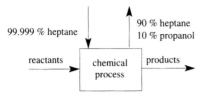

Of course, we should recycle *n*-heptane. But first we must purify the *n*-heptane. Our challenge is thus:

PROBLEM: "Separate *n*-heptane from *i*-propanol."

Note that this simple statement of the problem carries subtle meaning. Compare to "Separate *i*-propanol and *n*-heptane," which implies we wish to produce high purity *n*-heptane *and* high purity *i*-propanol. We only wish to produce a stream of high purity *n*-heptane; we do not care if the *i*-propanol is impure.

As we did several times in our design of the ammonia process, we investigate the possibility of separating the *n*-heptane by phase differences. We begin with the most efficient separation, gas/liquid or gas/solid, for which we need to know boiling points. Again we make a table of the relevant thermodynamic data.

Table 2.4. Boiling points for *n*-heptane purification

Compound	Boiling point at 1 atm, in °C
n-heptane	98.4
i-propanol	97.8

Evaporating *i*-propanol from an *n*-heptane liquid is feasible, but the boiling points are very close. When we explore the design of evaporators and condensers in Chapter 4 we will learn the implications of "the boiling points are very close." In theory any difference in boiling points can be exploited for a gas/liquid separation. But in practice the process may be too expensive. It is prudent to consider other methods.

Another method is solid/liquid separation. To explore this possibility, we add another column to our table.

Table 2.5. Melting points and boiling points for *n*-heptane purification

Compound	Boiling point at 1 atm, in °C	Melting point at 1 atm, in °C
n-heptane	98.4	−90.6
i-propanol	97.8	−126.2

There is a large difference in melting points, so solid/liquid separation seems feasible. Below $-90.6°C$ solid n-heptane will form in liquid i-propanol. (The melting point of a substance in a mixture is usually lower than the melting point of the pure substance.) The first unit in our process is thus a solid/liquid separator:

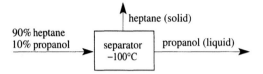

Because n-heptane is less dense than i-propanol, the solid floats on the liquid. We therefore show the solid effluent leaving the top of the separator.

But solid/liquid separation is more difficult than gas/liquid separation, for at least three reasons. Solid/liquid phases separate more slowly than gas/liquid phases. Because a gas is a factor of $\sim$1,000 less dense than a liquid, gas bubbles rise rapidly through a liquid. But an organic solid is only a factor of $\sim$1.2 or so different in density from an organic liquid. The solid particles settle (or rise) very slowly through the liquid. (In Chapter 5, we will calculate the velocities of rising bubbles and settling solids, which one needs to design a separator.) A second issue is the transport of the product streams. Both outputs from a gas/liquid separator are fluids, which are easily pumped through pipes. The solids from the solid/liquid separator will require special handling, such as an extruder or conveyor. But perhaps the most important issue is how "cleanly" the phases separate. Whereas it may be possible to produce a liquid free of solids (by filtration or settling, for example), it is usually not possible to produce a solid free of liquid.

So, given a choice, choose gas/liquid separation over solid/liquid separation. But we do not have a choice. The output of our solid/liquid separator is:

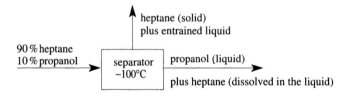

The impurity in the liquid stream is of little consequence. We regret losing some n-heptane, but it may not be feasible to recover *all* the n-heptane. The liquid entrained with the solid n-heptane poses a problem. Our new problem statement becomes:

PROBLEM: "Reduce (or eliminate) the liquid entrained by the solid n-heptane."

How do you remove liquid from a solid? What are some solids we dry routinely? How do you remove water from clothes after washing? The first step is the spin-dry, also known as a centrifuge. The next step is the air-dryer. Perhaps we could blow air (or some other gas) over the wet solid n-heptane at $-100°C$. When washing dishes, it was once common to dry the dishes with a towel and store the dishes in the cupboard. Perhaps we could design an absorbent conveyor belt to dry the solid n-heptane.

Centrifugation, air-drying, or toweling might work. However each is probably expensive, especially because the solid n-heptane must be dried at $-100°C$. And it is doubtful that any of the methods could remove all but 0.001% of the liquid, to yield 99.999% n-heptane.

Our ideas are unduly restricted by a poor statement of the problem. The problem is not that the n-heptane is wet with a liquid. The *real* problem is that the liquid is i-propanol; separating n-heptane and i-propanol is difficult. So we wash the solid n-heptane with a liquid that is easier to separate from n-heptane, either by distillation or perhaps liquid/liquid immiscibility.

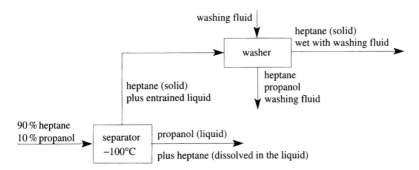

What is a suitable washing fluid? We could find a compound that is liquid at $-100°C$ and whose boiling point is different from n-heptane. But there are better criteria for the washing fluid. Wash with a liquid that need not be separated from n-heptane. We could review the chemistry in our process and perhaps find a compatible liquid. However, we already know a compatible liquid – n-heptane. Wash solid n-heptane with liquid n-heptane. The process to purify n-heptane becomes:

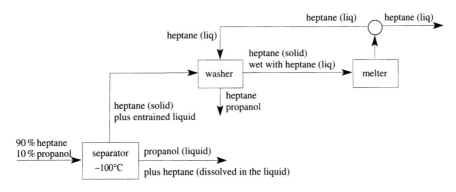

The temperature of liquid n-heptane recycled to wash the solid n-heptane should be close to its freezing point, to minimize melting the solid n-heptane.

Solid/liquid separation followed by washing should work. But the process involves some complicated units and n-heptane is lost in the wash. Note that recycling the

n-heptane/i-propanol wash to the separator will not help. The recycle will need a purge, to eliminate propanol from the loop. The purge will have the same n-heptane/i-propanol ratio as the stream leaving the washer. The purge will also have the same flow rate of i-propanol as the stream leaving the washer in the above design, and thus it will have the same total flow rate. The amount of n-heptane lost is the same, with or without a recycle.

There is a better design. We were restricted by the initial statement of the problem, "Separate n-heptane from the i-propanol," because it implied we needed to separate n-heptane from i-propanol. We do not care what happens to the i-propanol. It can be destroyed in the process of purifying the n-heptane. The *real* problem is:

PROBLEM: "Purify the n-heptane to 99.999%."

By changing the problem statement, we have more design options. For example, if n-heptane and i-propanol are difficult to separate, change i-propanol into something that you *can* separate from n-heptane. One might suspect that n-heptane and i-propanol have different reactivities. Why is n-heptane an acceptable solvent in this process? Among other properties, we can assume n-heptane does not react with any chemical in the process. Why is i-propanol unacceptable? We can assume i-propanol reacts with something in the process.

Chemical engineering students learn in organic chemistry that an alcohol is generally more reactive than a saturated hydrocarbon, such as n-heptane. One example is reactivity toward aluminum:

$$3 \begin{bmatrix} \text{CH}_3 \\ | \\ \text{CH}_3 - \text{C} - \text{OH} \\ | \\ \text{H} \end{bmatrix} + \text{Al} \rightarrow ((\text{CH}_3)_2\text{CHO})_3\text{Al} + \frac{3}{2}\text{H}_2, \tag{2.5}$$

$$\text{CH}_3(\text{CH}_2)_5\text{CH}_3 + \text{Al} \rightarrow \text{no reaction.} \tag{2.6}$$

The products of reaction (2.5) can be separated from n-heptane by distillation, as shown in Table 2.6.

Table 2.6. More melting points and boiling points for n-heptane purification

Compound	Boiling point at 1 atm, in °C	Melting point at 1 atm, in °C
n-heptane	98.4	−90.6
i-propanol	97.8	−126.2
Al propoxide	248	106
H_2	−253	−259

The chemistry in reactions (2.5) and (2.6) is used in the process below.

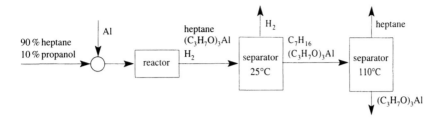

Is reacting i-propanol with aluminum a better process than solid/liquid separation of n-heptane and i-propanol? It is not possible to determine from the qualitative flowsheets developed here. Reacting i-propanol with aluminum has potential problems. The reaction to aluminum propoxide may not be sufficiently complete. The reactor may be expensive. In Chapters 3, 4, and 5 we will take up methods to analyze designs for product purity and cost, for example. During the design phase, it is essential to generate different possibilities, which can be analyzed later.

Proper definition of the problem statement is key to generating ideas. A quote from Albert Einstein eloquently makes this point:

> The mere formulation of a problem is far more often essential than its solution, which may be merely a matter of mathematical or experimental skill. To raise new questions, new possibilities, to regard old problems from a new angle requires creative imagination and makes real advances in science.

Proper definition of the problem statement also makes real advances in engineering.

Defining the problem is the first step in the McMaster problem-solving heuristic, which is described in two texts by chemical engineers: *Problem-Based Learning: How to Gain the Most from PBL* by D. R. Woods and *Strategies for Creative Problem Solving* by H. S. Fogler and S. E. LeBlanc. The McMaster five-step agenda for problem solving is:

Define the problem. Identify the true objective. List the constraints, criteria, and assumptions.
Generate ideas.
Choose a course of action.
Execute a plan.
Evaluate the plan.

This chapter (and especially the exercises that follow) provide experience in defining the problem and generating ideas. Chapters 3, 4, and 5 introduce three methods of analysis, the basis for the third step.

2.4 Flowsheet Conventions

Let's state some of the conventions for process flowsheets that we implicitly invoked in the preceding designs. First, each unit should represent only one physical or chemical operation. For example, unit 1 below is unacceptable. The flowsheet needs more details than unit 1 provides. In practice some units involve more than one operation. Some reactors are also mixers and some reactors are also separators. However, when designing a process here, draw an individual unit for each operation.

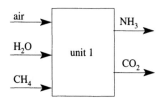

Second, process streams and units should be labeled. The chemical content of the streams and the physical properties of the unit should be included whenever possible. For example, a process stream might be labeled "1240 mol/min methane at 77°C," and a unit might be labeled "reactor, 500°C, 10 atm."

Third, streams should be drawn to and from units such that matter is neither created nor destroyed. Consider a washer for recycled glass bottles:

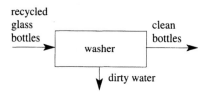

Water appears from nowhere in the diagram above. A proper flowsheet for the washer might be:

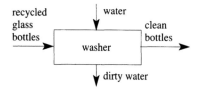

Reactors are sometimes a source of confusion. For example, a reactor may seem to destroy matter, such as nitrogen and hydrogen in the reactor below. A reactor may also seem to create matter. In this case the reactor seems to create ammonia.

Careful inspection of this unit, however, shows that NH_3 is created from its constituent atoms, H and N, which enter the unit as H_2 and N_2. Matter is not created. Similar reasoning reveals that although nitrogen and hydrogen do not appear in the effluent, matter is not destroyed. There is chemical change in the reactor, but matter is neither created nor destroyed.

Flowsheets are a graphical representation of a process. Creating a flowsheet is the first step toward producing architectural drawings for the process. As we shall learn, flowsheets are also used to *analyze* designs and compare competing designs.

2.5 The Production of Electronic-Grade Silicon

We will now create a flowsheet for a process to produce elemental silicon used in the manufacture of integrated circuits and other electronic devices. This seems easy – silicon is a nontoxic element, abundant in the Earth's crust, and it is stable at reasonable temperatures and pressures. However, the lack of a means to produce *electronic-grade* silicon once inhibited the mass production of solid-state electronic devices, such as transistors and diodes. The complication is the extraordinary purity needed.

PROBLEM: "Produce elemental silicon with less than 1 part-per-billion impurities."

To put this purity in perspective, consider that one part-per-billion (ppb) corresponds to one grain of salt in an entire railroad car of sugar.

Is this problem correctly defined? Must the impurity levels be so low? The answers to these questions are not trivial; approximately forty years of research and development in the field of condensed matter physics have been devoted to this issue. Without discussing the details of semiconductor device physics, let it suffice to say that the electrical properties of elements such as Si and Ge are extremely sensitive to impurities. Solid-state devices require ultra-pure materials that may be used in their pristine state or intentionally "doped" with impurities to tailor their electrical properties.

What is a good source of silicon? One need not look far. Silicon is the second most abundant element in the Earth's crust (oxygen is the most abundant), comprising about 25%. Virtually all of this silicon is found in various forms of sand, or silicon dioxide, SiO_2. The obvious route to preparing silicon, then, would be

$$SiO_2 + 911 \text{ kJ} \rightarrow Si + O_2. \tag{2.7}$$

However, this reaction proceeds only above about 10,000°C. What material could we use to construct the reactor? Most metals are molten at 10,000°C. And how would we heat the vessel? Recall that electricity is an expensive source of energy.

Recall the alternate source of energy we used in the ammonia synthesis. To produce hydrogen we burned water in methane, which transferred the oxygen atom from the hydrogen to carbon. For example,

$$CH_4 + \frac{1}{2}O_2 + H_2O \rightarrow CO_2 + 3H_2. \tag{2.8}$$

The analogous chemical reaction for silicon is

$$CH_4 + O_2 + SiO_2 \rightarrow Si + CO_2 + 2H_2O. \tag{2.9}$$

As before, the coefficients in reaction (2.9) will be determined by stoichiometry, the amount of energy needed, and the restrictions on the effluent. For example, is it acceptable to have carbon dioxide in the product? How about water? Methane is less expensive than electricity, but there is yet a cheaper source of energy – coal. We designate coal by its chief constituents as C_xH_y, where typically $x \approx 10y$, and the (unbalanced) reaction is

$$C_xH_y + O_2 + SiO_2 \rightarrow Si + CO + H_2O \tag{2.10}$$

Thus the first units in our silicon process are

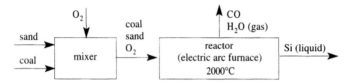

If coal is a cheaper energy source than methane, why don't we use coal in the ammonia synthesis as well? Coal contains higher concentrations of impurities, notably organic nitrogen, organic sulfur, and phosphorus, which would cause problems in the ammonia process. These impurities are not a problem in the silicon process because each reacts with oxygen and forms compounds that are gases at 2,000°C and thus are easily eliminated in the electric arc furnace.

The electric arc furnace violates the "one unit – one operation" principle – it reacts *and* separates. In a strict sense this is true. Such a unit would be unacceptable for the incomplete reaction in the ammonia reactor, because the products are all gases and their separation is not trivial.

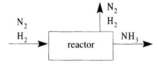

However, when the products are easily separated phases, such as the gas/liquid phases in the electric arc furnace, combining reaction and separation is acceptable.

The electric arc furnace produces liquid silicon that is about 98% pure. This "metallurgical" grade silicon is used chiefly to make silicone polymers. The principal impurities – aluminum and iron – derive from the sand. How do we purify this silicon further?

One possibility is to change SiO_2 into some other chemical that can be easily purified. This is impractical because SiO_2 is a very stable material and is inert to most simple chemical reactions. This is why glassware is so useful in chemistry. Another possibility is to remove Al and Fe from Si by sorting, perhaps with the assistance of magnets. But Al and Fe are not present in the Si as "lumps"; rather Fe and Al are atomically dispersed in the silicon.

Distillation is often a viable alternative. As before we consult a source of chemical reference data and create a table.

Table 2.7. Boiling points for silicon processing

Compound	Boiling point at 1 atm, in °C
CO	−191
H_2O	100
Si	2,355
Al	2,467
Fe	2,760

Clearly we may assume that CO and H_2O gases readily separate from the Si liquid. However, the boiling points for the two impurities of silicon – Al and Fe – are too close to silicon's for effective gas–liquid separation. Again, "too close" may not be obvious since the boiling points differ by at least 110°C. We only required a difference of 14°C to separate O_2 and N_2 previously. There are two effects here. First, the ratio of the absolute boiling points (2,628 K/2,740 K $= 0.96$) is very close to 1. Also, the required purity is one part per billion. This is too difficult to achieve with a boiling-point ratio of only 0.96.

How about a separation based on differences in melting points? Perhaps we can separate liquid silicon from solid Fe and solid Al. Again we look to the thermodynamic data.

Table 2.8. Melting points and boiling points for silicon processing

Compound	Boiling point at 1 atm, in °C	Melting point at 1 atm, in °C
CO	−191	−205
H_2O	100	0
Si	2,355	1,410
Al	2,467	660
Fe	2,760	1,535

Unfortunately, aluminum, not silicon, has the lower melting point. Such a process would be tenuous anyway since aluminum and iron are soluble in molten silicon.

Separation based on liquid–solid phase is thus impractical; how would we remove this liquid aluminum from the solid silicon? Furthermore, even if we could remove the liquid aluminum, how would we separate the iron liquid from the silicon?

The purification of Si is another example in which the problem statement needs to be examined carefully. We have been assuming that the problem is *remove the Al and Fe from Si*. In fact the real issue is to *produce pure Si*. This allows for other possibilities; consider using chemistry.

In the separation of *n*-heptane and *i*-propanol we used chemistry to transform the undesired component (*i*-propanol), because our only objective was to purify *n*-heptane. Selectively transforming the undesired components in Si (Al and Fe) is more difficult, if not impossible. Since Si, Al, and Fe are all metals, they have similar reactivities. More important, the Al and Fe atoms can be reached only by removing the surrounding Si atoms. We consider here a broader application of chemistry; transform everything, isolate the Si-containing chemicals, then reverse the chemistry to obtain Si.

The chlorides of silicon, iron, and aluminum are easily prepared from the elements. Consider the following reactions and boiling points of products (at 1 atm):

$$Si + HCl \rightarrow \quad SiH_4 \qquad (-112°C)$$
$$+ SiH_3Cl \qquad (-30°C)$$
$$+ SiH_2Cl_2 \qquad (8°C)$$
$$+ SiHCl_3 \qquad (33°C)$$
$$+ SiCl_4 \qquad (58°C) \tag{2.11}$$
$$Al + HCl \rightarrow \quad AlCl_3 \qquad (181°C) \tag{2.12}$$
$$Fe + HCl \rightarrow \quad FeCl_2 \qquad (1,024°C)$$
$$+ FeCl_3 \qquad (332°C) \tag{2.13}$$

At 35°C the chlorides of aluminum and iron will condense, and only $SiCl_4$ of the silicon chlorides will condense. We add two units to our silicon process:

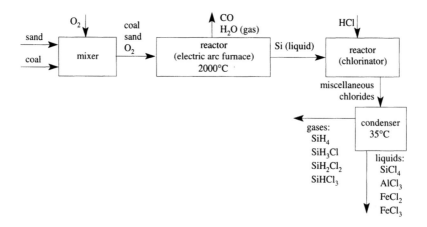

We now transform the silicon chlorides back into elemental silicon. For this we use the Siemans process:

$$SiHCl_3 + nH_{2(gas)} \rightarrow Si + HCl.$$ (2.14)

So we add units to our flowsheet to isolate $SiHCl_3$ and react with hydrogen to form silicon:

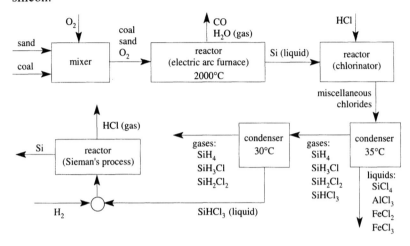

From this basic scheme for producing electronic-grade silicon, you might imagine ways to improve the process. Consider adding recycle and purge streams, as well as optimizing the use of energy, for example, by using hot exit streams to preheat incoming streams.

The silicon from this process is used to grow large Si crystals, called *ingots* or *boules*, typically six to twelve inches in diameter and over a meter long. The Si boule is sliced into wafers, which in turn are used to produce integrated circuits, also known as chips.

Chemical engineering is integral to the semiconductor industry. Andrew Grove, CEO of Intel and *Time* magazine's 1997 Man of the Year, earned a BS degree in chemical engineering from CCNY in 1960.

2.6 Generating Electrical Power with Fuel Cells

Chemical engineers play integral roles in the utility industries. A fuel-burning electricity generation plant converts chemical energy into electrical energy, usually by means of mechanical energy. The schematic below depicts the principal units in a coal-fired electricity generation plant. Chemical engineers design burners to efficiently burn coal in air. The burner is designed to minimize by-products such as carbon monoxide (CO) and nitrogen oxides, such as NO, NO_2, and N_2O, collectively indicated by NO_x. Chemical engineers also design units to collect and convert inevitable by-products including ash and sulfur oxides, SO_2 and SO_3, represented as SO_x. The heat from the

29

burner must be efficiently transferred to the water. The high-pressure steam drives turbines, which turn electricity generators.

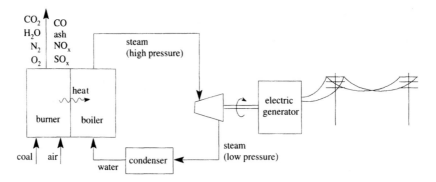

A coal-fired plant is large – a typical capacity is 1,000 MW, which can service about one million residences.

As you will learn in thermodynamics, the conversion from chemical energy to heat to mechanical energy and finally to electrical energy is accompanied by unavoidable energy losses. Because fuel cells convert chemical energy directly to electrical energy, fuel cells are potentially more efficient. Recent reports[1] indicate that fuel cells can replace burners and turbines in electricity generation plants. These plants would be much smaller, with capacity on the order of 1 MW.

In this section we design a chemical process to supply fuel to a fuel cell. This exercise offers more examples of problem definition and adds two unit operations to our repertoire: absorbers and adsorbers.

The energy source for the fuel cell is the reaction of hydrogen and oxygen, the reverse of reaction (2.2):

$$2H_2 + O_2 \rightarrow 2H_2O_{(gas)} + 484 \text{ kJ}. \tag{2.15}$$

If hydrogen and oxygen are mixed, the energy is released as heat. The key to producing electricity directly is to divide reaction (2.15) into three steps, as follows:

$$2H_2 \rightarrow 4H^+ + 4e^-, \tag{2.16}$$

$$O_2 + 4e^- \rightarrow 2O^-, \tag{2.17}$$

$$4H^+ + 2O^- \rightarrow 2H_2O_{(liquid)}. \tag{2.18}$$

The sum of reactions (2.16), (2.17), and (2.18) is reaction (2.15). The high-voltage electrons generated by reaction (2.16) are collected and sent down power lines. Utility customers use the high-voltage electrons to power various appliances, which reduces

[1] "Fuel-Cell Development Reaches Demonstration Stage," *Chemical and Engineering News*, August 7, 1995, p. 28; "Fuel Cells Poised to Provide Power," *Chemical Engineering Progress*, September 1996, p. 11.

the voltage of the electrons. Similarly, high-pressure steam drives a turbine in the power plant above, which reduces the pressure of the steam. The low-voltage electrons return to the plant and react with oxygen, as given in reaction (2.17). Finally, we combine the hydrogen and oxygen ions to make water, in a kinder, gentler fashion than reaction (2.15).

Reaction (2.16) poses an interesting separations problem. Assuming we could entice a hydrogen molecule to decompose into two protons and two electrons (as in a plasma) how do we separate the protons from the electrons? Protons and electrons are both gases and neither cares to be condensed separately. The key is to use a third player to ferry the electrons from the hydrogen to the oxygen. The fuel cell uses carbonate ions, as follows:

$$2H_2 + 2CO_3^{2-} \rightarrow 2CO_2 + 2H_2O + 4e^-, \tag{2.19}$$

$$4e^- + O_2 + 2CO_2 \rightarrow 2CO_3^{2-}. \tag{2.20}$$

As before, the sum of reactions (2.19) and (2.20) is reaction (2.15). The fuel cell in the simplified diagram below conducts reactions (2.19) and (2.20) in separate cells.

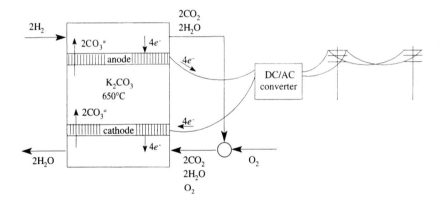

The K_2CO_3 (with secret additives) allows carbonate ions to diffuse between the anode and cathode. However, the K_2CO_3 is impermeable to the other, nonionic reactants. Our task as chemical engineers is:

PROBLEM: "Provide hydrogen and oxygen to power a fuel cell."

What are our options? We could purchase hydrogen and oxygen from a chemical manufacturer and have the reactants delivered by truck, or perhaps by rail. But transporting gases is expensive, especially an explosive gas (H_2) or a flammable gas (O_2). How did we supply H_2 and N_2 for the ammonia synthesis? We produced H_2 and N_2 on site. Let's investigate producing H_2 and O_2. What is a good source? Water is inexpensive and there are no by-products when it is decomposed. The decomposition reaction

$$2H_2O \rightarrow 2H_2 + O_2 \tag{2.21}$$

does, however, require energy. How much energy? Exactly the amount of energy generated by the fuel cell, if the fuel cell is 100% efficient.

We can obtain O_2 from the air, as we obtained N_2 (and O_2) from the air for the ammonia synthesis by condensation.

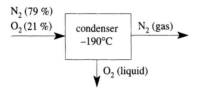

We generate H_2 also using a scheme from the ammonia plant; let's "burn" water. Analogous to reaction (2.4) we have the reaction

$$CH_4 + 2H_2O \rightarrow 4H_2 + CO_2. \tag{2.22}$$

CO_2 is an acceptable by-product because it is one of the components in the fuel cell cycle, reactions (2.19) and (2.20). A process to generate H_2 is

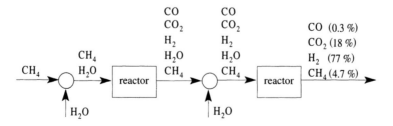

The actual process of producing H_2 from CH_4 is more complex than the reactor above and the similar reactor in the ammonia synthesis. Reaction (2.22) is the overall reaction. The actual process has an intermediate step that produces a mixture known as *synthesis gas*, or *syngas*:

$$CH_4 + H_2O \rightarrow 3H_2 + CO. \tag{2.23}$$

The CO is then reacted with H_2O,

$$CO + H_2O \rightarrow H_2 + CO_2. \tag{2.24}$$

The actual process is

Reaction (2.23) is catalyzed by NiO particles dispersed on aluminum oxide at 500 to 870°C and 20 atm. Reaction (2.24), also known as the water–gas shift reaction, is catalyzed by iron oxide, then copper-zinc oxide on aluminum oxide, at 220 to 250°C and 20 atm.

2.7 Desulfurization of Natural Gas – Absorbers and Adsorbers

Our process to produce H_2 from natural gas is idealized: We assumed that our natural gas source is 100% methane. Typically, natural gas has only 85 to 95% methane; the balance is chiefly ethane, propane, and propene. Natural gas also contains small concentrations (<1% total) of He, CO_2, and N_2. None of these constituents will affect the process diagram above and so we will ignore them. However, natural gas also contains H_2S. Up to 3 parts per million (ppm), or 0.0003%, H_2S is allowed. Although 3 ppm may seem like an inconsequential impurity level, it is a problem. A concentration of H_2S above 0.1 ppm will deactivate the catalyst in the first reactor. We need to remove H_2S from the natural gas.

Let's assume further that we are going to build our fuel-cell generator at the natural gas source, the well head. Instead of piping natural gas to consumers, we will convert it to electricity on site and "pump" electricity to consumers. But at the well head, the level of H_2S is even higher – 100 ppm. Our task is:

PROBLEM: "Reduce the H_2S level to less than 0.1 ppm in the natural gas."

Again, we first consider gas/liquid separation. We prepare a table of boiling points.

Table 2.9. Boiling points for desulfurization

Compound	Boiling point at 1 atm, in °C
CH_4	−164
H_2S	−61

Because there is a large difference in boiling points, we could condense H_2S from CH_4. But we would have to cool the natural gas to very low temperatures to remove all but 0.1 ppm (0.000001%) of the H_2S. That is, H_2S remains in the gas phase even at temperatures below its boiling point. Consider water, which boils at 100°C at 1 atm. There is still water vapor in the air at temperatures below 100°C. So H_2S might be removed by condensation, but it will be expensive. As a rule, cooling something 1°C costs four times as much as heating something 1°C. Engineering requires more than just a workable solution. Engineering also requires an economical solution. As stated by Arthur Mellen Wellington[2] (1847–1895),

> Engineering ... is the art of doing that well with one dollar, which any bungler can do with two after a fashion.

[2] *The Economic Theory of the Location of Railways*, 6th edition (1990).

We introduce here another type of separator, an *absorber*. Absorbers are prevalent in the chemical industry, second only to gas/liquid separation (distillation). An absorber extracts a substance from a gas by absorbing the substance into a liquid, such as water or oil. To explore the possibility of using a water absorber, we add a new column to our table of properties.

Table 2.10. Solubilities and boiling points for desulfurization

Compound	Boiling point at 1 atm, in °C	Solubility in water
CH_4	−164	slightly
H_2S	−61	very

Quantitatively, 100 mL of water will absorb 440 cm^3 of H_2S gas at 1 atm and 25°C. If water did not absorb H_2S sufficiently, or absorbed CH_4 too strongly, we could check the solubilities in a different liquid.

We bubble the natural gas through water to absorb H_2S. The absorber is shown below.

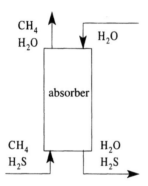

Absorbers are typically tall cylinders, also called *absorption columns*. An absorber is usually represented on a flowsheet by a tall, narrow rectangle. The liquid flows down the cylinder and gas bubbles upward. We will study the inner workings of absorbers in Chapter 4.

What shall we do with the water effluent from the absorber? Water is cheap, so there is little economic motivation to purify and recycle the water. Can we release the effluent into a nearby stream? H_2S smells like rotten eggs and at sufficient levels is flammable and poisonous; one or two inhalations of H_2S causes collapse, then coma, then death.[3] We must separate the H_2S from H_2O. Because the boiling point of H_2S is so low, it is easy to distill the H_2S from the water.

---

[3] *The Merck Index*, 12th edition (Merck & Co., Rahway, NJ, 1996).

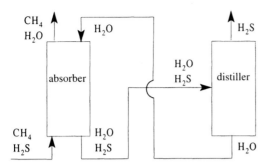

The H_2S can be sold; it is converted to H_2SO_4 by way of SO_2. In fact, because H_2S is so noxious, it is advisable to convert it to SO_2 immediately, and then store the SO_2. Eventually, we can ship the SO_2 to a sulfuric acid manufacturer.

The water in the absorber is thus recycled. What have we neglected to include with the recycle? A purge. If the natural gas contains impurities that are soluble in water, but have boiling points comparable to water, the impurities will accumulate in the absorber–distiller loop. Which stream should we purge? Obviously, it is easier to dispose of the stream after the distiller. And if we purge, we must replenish the loss of water in the loop. We add a combiner for the *make-up* stream.

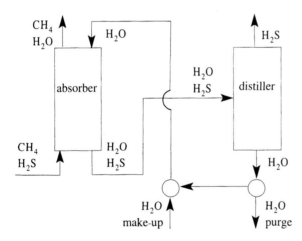

An absorber such as diagrammed can reduce the H_2S concentration to about 10 ppm. In practice, absorbers use liquids with higher capacity for absorbing H_2S. A good liquid for absorbing H_2S is monoethanolamine, which can reduce the H_2S to about 1 to 3 ppm. But even with monoethanolamine, the H_2S concentration is not low enough. We need to add another unit – an *adsorber*.

An adsorber extracts a substance from a gas (or liquid) by adsorbing onto a solid. The adsorbed molecule bonds to the surface of the solid. The stronger the bond, the more effectively the solid collects the substance. But a stronger bond also means it is harder to clean the surface to recycle the solid. Charcoal (activated carbon) is commonly used to adsorb impurities from drinking water. The packets of silica shipped with cameras and electronic equipment adsorb water from the air. On a macroscopic level, fly paper adsorbs insects from the air.

To reduce the H_2S concentration to less than 0.1 ppm, the solid must have a strong affinity for H_2S; the surface must form a chemical bond with H_2S. The preferred adsorbent for H_2S is ZnO, which adsorbs H_2S by the reaction

$$ZnO + H_2S \rightarrow ZnS + H_2O. \tag{2.25}$$

The surface of the ZnO adsorbs an H_2S molecule, an S atom exchanges with an O atom, and H_2O desorbs from the surface. The surface is rejuvenated by reaction with oxygen,

$$2ZnS + 3O_2 \rightarrow 2ZnO + 2SO_2. \tag{2.26}$$

We add units to our process to adsorb H_2S with ZnO, giving us

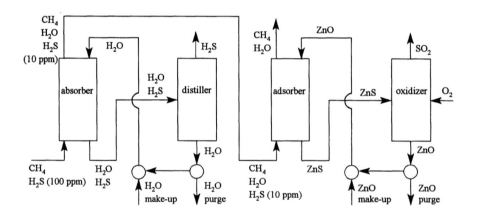

The stream leaving the ZnO adsorber has less than 0.1 ppm H_2S.

Recall that solids do not flow well. Solids are difficult to convey between units, and solids do not flow downward through cylinders as well as liquids. In practice the ZnO adsorber and oxidizer are batch processes rather than continuous processes as implied by the flowsheet above. The adsorber is initially packed with ZnO. After all the ZnO has converted to ZnS, the solids are dumped into the oxidizer and the adsorber is recharged with ZnO.

So why not eliminate the absorber and its attendant distiller from the process? Perhaps we should just feed the CH_4 with 100 ppm H_2S directly into the adsorber. The purity of the natural gas would be the same and we would simplify the design. How else would the process change if the absorber were eliminated? The adsorber would have to remove ten times as much H_2S, and therefore it would have to be emptied and refilled ten times as often (or you would need an adsorber ten times as large.) Although the design is more complex with an absorber *and* an adsorber, it is also less expensive.

This completes the chemical process to supply H_2 and O_2 to the fuel cell. We purify natural gas and then react it with water to produce H_2. We obtain O_2 from air by cryogenic distillation. In addition, we produce saleable by-products N_2, SO_2, and CO_2.

2.8 Summary

Chemical processes can be divided into discrete components known as units. A unit performs one operation in the process and is represented by a simple symbol, such as a square or circle, on a process flowsheet.

A complex design evolves incrementally from the key unit in the process. The key unit is often a reactor, but it can also be a specialized separation. Units are added by asking questions such as "How do I obtain the reactants needed in the reactor?" or "How do I isolate the product from the by-products of the reactor?" But more important than answering the question is verifying that you have identified the *real* problem to be solved at each stage in the design. Properly identifying the true goal allows for more possibilities for solving the problem. And a longer list of options increases the chance of finding the optimal solution.

One's skill at identifying the real problem improves with experience. Unfortunately, formal education does not exercise problem redefinition. In fact, formal education usually does not tolerate problem redefinition. When you are assigned exercises 1, 5, and 6 at the end of a chapter, you might perceive that the *real* problem is not to complete the assignment, but to learn the material in the chapter. Your redefinition of the problem would allow you to use means other than doing the homework to attain the true goal. Most likely the other means would not be tolerated. Rather, you are encouraged to practice identifying the real problems in everyday life.

REFERENCES

Fogler, H. S., and LeBlanc, S. E. 1995. *Strategies for Creative Problem Solving*, Prentice Hall, Upper Saddle River, NJ. (http://www.engin.umich.edu/~problemsolving)

Rudd, D. R., Powers, G. J., and Siirola, J. J. 1973. *Process Synthesis*, Prentice Hall, Upper Saddle River, NJ.

Woods, D. R. 1994. *Problem-Based Learning: How to Gain the Most from PBL*, Donald R. Woods, Watertown, Ontario, Canada. (http://chemeng.mcmaster.ca/pbl/pbl.htm)

EXERCISES

Process Analysis

2.1 A process for converting coal to diesel fuel is shown in the simplified flowsheet below. Coal has the generic formula $C_xH_yS_z$ where $x > y$ and $z \ll x, y$. Coal also contains some inorganic matter, called *ash*. Diesel fuel has the generic formula C_aH_b.

(A) Identify the units of this process. If a unit is not one described in this chapter, give it a name based on its function.

(B) Write the overall reaction for this process.

(C) Write the reactions that take place in each of the units you identify as reactors.

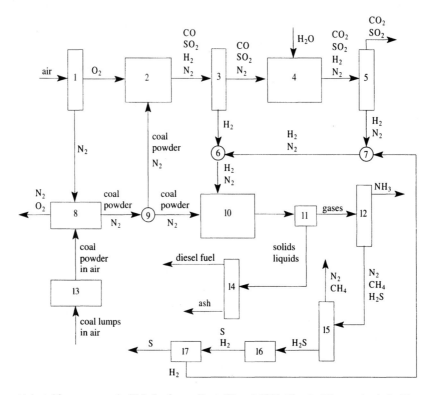

(Adapted from a process by W. L. Luyben and L. A. Wenzel, 1988. *Chemical Process Analysis: Mass and Energy Balances,* Prentice Hall, Upper Saddle River, NJ, pp. 46–9.)

2.2 A simplified flowsheet of the process to make ethylene oxide (C_2H_4O) from ethylene (C_2H_4) is shown below. Two reactions occur in the reactor:

$$2C_2H_4 + O_2 \rightarrow 2C_2H_4O,$$

$$C_2H_4 + 3O_2 \rightarrow 2CO_2 + 2H_2O.$$

Both reactions are incomplete; neither reactant is completely consumed in the reactor. Because ethylene oxide is very soluble in water, it can be extracted by bubbling the gaseous reactor effluent through water. Ethylene oxide is *absorbed* into the water. Ethylene oxide is then stripped from the water by steam.

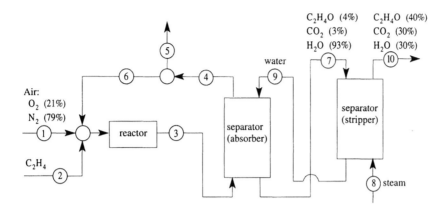

(A) What compounds constitute stream 3?

(B) What compounds constitute stream 4?

(C) What purpose is served by stream 5?

(D) Add one or more units to produce pure C_2H_4O from stream 10 in an energy-efficient manner.

Indicate the compounds present in each stream and label each unit.

Properties of some compounds at 1 atm

	H_2O	C_2H_4O	CO_2
melting point (°C)	0.	−111.	−57. (at 5.2 atm)
boiling point (°C)	100.	14.	−79. (sublimes)

(This exercise appeared on an exam. It was estimated that the exercise could be completed in 20 minutes.)

2.3 A process for producing ammonia from air, water, and methane was outlined in this chapter. Draw the flowsheet for the entire process. Label the units and list the components in each stream. Where appropriate, use the proper symbol to identify a unit operation. Finally, improve the energy efficiency of the process by adding heat exchangers before the air condenser and the ammonia reactor.

2.4 The process shown on the following page coats Si wafers with a thin polymer film, a crucial step in producing integrated circuits by photolithography. The process uses two reactants – monomer and Si wafers – and uses two solvents – air and acetone. Monomer molecules are linked into large polymer molecules in the reactor; an example is the reaction to form polyethylene, $n(CH_2=CH_2) \rightarrow (-CH_2-CH_2-)_n$, where $n > 1,000$. The Si wafers and the monomer used here are expensive. Because the monomer is nonvolatile, the air entering the dryer must be at least 200°C. The catalyst in the reactor cannot be exposed to air.

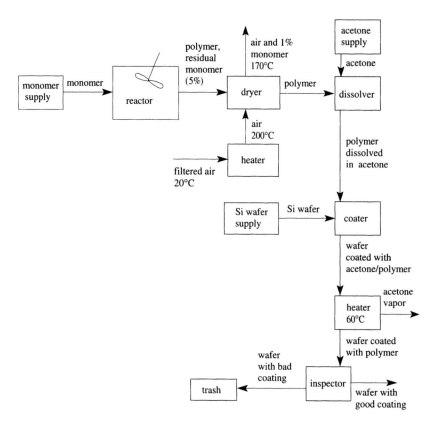

Improve the process with regard to pollution, economics, and energy efficiency. Your improvements must be in the form of unit operations, not a list of suggestions. Add to a copy of the flowsheet above and/or redraw portions of the process. Four significant improvements will suffice.

Note that the polymer is soluble in acetone. Also note that although air is a free raw material, the filtered air used in this process has been treated to remove particles larger than 1 μm and thus is valuable.

2.5 Sulfur dioxide can be produced by oxidizing sulfur:

$$S + O_2 \rightarrow SO_2.$$

Because the reaction is highly exothermic, the flow to the reactor must contain cool inert gas to maintain a low temperature in the reactor. Consider the two process schemes diagrammed on the following page. Scheme I uses N_2 as the inert gas in the reactor and Scheme II uses SO_2 as the inert gas in the reactor.

(A) In which sequence is the separation the easiest? State at least two reasons for your choice. Use the table of thermodynamic data on the following page.

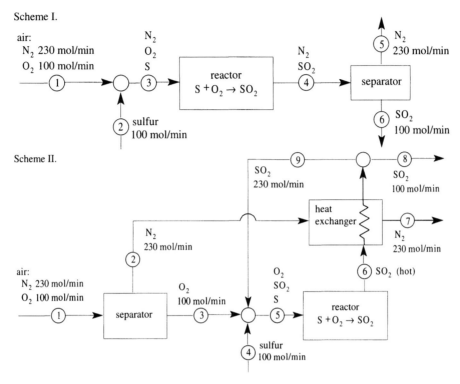

Scheme I.

Scheme II.

(Adapted from examples 1.3.2 and 3.4.1 by Rudd, D. F., Powers, G. J., and Siirola, J. J. 1973. *Process Synthesis*, Prentice Hall, Upper Saddle River, NJ, p. 12 and pp. 82–4.)

	N₂	O₂	S	SO₂
molecular wt (g/mol)	28	32	32	64
melting point (°C)	−210	−218	113	−73
boiling point (°C)	−196	−183	445	−10
heat capacity (joule (°C)⁻¹ mol⁻¹)	45 (gas)	30 (gas)	32 (solid)	43 (gas)
			34 (liq)	
heat of melting (joule mol⁻¹)	7.2×10^2	4.4×10^2	1.7×10^3	7.4×10^3
heat of vaporization (joule mol⁻¹)	5.6×10^3	6.8×10^3	9.6×10^3	2.5×10^4

(B) Schemes I and II assume that O_2 and S enter the reactor in exact stoichiometric ratio. Consider the consequences of a nonstoichiometric ratio of O_2 and S. In which scheme is the product purity least sensitive to an excess of air? Explain your choice.

Process Design

2.6 Sketch a flowsheet for a process to wash clothes. Your process should include units to perform the following operations:

(A) Sort the laundry into three types: white items, colored items, and delicate items.

(B) Wash laundry types separately.

(C) Spin dry.

(D) Air dry.

Label all unit operations in general terms, such as mixer, separator, heater, and cooler. Each unit on your flowsheet should perform only one operation, even though in actual practice one machine may perform multiple operations in sequence. Add the necessary streams for water, soap, bleach, and air and list the components in all streams.

> Laundry is an extremely complex system. Those of us who do the laundry don't get enough credit for the technical decisions we make every day. Some very serious science goes into doing the laundry.
>
> K. Obendorf, Professor of Textiles and Apparel, Cornell University, in *Smithsonian Magazine*, September 1997.

2.7 Consider the following recipe for pizza dough:

> Mix 4 packages of Baker's yeast with 1/4 cup of milk and 1/2 cup of warm water. Gently warm and stir for 15 minutes.
>
> Prepare two cups of "scalded milk," that is, milk that has been heated to a boil for about 1 minute. Add 1 tablespoon of molasses and 1 tablespoon of salt to the milk after scalding.
>
> Make the dough by combining the yeast mixture and the scalded milk mixture with 6 cups of flour. Gently knead the dough for about 5 minutes. Let the dough rise in warm, moist air for 15 minutes. The risen dough should be about double its initial size. There should be sufficient dough to make 4 pizzas.
>
> Roll one quarter of the dough into a flat circle, 15 inches in diameter. Lightly coat the dough with olive oil. Bake the dough at 535°F for 5 to 6 minutes.

Design a process to continuously produce prebaked pizza dough for a major grocery chain. Your flowsheet should have labels on all units and qualitative compositions of all streams. For each unit list key quantities such as temperature and residence time of material through the unit.

2.8 Your company manufactures specialty filters sketched below. The filter consists of an aluminum container packed with threads of polymer Q. The threads are coated with an active ingredient, chemical Z. The threads are approximately 1 mm diameter and 10 cm long and are wadded into entangled balls, which prevents the loss of any threads as fluid passes through the filter. The filters are produced by tamping a thread ball into an aluminum container open at one end, and then crimping the cap over the open end.

Over 10% of the filters fail to meet specifications. Filters fail for myriad reasons. Typical flaws include containers that leak, filters with too much (or too little) Q threads, and Q threads with too much (or too little) Z coating. We wish to recycle the rejected filters.

Sketch the unit operations of a process to recover chemical Z, polymer Q, and aluminum scrap. Indicate the species in each stream. The simplest, safest process is often the best process.

Component	Weight %	Soluble in acetone?	Soluble in water?	Melting point (°C)	Boiling point (°C)	Density (kg/m³)
chemical Z	5	yes	yes	80	350	960
polymer Q	55	yes	no	250	decomposes at 450	870
aluminum	40	no	no	660	2,467	2,700

Listed below are properties of some solvents at 1 atm. (Note that acetone is volatile and highly flammable. Inhalation may cause headaches, fatigue, and bronchial irritation.)

	Water	Acetone
melting point (°C)	0.	−95.
boiling point (°C)	100.	56.
liquid density (kg/m³)	1,000.	790.

(This exercise appeared on an exam. It was estimated that it could be completed in 30 minutes.)

2.9 Design a process to recycle all the components in a pharmaceutical tablet. Each tablet is chiefly sugar. Interspersed in the sugar are tiny (smaller than 0.1 mm diameter), hard capsules that contain a drug. The capsules are impermeable to water, which protects the drug. The capsules melt at 95°F (35°C) to release the drug in one's stomach. The tablets are to be recycled because the dosage is incorrect; there are too many capsules per tablet.

The tablets are sealed in individual plastic pouches. An unopened pouch floats on water and dissolves in acetone. The external surfaces of the recycled pouches are dirty.

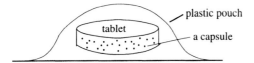

Cross-sectional side view of a tablet in pouch.

Your process should be functional, simple, energy efficient, and environmentally benign. Label the components in each stream of your process. Note that acetone is volatile and highly flammable. Inhalation may cause headaches, fatigue, and bronchial irritation.

Component	Weight %	Soluble in acetone?	Soluble in water?	Melting point (°C)	Boiling point (°C)	Density (kg/m³)
plastic pouch	42%	yes	no	185	decomposes at 450	860
dirt	~0.3%	yes	yes	—	—	~2,500
sugar	52%	no	yes	146	decomposes	1,560
capsule	2%	yes	no	35	85	1,140
drug	1%	yes	yes	5	decomposes	1,420

(This exercise appeared on an exam. It was estimated that it could be completed in 30 minutes.)

2.10 Soda ash (sodium carbonate, Na_2CO_3) is a key reactant in the production of glass and paper. Soda ash is produced from limestone and salt via the reaction

$$CaCO_3 + 2NaCl \rightarrow Na_2CO_3 + CaCl_2.$$

But this reaction does not proceed as written above. Soda ash is instead produced indirectly by the ingenious route invented by Ernst Solvay (1838–1922). There are five salient reactions in this process, beginning with the decomposition of limestone into lime and carbon dioxide at 1,000°C:

decomposition (1,000°C):	$CaCO_3 \rightarrow CaO + CO_2,$
slaking:	$CaO + H_2O \rightarrow Ca(OH)_2,$
chlorination:	$Ca(OH)_2 + 2NH_4Cl \rightarrow 2NH_4OH + CaCl_2,$
carbonation:	$NH_4OH + CO_2 + NaCl \rightarrow NH_4Cl + NaHCO_3,$
calcination (300°C):	$2NaHCO_3 \rightarrow Na_2CO_3 + CO_2 + H_2O.$

Design a chemical process to produce soda ash by the Solvay route. Your process should minimize waste and reactants. Label the units and list the components in all streams.

You may assume that each reaction goes to completion; reactants are entirely converted to products. Slaking, chlorination, and carbonation occur in aqueous solutions. $CaCl_2$, NH_4OH, and NH_4Cl are highly soluble in water, whereas $NaHCO_3$ is only slightly soluble. $CaCO_3$, $Ca(OH)_2$, $NaHCO_3$, Na_2CO_3, and NH_4Cl are solids at 20°C; NH_4OH exists only in aqueous solution. Ionic solids decompose when heated: The boiling point of an ionic solid is undefined. Finally, you may assume $CaCl_2$ does not react with CO_2.

(Adapted from descriptions of the Solvay process by Rudd, D. F., Powers, G. J., and Siirola, J. J. 1973. *Process Synthesis*, Prentice Hall, Upper Saddle River, NJ, example 2.4.1, pp. 42–5, and by Reklaitis, G. V. 1983. *Introduction to Material and Energy Balances*, Wiley, New York, pp. 182–4, exercise 3.33.)

2.11 Newly formed steel has an oxide crust that may be removed by washing in sulfuric acid:

$$\text{Fe with FeO crust} + H_2SO_4 \rightarrow Fe + FeSO_4 + H_2O.$$

The H_2SO_4 is dissolved in water and is in excess. $FeSO_4$ is soluble in water and soluble in acidic solution. We wish to recover the Fe and the H_2SO_4 with the following reactions:

$$FeSO_4 + 2HCN \rightarrow Fe(CN)_2 + H_2SO_4.$$

The HCN is bubbled into the $FeSO_4$ solution and is in excess. Note that HCN is highly toxic. Assume the $FeSO_4$ is entirely converted to $Fe(CN)_2$. $Fe(CN)_2$ is insoluble in water and insoluble in acidic solution. In water, $Fe(CN)_2$ is entirely converted to FeO,

$$Fe(CN)_2 + H_2O \xrightarrow{500°C} 2HCN + FeO.$$

Solid FeO may be recycled to the smelting process, which converts iron oxides to iron metal.

(A) Design a process based on the above reactions. Indicate the compounds present in each stream of your process and label each unit.

(B) Assume further that the steel is contaminated with grease. The generic reaction of grease with sulfuric acid is

$$grease + H_2SO_4 \rightarrow CO_2 + H_2O + SO_2.$$

CO_2 and SO_2 do not react with HCN. Modify your process to accommodate grease on the steel. Draw only the modified portion of your process; you need not redraw the entire process.

(C) Assume the steel contains no grease but is contaminated with calcium oxide. The pertinent reactions are

$$CaO + H_2SO_4 \rightarrow CaSO_4 + H_2O,$$
$$CaSO_4 + 2HCN \rightarrow Ca(CN)_2 + H_2SO_4,$$
$$Ca(CN)_2 + H_2O \xrightarrow{500^\circ C} CaO + 2HCN.$$

Both $CaSO_4$ and $Ca(CN)_2$ salts are soluble in water and soluble in acidic solutions. Modify your process to accommodate CaO on the steel. Draw only the modified portion of your process; you need not redraw the entire process.

Properties of some compounds at 1 atm

	H_2O	H_2SO_4	HCN	SO_2
melting point (°C)	0	11	−14	−73
boiling point (°C)	100	338	26	−10

(Adapted from Rudd, D. F., Powers, G. J., and Siirola, J. J. 1973. *Process Synthesis*, Prentice Hall, NJ, p. 49. This exercise appeared on an exam. It was estimated that it could be completed in 40 minutes.)

2.12 Styrene (a commodity chemical) is produced in two steps. First, benzene and ethylene react at 450°C in the presence of catalyst 1 to form ethylbenzene:

$$alkylation~(450^\circ C): C_6H_6 + CH_2{=}CH_2 \rightarrow C_6H_5{-}CH_2CH_3. \qquad (rxn~1)$$

The ethylbenzene is then dehydrogenated at 620°C in the presence of catalyst 2 to produce styrene:

$$dehydrogenation~(620^\circ C): C_6H_5{-}CH_2CH_3 \rightarrow C_6H_5{-}CH{=}CH_2 + H_2. \qquad (rxn~2)$$

Unfortunately, side reactions occur in both reactors. In the alkylation reactor, ethylbenzene reacts with ethylene to form diethylbenzene:

$$alkylation~(450^\circ C): C_6H_5{-}CH_2CH_3 + CH_2{=}CH_2 \rightarrow C_6H_4(CH_2CH_3)_2. \qquad (rxn~1a)$$

However, the effects of side reaction 1a can be minimized (but not eliminated entirely) if an excess of benzene is present, by reaction 1b:

$$alkylation~(450^\circ C): C_6H_4(CH_2CH_3)_2 + nC_6H_6 \rightarrow 2C_6H_5{-}CH_2CH_3 + (n-1)C_6H_6.$$
$$(rxn~1b)$$

A nuisance in the dehydrogenation reactor is the reversal of reaction 1:

$$\text{dealkylation (620°C): } C_6H_5\text{–}CH_2CH_3 \rightarrow C_6H_6 + CH_2\text{=}CH_2. \qquad \text{(rxn 2a)}$$

None of the reactions go to completion. Reaction 1 is exothermic (releases heat) whereas reaction 2 is endothermic (requires heat). The energy to drive reaction 2 is conveniently supplied by injecting super-heated steam with the reactants. The steam does not affect any of the reactions and improves the catalyst performance by removing residue.

Rather than use the chemical formulae given in reactions 1 and 2, you may wish to use the more expedient (although less informative) nomenclature of benzene $\equiv B$, ethylene $\equiv E$, styrene $\equiv S$, ethylbenzene $\equiv EB$, diethylbenzene $\equiv DEB$, superheated steam $\equiv SHS$, and water $\equiv W$.

Reactions 1 through 2 are thus:

alkylation (450°C):	$B + E \rightarrow EB$,	(rxn 1)
dehydrogenation (620°C):	$EB \rightarrow S + H_2$,	(rxn 2)
alkylation (450°C):	$EB + E \rightarrow DEB$,	(rxn 1a)
alkylation (450°C):	$DEB + nB \rightarrow 2EB + (n-1)B$,	(rxn 1b)
dealkylation (620°C):	$EB \rightarrow B + E$.	(rxn 2a)

Design a process to produce styrene.

- Assume no other chemistry than the reactions given.
- Maximize the product and minimize by-products.
- Minimize the number of process units; then minimize the sizes of the process units.
- List the substances in every stream. You need not specify flow rates or compositions.
- Label all units and the typical temperatures in each unit.
- You may neglect pumps, heat exchangers, heaters, and coolers.

Physical properties at 1 atm

		Melting pt (°C)	Boiling pt (°C)
H_2	H_2	-259	-253
$CH_2\text{=}CH_2$	E	-169	-104
C_6H_6	B	6	80
H_2O	W	0	100
$C_6H_5\text{–}CH_2CH_3$	EB	-95	136
$C_6H_5\text{–}CH\text{=}CH_2$	S	-31	145
$C_6H_4(CH_2CH_3)_2$	DEB	-84	182

You may assume that all species (B, E, EB, DEB, S, and H_2) are insoluble in water and that water does not dissolve in any of the species encountered in this process.

(Professor C. Cohen (Cornell) created this exercise for an exam. He estimated it could be completed in 25 minutes.)

2.13 **(A)** Design a process to produce vinyl chloride ($CH_2{=}CHCl$) from ethylene ($CH_2{=}CH_2$) and chlorine (Cl_2). Your process should use chlorination followed by pyrolysis.

chlorination (60°C):	$CH_2{=}CH_2 + Cl_2 \rightarrow CH_2Cl{-}CH_2Cl$	(rxn 1)
pyrolysis (500°C):	$CH_2Cl{-}CH_2Cl \rightarrow CH_2{=}CHCl + HCl$	(rxn 2a)
	$CH_2Cl{-}CH_2Cl \rightarrow CH_2{=}CH_2 + Cl_2$	(rxn 2b)
net reaction	$CH_2{=}CH_2 + Cl_2 \rightarrow CH_2{=}CHCl + HCl$	

A border drawn around your entire process must agree with the *net reaction*.

These compounds undergo additional reactions, which one may wish to exploit or avoid:

hydrochlorination (150°C):	$CH_2{=}CH_2 + HCl \rightarrow CH_3{-}CH_2Cl,$	(rxn 3)
pyrolysis (500°C):	$CH_3{-}CH_2Cl \rightarrow CH_2{=}CH_2 + HCl.$	(rxn 2c)

Reactions 1 and 3 each require a special catalyst and do not proceed in the absence of their respective catalysts. Reactions 1 and 3 have 100% conversion of reactants into products. Reactions 2a, 2b, and 2c have about 35% conversion.

- Assume no other chemistry than the reactions given.
- Maximize product and minimize by-products.
- Minimize the number of process units; then minimize the sizes of the process units.
- List the substances in every stream. You need not specify flow rates or compositions.
- Label all units and the typical temperatures in each unit.
- You may neglect pumps, heat exchangers, heaters, and coolers.

(A table of melting points and boiling points appears at the end of this exercise.)

(B) Early processes used the HCl by-product from the net reaction to hydrochlorinate acetylene ($CH{\equiv}CH$), as follows:

hydrochlorination (150°C):	$CH{\equiv}CH + HCl \rightarrow CH_2{=}CHCl.$	(rxn 4)

Reaction 4 has about 90% conversion and requires a special catalyst: the same catalyst as reaction 3. Unfortunately, about 10% of the product $CH_2{=}CHCl$ reacts with HCl in the hydrochlorination reactor.

hydrochlorination (150°C):	$CH_2{=}CHCl + HCl \rightarrow CH_2Cl{-}CH_2Cl$	(rxn 5a)
	$CH_2{=}CHCl + HCl \rightarrow CH_3{-}CHCl_2$	(rxn 5b)
pyrolysis (500°C):	$CH_3{-}CHCl_2 \rightarrow CH_2{=}CHCl + HCl$	(rxn 2d)
(see also rxns 2a–c)	$CH_3{-}CHCl_2 \rightarrow CH_2{=}CH_2 + Cl_2$	(rxn 2e)
net reaction (rxns 1–5)	$CH_2{=}CH_2 + CH{\equiv}CH + Cl_2 \rightarrow 2CH_2{=}CHCl$	

The net reaction in (B) is superior to the net reaction in (A) because there is no HCl by-product. Reactions 1 and 4 must be conducted in different reactors (different temperatures, different catalysts). Reactions 2d and 2e have about 35% conversion.

Design a process to produce CH_2=CHCl using the chemistry in parts (A) and (B). Use the guidelines and rules stated in part (A). You do not need to redraw your entire flowsheet from part (A). Integrate these new reactions into the process and redraw only the portions that change.

(C) Processes that used the chemistry in part (B) were made obsolete in the early 1950s by technology for the following reaction:

$$\text{oxychlorination } (275^\circ C): CH_2=CH_2 + 2HCl + \frac{1}{2}O_2 \rightarrow CH_2Cl-CH_2Cl + H_2O.$$

$$(\text{rxn } 6)$$

Reaction 3 also occurs in the oxychlorination reactor. The chemistry in (A) and reaction 6 yield the *net reaction:*

$$2CH_2=CH_2 + Cl_2 + \frac{1}{2}O_2 \rightarrow 2CH_2=CHCl + H_2O.$$

Reactions 1 and 6 must be conducted in separate reactors. Reaction 6 has 100% conversion.

Design a process to produce CH_2=CHCl using the chemistry in parts (A) and (C). Use the rules and guidelines stated in part (A).

(D) Technology invented in the early 1990s may render obsolete processes that use the oxychlorination reaction (rxn 6) in part (C) (see *Chemical Engineering Progress*, April 1993, p. 16). The new technology converts HCl into Cl_2 as follows:

$$\text{copper chlorination } (260^\circ C): \quad 2HCl + CuO \rightarrow CuCl_2 + H_2O, \qquad (\text{rxn } 7)$$

$$\text{copper oxidation } (350^\circ C): \quad CuCl_2 + \frac{1}{2}O_2 \rightarrow CuO + Cl_2. \qquad (\text{rxn } 8)$$

Reactions 7 and 8 have 100% conversion. The chemistry in (A) with reactions 7 and 8 yields the *net reaction:*

$$2CH_2=CH_2 + Cl_2 + \frac{1}{2}O_2 \rightarrow 2CH_2=CHCl + H_2O.$$

Design a process to produce CH_2=CHCl using the chemistry in parts (A) and (D). Use the rules and guidelines stated in (A).

(Parts (A) and (B) appeared on an exam. It was estimated that (A) and (B) could be completed in 55 minutes.)

Some properties (at 1 atm unless noted)

	Melting pt (°C)	Boiling pt (°C)
$CH_2=CH_2$	−169	−104
HCl	−115	−85
$CH\equiv CH$	−81[1]	−84[2]
Cl_2	−101	−35
$CH_2=CHCl$	−154	−13
CH_3-CH_2Cl	−136	12
CH_3-CHCl_2	−97	57
CH_2Cl-CH_2Cl	−35	83
O_2	−218	−182
H_2O	0	100
CuO	1,326	—
$CuCl_2$	620	993[3]

Notes: [1] Sublimation of solid. [2] At 1.2 atm. [3] Decomposes.

2.14

(A) Design a process to produce a nitric acid–water solution (HNO_3/H_2O) from ammonia (NH_3) and air (N_2 and O_2), using the following reaction

$$NH_3 + 2O_2 \xrightarrow[\substack{750°C}]{\text{Pt/Rh catalyst}} HNO_3 + H_2O.$$

Assume that using air is preferable to using O_2 obtained by cryogenically separating N_2 and O_2. Cryogenic separation is expensive. N_2 in the reactor yields no by-products, although it does increase the size of the reactor.

The reaction conversion is less than 100%: The effluent will contain one or more reactant, as follows.

- If a stoichiometric mixture (1 mole NH_3 + 2 moles O_2) enters the reactor, the effluent will contain both NH_3 and O_2 (and N_2, HNO_3, and H_2O).
- If the reactant mixture has excess NH_3, the effluent will contain no O_2.
- If the reactant mixture has excess O_2, the effluent will contain no NH_3.

Design guidelines:

- Assume no other chemistry than the reaction given.
- Minimize the number of process units; then minimize the sizes of the process units.
- List the substances in every stream. You need not specify flow rates or compositions, but indicate if a reactant is in excess.
- Label all units. If a unit is a separator, specify the basis for the separation (*e.g.*, liquid/solid separator) and the typical temperature in the separator.
- You may neglect heat exchangers.

(B) Design a process to produce a HNO_3/H_2O solution from methane, water, and air. That is, design a process to produce ammonia, and then use the ammonia to produce nitric acid. *Do not simply copy the ammonia process developed in Chapter 2. The ammonia process can be improved when integrated with the nitric acid synthesis. Specifically,* *you can avoid the costly cryogenic separation* of N_2 and O_2. Hint: Use excess NH_3 in the feed to the nitric acid reactor. (Note: Excess NH_3 is not necessarily best for part (A).)

Recall that ammonia is synthesized by the reaction:

$$N_2 + 3H_2 \xrightarrow[\quad 500°C \quad]{\text{Fe catalyst}} 2NH_3.$$

The conversion to ammonia is less than 100%: The effluent will contain N_2 and H_2.

Generate H_2 by the method used for the fuel cell. You may assume methane can be "burned" in water in one step:

$$CH_4 + 2H_2O \xrightarrow[\quad 450°C \quad]{\text{catalyst X}} CO_2 + 4H_2.$$

Furthermore, *you may assume* that excess H_2O will consume *all* the CH_4.

Below is a table of melting points, boiling points, and a summary of substances required, tolerated, and forbidden in the three reactors.

				Reactor	
Substance	Melting pt (°C)	Boiling pt (°C)	Methane burner	Ammonia synthesis	Nitric acid synthesis
H_2	-259	-253	product	reactant	ok
N_2	-210	-196	ok	reactant	ok
O_2	-218	-183	ok	**forbidden**	reactant
CH_4	-182	-164	reactant	ok	**forbidden**
CO_2	-56[1]	-79[2]	by-product	**forbidden**	ok
NH_3	-78	-33	**forbidden**	product	reactant
HNO_3	-42	83	**forbidden**	**forbidden**	product
H_2O	0	100	reactant	**forbidden**	by-product

Notes: [1] At 5. 2 atm. [2] Sublimation point; gas–solid transition.

Although O_2 is allowed in the methane burner, it will consume some of the product, H_2, by the following reaction:

$$2H_2 + O_2 \rightarrow 2H_2O,$$

and some of the reactant CH_4 by the reaction

$$CH_4 + 2O_2 \rightarrow CO_2 + 2H_2O.$$

Similarly, H_2 in the nitric acid synthesis reactor will be consumed by reaction with O_2. Thus it is wasteful to admit O_2 to the methane burner or H_2 to the nitric acid synthesis reactor. If you choose to do so, be sure the waste can be justified, for example, because doing so eliminates a process unit.

You need only specify the steady-state conditions in your process. You need not worry about how to start your process, which might require intermediates that are not present until the process is running. For example, for the steady-state process "chicken yields egg yields chicken yields egg, and so on" it would not be necessary to specify which came first. That is, how the process starts is not important here.

(This exercise appeared on an exam. It was estimated that part (A) could be completed in 20 minutes and part (B) could be completed in 30 minutes.)

2.15

(A) Air contaminated with SO_2 and NO_2 can be cleaned by bubbling the air through a suspension of $Mg(OH)_2$ in water. Two reactions occur in the suspension:

$$SO_2 + Mg(OH)_2 \rightarrow MgSO_3 + H_2O,$$

$$2NO_2 + Mg(OH)_2 \rightarrow Mg(NO_2)_2 + H_2O + \frac{1}{2}O_2.$$

Excess $Mg(OH)_2$ removes all the SO_2 and NO_2 from the air. The magnesium solids decompose upon heating:

$$MgSO_3 \xrightarrow{200°C} MgO + SO_2,$$

$$\frac{1}{2}O_2 + Mg(NO_2)_2 \xrightarrow{250°C} MgO + 2NO_2,$$

$$Mg(OH)_2 \xrightarrow{350°C} MgO + H_2O.$$

The second reaction above uses excess oxygen. $Mg(OH)_2$ can be regenerated by the following reaction:

$$MgO + H_2O \xrightarrow{20°C} Mg(OH)_2.$$

SO_2 and NO_2 are collected separately and later converted to H_2SO_4 and HNO_3, respectively.

Design a process to produce clean air, SO_2, and NO_2 from air polluted with SO_2 and NO_2. You may use MgO, H_2O, and air as reactants. You need not isolate pure SO_2 and NO_2; you may produce water solutions of each (to be used in parts (B) and (C) below).

	Melting pt (°C)	Boiling pt (°C)	Solubility in water
N_2	−210	−196	
O_2	−218	−183	
SO_2	−73	−10	
NO_2	−11	21	
H_2O	0	100	
$MgSO_3$	200[1]		insoluble
$Mg(NO_2)_2$	250[1]		highly soluble
$Mg(OH)_2$	350[1]		insoluble
MgO	2,852	3,600	insoluble

Note: [1] Decomposes.

(B) Convert the NO_2 (or NO_2/water solution) to a HNO_3/water solution using the following reactions:

$$3NO_2 + H_2O \xrightarrow{20°C} 2HNO_3 + NO,$$

$$NO + \frac{1}{2}O_2 \xrightarrow[\substack{\text{Pt catalyst} \\ 250°C}]{} NO_2.$$

The first reaction occurs only at significant concentrations of NO_2 (>10%). The reaction does not occur in the scrubber in part (A). Neither reaction goes to completion. You may use O_2 as a reactant in part (B). H_2O and O_2 are permitted in both reactors.

	Melting pt (°C)	Boiling pt (°C)	Solubility in water
NO	−164	−152	slightly soluble
HNO_3	−42	83	highly soluble

(C) Convert the SO_2 (or SO_2/water solution) to a H_2SO_4/water solution:

$$SO_2 + \frac{1}{2}O_2 \xrightarrow{300°C} SO_3,$$

$$SO_3 + H_2O \xrightarrow[\substack{\text{Pt catalyst} \\ 250°C}]{} H_2SO_4.$$

The first reaction does not go to completion. The second reaction uses excess water and completely consumes the SO_3. You may use O_2 as a reactant in part (C). H_2O and O_2 are permitted in both reactors.

	Melting pt (°C)	Boiling pt (°C)	Solubility in water
SO_3	−17	45	reacts with water
H_2SO_4	10	338	highly soluble

2.16 Chemicals A and X react to form a valuable chemical P:

$$X + A \rightarrow P.$$

Chemical P commands a high price because the conversion of the above reaction is low. The optimal performance of the reactor is shown below.

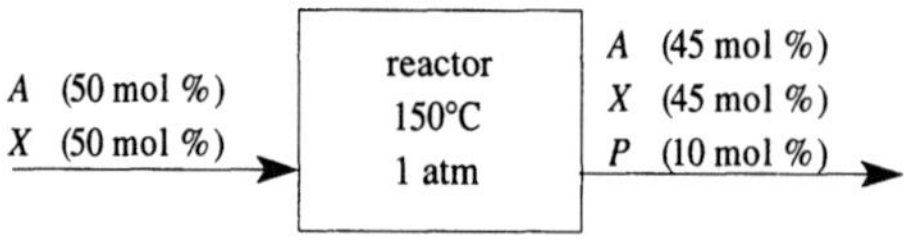

A (50 mol %)
X (50 mol %)

reactor
150°C
1 atm

A (45 mol %)
X (45 mol %)
P (10 mol %)

Chemical X is available commercially. Chemical A is available only in a mixture with chemicals B and C. Chemicals A, B, and C do not react with each other. The chemistry with chemical X is

$X + B \rightarrow$ no reaction,

$X + C \rightarrow Q$.

Moreover, Q forms readily under optimal conditions for forming P:

Finally, P does not react with any of the chemicals in this process.

Design a process to produce P from two raw materials: (1) pure X and (2) a mixture of A (30 mol%), B (5 mol%), and C (65 mol%).

· Label the function of each unit.
· Indicate *qualitatively* the chemicals in each stream.
· Indicate the temperature of each stream.
· Indicate the destination of any stream that leaves the process (for example, "vent to air," "to landfill," "to market").

Many designs are workable. Explore different options and choose your best design.

Some guidelines:

· Minimize the number of units in your process.
· Minimize waste of material and energy in the process.
· Maximize the yield of product P *and* any other marketable by-product.
· Your process should be safe and environmentally benign.

Some physical parameters at 1 atm

	Melting pt (°C)	Boiling pt (°C)	Soluble in benzene[1]?	Soluble in water?	Market value ($/kg)	Comments
A	60	235	yes	no	10	nontoxic, suitable for landfill disposal
B	32	232	no	yes	30	nontoxic, suitable for landfill disposal
C	28	135	no	no	160	nontoxic, suitable for landfill disposal
P	86	120	yes	no	400	mild irritant
Q	230	800	yes	yes	20	toxic
X	120	320	yes	no	150	flammable

Note: [1] Human toxicity of benzene is acute; irritates mucous membranes; death may follow respiratory failure.

(This exercise appeared on an exam. It was estimated that it could be completed in 35 minutes.)

2.17 Chemicals F and G react to form a valuable product P. The reaction

$F + G \rightarrow P$

goes to 90% conversion at optimal conditions on the following page.

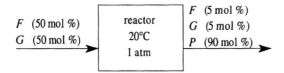

The reaction to form P occurs in a solution of F and G or in a mixture of F and G. Pure G is cheap. F is available only in a dilute aqueous solution. F is extracted from fish scales, a by-product of the halibut industry. Consequently a dilute solution of F is cheap on a basis of $/(kg F), but it contains an impurity, I. The chemical I does not react with F but reacts with G to form a less valuable by-product, B. The reaction

$$I + G \rightarrow B$$

goes to 100% conversion at the optimal conditions for producing P, as shown below.

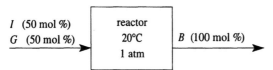

The reaction to form B occurs in a solution of I and G or in a mixture of I and G. Chemical G reacts preferentially with I over F. Chemical I reacts only with G; I does not react with F, P, B, or water. Likewise, chemical F reacts only with G; F does not react with P, I, B, or water.

Some physical parameters at 1 atm

		Melting pt (°C)	Boiling pt (°C)	Soluble in water?	Market price ($/kg)	Comments
F	reactant	35	125	yes	30[1]	nontoxic, may be disposed in landfill or waste water
G	reactant	18	112	yes	5	nontoxic, may be disposed in landfill or waste water
I	impurity	62	126	yes	10	nontoxic, may be disposed in landfill or waste water
P	product	−28	57	no	200	mild irritant
B	by-product	7	78	yes	5	nontoxic, may be disposed in landfill or waste water
H_2O	solvent	0	100		0.002	

Note: [1]Cost of 1 kg of F in 350 kg of water, with 0.2 mol% I impurity.

Design a process to produce P from two raw materials: (1) pure G and (2) a solution containing H_2O (99.7 mol%), F (0.1 mol%), and I (0.2 mol%).

· Label the function of each unit.
· Indicate *qualitatively* the chemicals in each stream. For the stream entering the reactor, indicate the approximate relative flow rates of all *reacting* chemicals.
· Indicate the temperature of each stream.
· Indicate the destination of any stream that leaves the process (for example, "vent to air," "to landfill," or "to market").

Many designs are workable. Explore different options and choose your best design. A good design is simple, optimizes the amount of product, and employs easy separations.

(This exercise appeared on an exam. It was estimated that it could be completed in 30 minutes.)

2.18 Design a process to produce G via the reaction

$$M + E \rightarrow 2G \quad \text{(Good product)}. \tag{rxn 1}$$

Reactants available: a liquid mixture of $M(50 \text{ mol\%})$ and $X(50 \text{ mol\%})$,

a liquid mixture of $E(50 \text{ mol\%})$ and $I(50 \text{ mol\%})$.

Compound I is inert. In a mixture of M, X, E, and I, there is only one side reaction,

$$X + E \rightarrow 2B \quad \text{(By-product)}. \tag{rxn 2}$$

Reactions 1 and 2 are highly exothermic. A mixture of pure M and pure E will explode. Likewise for X and E. To avoid explosion, the concentration of E must not exceed 1 mol% if M and/or X are present.

All reactants and products are soluble in water. Water is inert to all reactants and products. Reactions 1 and 2 do not go to completion at the safe reaction temperature, 20°C. That is, if M and E enter a reactor, the mol fractions in the stream leaving the reactor will be such that

$$\frac{(\text{mol\% } G)^2}{(\text{mol\% } M)(\text{mol\% } E)} = 60.$$

Two examples of reactor effluents are shown below. The first entails a stoichiometric mixture of M and E.

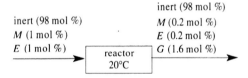

The second example shows a result when M is in excess.

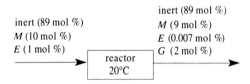

Likewise if X and E enter a reactor, the concentrations in the stream leaving the reactor will be such that

$$\frac{(\text{mol\% } B)^2}{(\text{mol\% } X)(\text{mol\% } E)} = 60.$$

Reactions 1 and 2 proceed at the same rate.

Some properties at 1 atm

		Melting pt (°C)	Boiling pt (°C)	Market value ($/mol)	Comments
M	reactant	2	78	50	nontoxic
X	reactive impurity	−40	42	50	mild irritant
E	reactant	−28	47	100	nontoxic
I	inert impurity	−30	47	2	nontoxic
G	product	−54	19	300	nontoxic
B	by-product	10	92	10	nontoxic
water	solvent	0	100	0.00001	

- Label the function of each unit.
- Indicate *approximate* compositions for each stream. Estimates with accuracy 10 mol%, 1 mol%, 0.1 mol% are sufficient.
- Indicate the temperature of all separations.
- Indicate the destination of any stream that leaves the process (for example, "vent to air," "to disposal," "to market").

A good design is simple, optimizes the amount of product, and employs easy separations. Many designs will work. Explore different options and choose your best design.

(This exercise appeared on an exam. It was estimated that it could be completed in 30 minutes.)

Problem Redefinition

Exercises 2.19 through 2.30 describe scenarios in which the problem is poorly defined. Propose an improved definition of the problem for each scenario. Here are three example scenarios.

Scenario 1: Two hikers are confronted by a mountain lion.

PROBLEM: Devise a plan to prevent the lion from eating the hikers.

One hiker quickly discards her pack, removes her boots, and puts on her running shoes. "What are you doing?" asks the second hiker. "You can't outrun a mountain lion!" The first hiker responds, "I don't have to outrun the lion – I only have to outrun you."

REAL PROBLEM: Devise a way to avoid being eaten.

(Adapted from Fogler, H. S. and LeBlanc, S. E. 1995. *Strategies for Creative Problem Solving*, Prentice Hall, Upper Saddle River, NJ, attributed to Professor J. Falconer, Department of Chemical Engineering, University of Colorado, Boulder.)

Scenario 2: An issue arose during the development of the manned space program in the early 1960s. A space capsule reentering the Earth's atmosphere is heated by air friction to thousands of degrees. This is hazardous to the capsule contents, such as its human passengers.

PROBLEM: Find a material that can withstand the extreme heat of reentry.

This problem was eventually solved, but not until twenty years of manned space flight and the construction of the space shuttle. This problem could not be solved with 1960s technology.

REAL PROBLEM: Protect the astronauts.

One scientist considered how meteorites survive entry into the Earth's atmosphere – the meteorite surface is ablated by the heat, which protects the core. The same was done for the early space capsules – the entry side of the capsule was covered with an ablative material that absorbed the heat and vaporized.

(Adapted from Fogler, H. S. and LeBlanc, S. E. 1995. *Strategies for Creative Problem Solving*, Prentice Hall, Upper Saddle River, NJ.)

Scenario 3: The waste water from your chemical production facility contains a infinitesimal amount of dioxin (2,3,7,8-tetrachlorodibenzo-p-dioxane or TCDD). Although the amount of dioxin was below the allowable level, the allowable level was decreased and your facility is now in violation.

Dioxin: teratogenic and highly toxic.

PROBLEM: Decrease the amount of dioxin in the waste water to an allowable level.

Dioxin can be extracted by treating the waste water with charcoal. However, this process is prohibitively expensive and decreases the dioxin level to only slightly below the allowable level.

REAL PROBLEM: Alter the process so dioxin is not a by-product.

The dioxin was traced to a solvent that was reacting to form dioxin. The solvent was replaced with an equally effective solvent and no dioxin was produced. This solution is superior to treating the waste water. The solution also represents a canon in the chemical process industries: Design for "zero" emissions, rather than design to comply with regulations. In this way you anticipate new regulations. Instead of working to remedy environmental hazards, don't contaminate in the first place.

2.19   A shooting range adjacent to the San Francisco bay placed the targets at the water's edge so that stray bullets fell harmlessly into the bay. Normally shooting ranges recover spent bullets as metal scrap. However, recovery was impractical at the bay because the bullets sank into the muddy bottom, estimated to be 3 m of silt below 10 m of water. The lead content in the water of the bay was recently found to be abnormally high. Initially a nearby chemical plant was blamed, but close examination revealed that the plant effluent contained no lead and that no process at the plant ever involved lead.

PROBLEM: Devise a means to remove the bullets from the bay.

2.20   A pneumatic system is used to transport powdered solids from a storage bin to a shipping container, as on the following page.

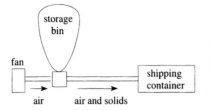

The pipe conducting the air and powdered solids plugs frequently.

PROBLEM: Devise a means to unplug the pipe.

(Adapted from Fogler, H. S. and LeBlanc, S. E. 1995. *Strategies for Creative Problem Solving*, Prentice Hall, Upper Saddle River, NJ.)

2.21 You construct your new restaurant on a building lot in a completely developed commercial zone. After a few months of operation the restaurant is still not showing a profit. The cost of paying off the new kitchen equipment and the salaries of the kitchen employees exceeds the dining revenues. The chief reason is that not enough customers are dining at your restaurant. The customers are pleased with the food, service, and price but complain that they have trouble finding a place to park; the restaurant parking lot is full every evening, although the dining room has empty tables. A quick study reveals that all the cars in the parking lot belong to your customers; there are no illegal parkers taking up precious spaces.

PROBLEM: Devise a plan to accommodate more cars in the parking lot.

2.22 A hotel remodels its upper floors to convert its penthouse suites into dozens of small rooms. Although the daily rate on the smaller rooms is less, there are many more guests, which increases the hotel's revenues. However, the guests on the remodeled upper floors complain that they must wait a seemingly interminable time for an elevator. The elevators are operating at peak efficiency. It's just that the number of elevators originally installed was based on a lower occupancy of the upper floors.

PROBLEM: Devise a plan to add a new elevator to the hotel.

Ultimately the *real* problem was solved without adding a new elevator.

(Adapted from Fogler, H. S. and LeBlanc, S. E. 1995. *Strategies for Creative Problem Solving*, Prentice Hall, Upper Saddle River, NJ.)

2.23 An airline sought to improve service by initiating a policy of docking incoming flights at gates close to the baggage claim. This caused two effects: Travelers arrived in the baggage claim area much faster and travelers complained about delays in receiving baggage. Before the new policy there were seldom complaints about baggage delay or any aspects of arrival.

(*The Washington Post*, A3, December 14, 1992.)

PROBLEM: Devise a means to deliver baggage to travelers more rapidly.

(Adapted from Fogler, H. S. and LeBlanc, S. E. 1995. *Strategies for Creative Problem Solving*, Prentice Hall, Upper Saddle River, NJ.)

2.24 A stream of air contains chlorosilanes and other impurities, which must be removed before the air is released. As shown in the diagram below, the stream is first passed through a condenser, which removes the chlorosilanes. The remaining impurities are removed with a scrubber, in which air is bubbled through water. The water effluent from the scrubber was frequently plugged by a highly viscous liquid. It was determined that the viscous liquid resulted from the reaction of water and chlorosilanes.

> **PROBLEM:** Modify the scrubber to accommodate the chlorosilanes that escape the condenser.

(From Elizabeth Lim, Cornell Class of '94.)

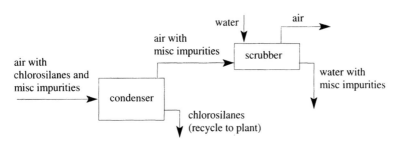

2.25 One of the dozens of employees that you supervise is best described as a "loner" but otherwise performs his tasks expertly. Lately he carries a briefcase with him *everywhere*, although no facet of his job requires items that might be kept in a briefcase. The other employees are reacting badly to the briefcase and have begun to complain.

> **PROBLEM:** Devise a means to eliminate the offending briefcase.

2.26 One step in a process to manufacture integrated circuits on silicon wafers (computer chips) involves etching the surface with a plasma. The metal layer on the silicon wafer is coated with a polymer template, such that portions of the metal are exposed. A plasma above the silicon wafer selectively etches the exposed metal. When the polymer is dissolved away, a pattern of the protected metal remains. This metal pattern connects electronic devices on the silicon surface, such as transistors and diodes.

   The particle concentration in the etching chamber is monitored optically because it is crucial that no particles deposit on the silicon wafer. The particle concentration is measured by a laser beam passed through the chamber to a detector. If a particle crosses the beam, the detector sends a signal. The signals from the detectors in several chambers are collected and analyzed by a computer, which displays the particle concentrations in each chamber. The software is updated frequently. Shortly after an update, the displays of particle levels in each chamber show no signals. The attending engineer, a Cornell co-op student, was given the following problem.

> **PROBLEM:** Find the error in the software.

(From Tom Sequist, Cornell Class of '95.)

2.27 Titanium dioxide is a common whitener that appears in foods, paint, paper, and health care products such as toothpaste. A process that manufactures titanium dioxide yields an effluent of dirty water. The dirty water is treated in a clarifier, which allows the solid particles to flocculate and settle to the bottom. The clarifier was plagued by a strong odor of chlorine. An engineer assigned to the process was assigned the following problem.

PROBLEM: Prevent chlorine from leaving the clarifier.

(From Vivian Tso, Cornell Class of '95.)

2.28 Chemical plants are constructed on large slabs of concrete that typically cover dozens of acres. During normal operation oil and other liquids leak onto the concrete and accumulate in various ditches. When it rains, the water washes the oil off the concrete. The runoff from the plant is therefore polluted and cannot be released to a river or lake.

PROBLEM: Construct a holding tank to catch the water so it can later be treated and then released. Hint: What volume of holding tank is needed to collect rain at the rate of 1 inch per hour, falling for 8 hours on a plant that covers 100 acres?

(From Alfred Center, Cornell Class of '64, '65.)

2.29 Your chemical plant ships product via a fleet of tankers. Your company has leased a fleet of tankers with capacity of 50,000 barrels each. These are modest tankers, neither large nor small. Your wharf has three loading docks. Each day you must ship 150,000 barrels. A tanker must be in port one day. Much of the day in port is spent docking, connecting to the transfer lines, and disconnecting from the transfer lines. Under normal conditions, the system works well – your wharf accommodates three tankers per day, each has a capacity of 50,000 barrels, and so you ship 150,000 barrels every day.

But sometimes bad weather or shipping schedules do not allow you to have three ships in port. The shipping falls behind on those days. The deficit can never be made up because your maximum shipping rate is the shipping rate on a normal day.

PROBLEM: Enlarge the wharf to accommodate five tankers.

(From Alfred Center, Cornell Class of '64, '65.)

2.30 In a large chemical process, energy to run the units comes from steam. (Steam is preferable to electricity because it is less likely to spark an explosion.) Your plant has several steam-generating units in a location at the periphery. Most of the units burn natural gas (methane) but one burns a high-sulfur oil. Whereas there are negligible emissions of SO_x from the methane-burning units, the emissions of SO_x from the stack of the oil-burning unit are too high.

PROBLEM: Install a scrubber to remove the SO_x from the offending stack.

(From Alfred Center, Cornell Class of '64, '65.)

3

Models Derived from Laws and Mathematical Analysis

HAVE YOU EVER MARVELED at the beauty of an exquisite design? How did the engineers devise such a complex device or system? Surely they must have been geniuses. Although this may be true in some cases, good engineering generally does not require genius (although it certainly helps). Complex designs usually evolve from simple designs. It is important to be able to analyze and evaluate the effects of each evolutionary step in a design.

How does a good design begin? Linus Pauling, who earned his B.S. degree in chemical engineering (Oregon State, 1922), and received a Nobel Prize in Chemistry (1954) and a Nobel Peace Prize (1962), recommended "The best way to get a good idea is to get a lot of ideas." We modify and extend this useful advice to "The best way to get a good design is to create many designs and choose the best." To choose the best design, one must be able to *analyze and evaluate* the designs.

How does one analyze and evaluate a design? One option is to build the system and test it. This is appropriate when the design is for a small item, such as a can opener. This approach is also appropriate for a medium-sized item, such as an automobile, if one expects to manufacture thousands of units. But clearly it is impractical to build a full-sized system when the system is very large and only a few, perhaps only one, will be built. Most chemical processes are too large and too expensive to allow one to build a full-sized model. Chemical engineers must use other means to analyze and evaluate a design. We introduce three methods in this text: mathematical modeling (this chapter), graphical analysis of empirical data (Chapter 4), and dimensional analysis/dynamic scaling (Chapter 5).

Dimensional analysis and dynamic scaling allow one to predict, for example, the performance of a full-sized system based on the performance of a small prototype of the system. One constructs a model of the system, tests the model under various operating conditions, and then applies dynamic scaling to predict how the actual system would behave, with the operating conditions scaled as well.

Rather than build a physical model of the system, one can construct a mathematical model. Rather than translate the design to a physical device, one can translate the design to equations. The equations can then be used to predict the performance of the system under various operating conditions. To translate from the description of the system (the design) to the mathematical model (the equations) one invokes various laws of nature, such as Newton's second law, $F = ma$. If our design involved a device exerting a force on a mass, we could predict the acceleration of the mass. And with a little calculus, we could predict the velocity and position of the mass, given the duration of the force (an operating condition in this case). We knew qualitatively that the force would accelerate the mass. Our model would provide a quantitative description, necessary to evaluate various alternatives. As eloquently stated by Richard Alkire (University of Illinois), who was recognized with the Professional Progress Award of the AIChE in 1985, "A mathematical model articulates a qualitative description." A key component of chemical engineering is developing mathematical descriptions of unit operations and processes.

The third alternative, graphical analysis of empirical data, is used when the complexity of the system prohibits mathematical analysis. This alternative is usually simpler than building a prototype.

In this chapter we will introduce mathematical modeling in the context of desalinating seawater.

3.1 Desalination of Seawater

Desalination of seawater to water suitable for human consumption and/or agricultural uses is a vital operation in many parts of the world. Table 3.1 gives the typical salt content of seawater and the maximum salt content of drinking water. The chief component of sea salt is NaCl; seawater contains 1.1 wt% Na and 1.9 wt% Cl.

Table 3.1. The composition of seawater

	H_2O (wt%)	Salt (wt%)
seawater	96.5	3.5
potable water	>99.95	<0.05

Consider an ideal process to convert seawater to pure water, diagrammed in Figure 3.1.

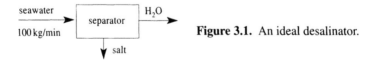

Figure 3.1. An ideal desalinator.

What are the flow rates of the water and salt streams that leave the ideal desalinator? The composition given in Table 3.1 tells us that the 100 kg of seawater that enters the process each minute contains 96.5 kg of water and 3.5 kg of salt. By inspection,

96.5 kg/min of water and 3.5 kg/min of salt leave the separator. How did you determine these flow rates? Probably your intuition told you that everything that enters the separator must leave the separator.

Consider a more realistic separator, shown in Figure 3.2.

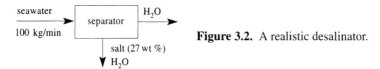

Figure 3.2. A realistic desalinator.

What are the flow rates of the water and salt streams that leave the realistic desalinator? The answers are less intuitive. We need to calculate. As such, we need equations to describe the separator. To obtain these equations we apply a physical law to the separator.

3.2 The Conservation Principles

We now consider the first class of Great Physical Laws – the Conservation Principles:

Certain quantities are invariant – their amount is constant.

The following quantities are conserved in any process: mass,[1] energy, linear momentum, angular momentum, electric charge, baryon number, and strangeness.

To model the separator we will apply the conservation of mass. Later we will appeal to the conservation of energy. As chemical engineers you will also use the conservation of linear momentum, angular momentum, and electric charge. Most likely you will never be concerned with the conservation of baryon number and strangeness, quantities encountered in particle physics.

The key to applying a conservation principle is careful accounting. Define a system by enclosing a volume of space with borders (Figure 3.3).

system   surroundings   **Figure 3.3.** A system, its borders, and its surroundings.

If nothing crosses the borders, conserved quantities are constant within the system. This conservation principle is translated into mathematics in Eqs. (3.1) and (3.2). This is a *closed system*. We can model a closed system, which is usually a trivial system, with Eqs. (3.1) and (3.2):

$$mass\ in\ system\ at\ time\ 1 = mass\ in\ system\ at\ time\ 2, \tag{3.1}$$

$$mass(t_1) = mass(t_2). \tag{3.2}$$

---

[1] Strictly speaking, mass is not conserved; it can be converted to energy. We will ignore nuclear processes in this text.

But our separator is not a closed system. Things enter and exit the system, crossing the borders (see Figure 3.4).

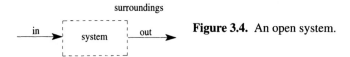

surroundings

Figure 3.4. An open system.

For open systems, the conservation of mass yields

$$mass(t_2) = mass(t_1) + (mass\ entering\ from\ t_1\ until\ t_2)$$
$$- (mass\ exiting\ from\ t_1\ until\ t_2). \qquad (3.3)$$

Assume that the rate of mass flow into and out of the system is constant over the epoch t_1 to t_2. The amount of mass entering is the flow rate times the time evolved:

$$mass(t_2) = mass(t_1) + (rate\ of\ mass\ flow\ in)(t_2 - t_1)$$
$$- (rate\ of\ mass\ flow\ out)(t_2 - t_1). \qquad (3.4)$$

This expression can be rearranged as follows:

$$\frac{mass(t_2) - mass(t_1)}{t_2 - t_1} = rate\ of\ mass\ flow\ in - rate\ of\ mass\ flow\ out. \qquad (3.5)$$

We now take the limit of Eq. (3.5) as the time between t_1 and t_2 becomes infinitely short, that is, in the limit that $t_2 - t_1 \rightarrow 0$:

$$\frac{d(mass)}{dt} = rate\ of\ mass\ flow\ in - rate\ of\ mass\ flow\ out. \qquad (3.6)$$

The assumption that the flow rates are constant is no longer restrictive since all flow rates are constant over an infinitely short time period. Equation (3.6) is commonly written as

$$rate\ of\ accumulation = rate\ in - rate\ out. \qquad (3.7)$$

A negative rate of accumulation means mass is being depleted from the system.

We now assume that there is no accumulation in the system. That is, the system is assumed to be at *steady state*. For systems at steady state, the conservation of mass translates into the following:

$$0 = rate\ in - rate\ out, \qquad (3.8)$$

$$rate\ in = rate\ out. \qquad (3.9)$$

We assume steady-state operation for two reasons: efficient operation and simplified analysis. Steady-state processes are easier to control and do not require vessels to accumulate materials between process units. Steady-state processes also yield products with uniform quality. Assuming steady state simplifies the analysis because steady-state processes are described by algebraic equations. Systems not at steady state, called *transient systems*, require calculus. We will study transient systems in Chapter 6, by which time you will have learned differentiation and integration in your calculus class.

3.3 Modeling a Desalinator

We now return to the realistic desalinator in Figure 3.2 and apply the conservation of mass. We begin by defining our system. We draw system borders around the separator, a trivial step with this system. You will be tempted to omit this step. Don't. Recall that applying conservation laws requires careful accounting, which begins with drawing the system borders (Figure 3.5).

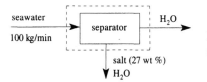

Figure 3.5. System borders for mass balances on a desalinator.

Identifying the streams with narration such as "the pure-water stream exiting the separator to the right" is impractical. It is useful to label the streams. Here we will use numbers (Figure 3.6).

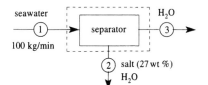

Figure 3.6. System borders and stream labels for mass balances on a desalinator.

It is useful to define some variables to represent quantities such as "the flow rate of salt in stream 1." First, we assign letters for the principal components. Let water $\equiv W$, salt $\equiv S$, and total $\equiv T$. It is prudent to choose obvious and mnemonic letters, such as W for water, so that when you see "W" in an equation you can immediately translate to "water." Labeling water as some obtuse Greek character such as ζ will complicate your analysis unnecessarily. Finally, we will use the nomenclature that $F_{i,n}$ represents the flow rate of component i in stream n. Thus, "the flow rate of salt in stream 1" translates to $F_{S,1}$.

We now translate the design into mathematics using the conservation of mass given in Eq. (3.9). We start with "The total flow rate of stream 1 entering equals the total flow rate of stream 2 exiting plus the total flow rate of stream 3 exiting." With our nomenclature, this translates to

$$\text{total:} \quad F_{T,1} = F_{T,2} + F_{T,3}. \tag{3.10}$$

Similarly we can write mass balances on water and salt:

water: $F_{W,1} = F_{W,2} + F_{W,3}$, $\qquad$ (3.11)

salt: $\quad F_{S,1} = F_{S,2} + F_{S,3}$. $\qquad$ (3.12)

Equations (3.10), (3.11), and (3.12) are related because the sum of the individual components equals the total. The total flow rate of a stream is the sum of the component flow rates. For each stream we can write

stream 1: $\quad F_{T,1} = F_{W,1} + F_{S,1}$, $\qquad$ (3.13)

stream 2: $\quad F_{T,2} = F_{W,2} + F_{S,2}$, $\qquad$ (3.14)

stream 3: $\quad F_{T,3} = F_{W,3} + F_{S,3}$. $\qquad$ (3.15)

In other words, mass is an extensive property. This is not true for the total temperature, an intensive property; $T_T \neq T_S + T_W$.

Another clue that Eqs. (3.10)–(3.15) are not independent is that we now have six equations for six unknowns. If the equations were independent *we could solve the problem with no further information*. We would not need the inlet flow rate or the composition of seawater, for example. Obviously this is not true. The output must depend on the input.

Let's translate the information given into the nomenclature for this model. The total flow rate entering via stream 1 is 100 kg/min. Thus

$$F_{T,1} = 100. \, \text{kg/min}. \qquad (3.16)$$

We are given the compositions of all the streams. Stream 1 is 96.5 wt% water. Translating to our nomenclature we get

$$F_{W,1} = 0.965 F_{T,1} = (0.965)100. = 96.5 \, \text{kg/min}. \qquad (3.17)$$

Similarly,

$$F_{S,1} = 0.035 F_{T,1} = (0.035)100. = 3.5 \, \text{kg/min}. \qquad (3.18)$$

The flow rate of salt in stream 1 could have been obtained alternatively from Eq. (3.13). Stream 3 is 100% water, so

$$F_{W,3} = F_{T,3} \quad \text{and} \quad F_{S,3} = 0. \qquad (3.19)$$

And finally, stream 2 is 27 wt% salt:

$$F_{S,2} = 0.27 F_{T,2}. \qquad (3.20)$$

The system of equations presented here could be solved simultaneously with methods of linear algebra. However, methods of successive substitution will be sufficient for the applications you will encounter in this text. Indeed, successive substitution is a useful means of guiding one's thinking through derivations of this type. We seek the flow rate of pure water. Because the total flow out must be 100 kg/min, if we knew the flow out of stream 2 we would know the flow rate of stream 3. Because we know the composition of stream 2, if we knew the salt flow rate we would know the total flow rate. But salt leaves the separator only by stream 2, so we do know the flow rate of salt in stream 2. Let's follow this logic backward from the known to the desired. Start with the salt balance, Eq. (3.12), and substitute Eqs. (3.18) and (3.19). In summary, we started at the objective and traced backward to something we knew:

To find $F_{W,3}$, we could use Eq. (3.11). We know $F_{W,1}$, but we need $F_{W,2}$.
To find $F_{W,2}$, we could use Eq. (3.14), but we need $F_{T,2}$ and Eq. (3.20).
To find $F_{T,2}$, we could use Eq. (3.20), but we need $F_{S,2}$.

To find the salt flow in stream 2 use

$$\text{salt:} \quad F_{S,1} = F_{S,2} + F_{S,3}, \tag{3.12}$$

$$F_{S,2} = F_{S,1} - F_{S,3} = 3.5 - 0.0 = 3.5 \text{ kg/min.} \tag{3.21}$$

From the salt flow in stream 2 we can use Eq. (3.20) to obtain the total flow rate of stream 2. Rearrange Eq. (3.20) to solve for $F_{T,2}$ and substitute from Eq. (3.21) to get

$$F_{T,2} = F_{S,2}/0.27 = 3.5/0.27 = 13. \text{ kg/min.} \tag{3.22}$$

Finally, we substitute for $F_{T,2}$ in the total mass balance, Eq. (3.10),

$$\text{total:} \quad F_{T,1} = F_{T,2} + F_{T,3}, \tag{3.10}$$

$$100. = 13. + F_{T,3}, \tag{3.23}$$

$$F_{T,3} = 87. \text{ kg/min.} \tag{3.24}$$

We have our answers: 87 kg/min of pure water and 13 kg/min of salty water. But we are not done until we check the answer. First, a qualitative check – does 87 kg/min seem reasonable? Sure – anything between 0.0 and 96.5 kg/min is reasonable. How much water leaves via stream 2? $96.5 - 87.0 = 9.5$ kg/min. Is this consistent with the composition of stream 2? $9.5/13 = 0.73$, which is 27% salt. Our answers check!

Compare the realistic separator to the ideal separator. As the weight fraction of salt in stream 2 decreases, or as the separator works less ideally, the output of pure water decreases. In these two examples, the salt content decreased from 100% to 27% and the output of pure water decreased from 96.5 to 87 kg/min. Often one wishes to predict the performance of a process given different operating conditions. In this case, one might wish to know the output of pure water given the salt content of stream 2, for example. Rather than repeat the above process of successive substitution for a different operating condition, it would be useful to have an equation that yields the flow rate of pure water given the salt content of stream 2.

If one defines the weight fraction of salt in stream 2 to be σ ("sigma" is the Greek symbol[2] for "s" and "s" is a good mnemonic for "salt"),

$$\sigma = \frac{F_{S,2}}{F_{T,2}},$$
(3.25)

one arrives at the following equation:

$$F_{W,3} = F_{T,1}\left[1 - \frac{0.035}{\sigma}\right].$$
(3.26)

Using Eq. (3.26) one obtains σ by successive substitution following the same strategy used to calculate the specific result for $\sigma = 0.27$. We recommend that you verify you can derive Eq. (3.26). Again, we are not finished until we check the result. We substitute $\sigma = 0.27$ into Eq. (3.26) and calculate $F_{W,3} = 87.$ kg/min, which agrees with the previous result.[3] It is wise to test the equation at a few other values for σ. What other values should yield obvious results? When stream 2 is entirely salt, $\sigma = 1$, all the water entering in stream 1 (96.5 kg/min) must exit via stream 3. When one substitutes $\sigma = 1$ into Eq. (3.26) one calculates $F_{W,3} = 96.5$ kg/min, as predicted. One final check – what is the minimum value for σ? If no pure water is produced, then stream 1 = stream 2, and $\sigma = 0.035$, and stream 3 should have no flow. Substituting $\sigma = 0.035$ again yields the expected result, $F_{W,3} = 0$. Our answer checks.

We have analyzed a simple desalinator by translating the physical description to a mathematical description. The basis for our analysis was a concept from physics – the conservation of mass.

3.4   A Better Desalinator

What are our options for producing potable water from salt water? Perhaps the most obvious design is an evaporator. The process separates vapor from liquid, which is easy and efficient. And because water boils at a temperature much lower than salt sublimes, the vapor phase will be essentially pure H_2O. However, evaporation may require too much energy to be cost effective. Even if one makes liberal use of heat exchangers and condensers to reclaim the heat from the steam and the hot brine, the process still may be too expensive – although the energy costs will be reduced, the

---

[2] Greek symbols are common in mathematical modeling, especially in chemical engineering modeling. Experience shows that Greek symbols cause anxiety for some first-year engineers. Mathematical manipulations easily done with x and y become scary with η and θ. But avoiding Greek symbols in this text would only postpone the inevitable. Recognize this potential source of trepidation and deal with it.

[3] Actually we calculate $F_{W,3} = 87.037037$, but we retain only two significant figures. If it is not clear why only two figures are significant, you should study Appendix C.

cost of the equipment may be too high. Alternatively, one could use an inexpensive source of energy to evaporate the water, such as solar energy. Or, one can avoid the energy-intensive step of evaporating water. One could use membranes that allow water to pass but exclude salt. Solar heating and membrane separation, however, are low-yield processes and are also expensive owing to the cost of the equipment. Let's consider another option – freezing. Ice formed in salt water contains negligible salt. And the energy required to freeze water (334 kJ/kg) is substantially less than required to evaporate water (2,260 kJ/kg).

Consider the design shown in Figure 3.7 to produce pure water from seawater by freezing.

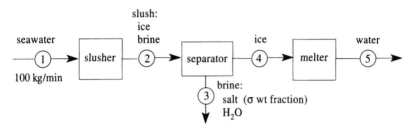

Figure 3.7. A desalination scheme based on freezing.

Our initial design divides the process into the three unit operations: cooling the seawater to form an ice/brine[4] slush, separating the ice from the brine, and melting the ice to yield water. In practice the process is not as simple. What have we trivialized? Have you ever tried to separate ice from a water solution? The ice retains water. Solids will retain liquids in most solid/liquid separations, unless special efforts are made. Therefore, in contrast to the vapor/liquid separator used in the process that evaporated the water, we must allow for the nonideality of the ice/brine separator. Assume that the slush formed in our process retains 1.0 wt% liquid. Our process is now as depicted in Figure 3.8.

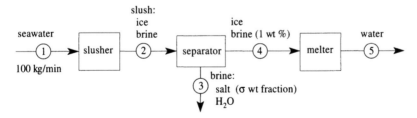

Figure 3.8. A more realistic desalination scheme based on freezing.

Because the operation of the separator is crucial to the performance of the desalinator its mechanism bears examination. As diagrammed in Figure 3.9, our separator

---

[4] In this text "brine" will imply water with dissolved salt, not necessarily the same concentration as seawater.

allows the ice to rise to the top of the vessel and a skimmer pushes the top layer out a chute. What is the rate of production of desalinated water with this process? Our freezing process will have some salt in the desalinated water. Recall that potable water may have no more than 0.05 wt% salt. How much salt will be in the water produced by this process? Is this process capable of producing potable water?

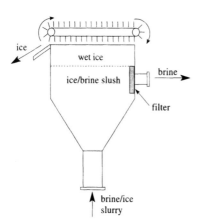

Figure 3.9. The mechanism of an ice/brine separator. A piston of ice/brine is pushed upward from the bottom and brine is filtered off a side. A belt with paddles pushes wet ice out a chute on the other side. Adapted from Rudd, D. R., Powers, G. J., and Siirola, J. J. 1973. *Process Synthesis*, Prentice Hall, Upper Saddle River, NJ.

Let's estimate the salt content in stream 5. Again, working backward from the objective, to know the concentration of salt in stream 5, we need to know the salt content of the brine retained by the ice in stream 4. The minimum salt content of the retained brine is that of the input, 3.5 wt%. The brine is 1 wt% of stream 4, so the salt is about $3.5/100 = 0.035$ wt%, which is acceptable. However, the salt content of the brine equals that of seawater only when no ice is formed, thus only when *all* the water leaves by stream 3. Hence there is no water flowing out stream 5. By similar logic, the salt content in stream 5 would be about 0.05 wt% when the salt content in stream 3 is 5 wt%. What is the flow rate of desalinated water? Or, to pose a more general question: What is the flow rate of desalinated water, given the salt content of the brine? We need an operating equation that expresses the flow rate of desalinated water in terms of σ, the salt content of stream 3, analogous to the operating equation for the evaporation process, Eq. (3.26).

We use the same nomenclature as before; water $\equiv W$, salt $\equiv S$, and total amount $\equiv T$. But now we have an additional component, ice. Do we need to define ice = I? We could, but that would require a source for the ice, such as the physical reaction

$$water \rightarrow ice. \tag{3.27}$$

It is not necessary to label ice as a pure component. Instead we will lump the ice and water into W; thus $H_2O = W =$ water plus ice. Likewise we do not need to create a label for brine.

Because we seek information only on streams leaving the process, streams 3 and 5, perhaps it is not necessary to calculate the details of the intermediate streams. We can avoid the details of the intermediate streams by applying mass balances to the entire process, which we start by drawing system boundaries that enclose all three unit operations (Figure 3.10).

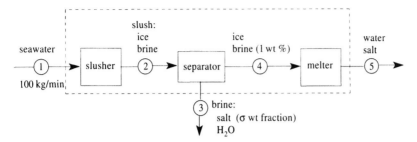

Figure 3.10. System borders enclosing entire freezer–desalinator.

Write mass balances on salt and water:

salt: $F_{S,1} = F_{S,3} + F_{S,5}$,   (3.28)

water: $F_{W,1} = F_{W,3} + F_{W,5}$.   (3.29)

Recall that writing a total mass balance provides no additional information. The total mass balance,

total: $F_{T,1} = F_{T,3} + F_{T,5}$,   (3.30)

can be obtained from the sum of Eqs. (3.28) and (3.29) and the definition of the total, $F_{T,i} = F_{W,i} + F_{S,i}$. As in the first design, the input flow rates of salt and water are known from the total flow rate and the composition of seawater,

$$F_{W,1} = 0.965 F_{T,1} = (0.965)100. = 96.5 \text{ kg/min},$$ (3.17)

$$F_{S,1} = 0.035 F_{T,1} = (0.035)100. = 3.5 \text{ kg/min}.$$ (3.18)

We are also given that the weight fraction of salt in the brine (stream 3) is σ. We translate this into mathematics as

$$\sigma = \frac{F_{S,3}}{F_{W,3} + F_{S,3}}.$$ (3.31)

This is all we know about streams 1, 3, and 5. Do we have enough information? We have six unknowns: $F_{W,i}$ and $F_{S,i}$ for $i = 1, 3, 5$. But we have only five equations: (3.28), (3.29), (3.17), (3.18), and (3.31). We need more information. What other facts do we have? We are given that the ice in stream 4 retains 1 wt% brine. We will assume that the retained brine has the same composition as the brine in stream 3. Consequently, we must delve into the details of the process. We will calculate the composition of stream 4, from which we can calculate the composition of stream 5. We redraw the system borders (Figure 3.11)

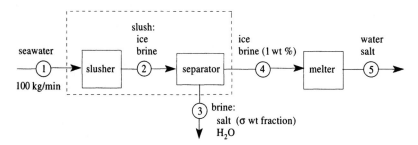

Figure 3.11. System borders enclosing slusher and separator of freezer–desalinator.

and write mass balances on salt and water, analogous to Eqs. (3.28) and (3.29):

salt: $F_{S,1} = F_{S,3} + F_{S,4}$,   (3.32)

water: $F_{W,1} = F_{W,3} + F_{W,4}$.   (3.33)

We need to translate into mathematics the information that stream 4 is 1% brine, which contains σ wt fraction salt. That is,

retained brine in stream $4 = 0.01$ (*stream 4*) $= 0.01(F_{W,4} + F_{S,4})$.   (3.34)

And

σ(*retained brine in stream* 4) $=$ (*salt in stream* 4) $= F_{S,4}$.   (3.35)

Substituting Eq. (3.34) into (3.35) yields

$\sigma 0.01(F_{W,4} + F_{S,4}) = F_{S,4}$.   (3.36)

We have six unknowns: $F_{W,i}$ and $F_{S,i}$ for $i = 1, 3, 4$ and six independent equations: (3.17), (3.18), (3.31), (3.32), (3.33), and (3.36). We are assured the problem has a unique solution. We proceed by successive substitution, which eliminates an unknown with each substitution. Substitute Eqs. (3.17) and (3.18) into Eqs. (3.32) and (3.33) respectively, and solve for $F_{W,4}$ and $F_{S,4}$.

salt: $F_{S,3} = 3.5 - F_{S,4}$,   (3.37)

water: $F_{W,3} = 96.5 - F_{W,4}$.   (3.38)

Substitute Eqs. (3.37) and (3.38) into Eq. (3.31) to get

$$\sigma = \frac{F_{S,3}}{F_{W,3} + F_{S,3}} = \frac{3.5 - F_{S,4}}{96.5 - F_{W,4} + 3.5 - F_{S,4}} = \frac{3.5 - F_{S,4}}{100. - F_{W,4} - F_{S,4}}. \qquad (3.39)$$

Equations (3.39) and (3.36) have reduced the problem to two equations and two unknowns. We may now solve either equation for either of the unknowns, and then substitute into the other equation. Let's solve Eq. (3.36) for the salt in stream 4:

$$\sigma 0.01(F_{W,4} + F_{S,4}) = F_{S,4}, \tag{3.36}$$

$$F_{W,4} + F_{S,4} = \frac{100 F_{S,4}}{\sigma}, \tag{3.40}$$

$$F_{S,4} = \frac{\sigma}{100 - \sigma} F_{W,4}. \tag{3.41}$$

Substitute Eq. (3.41) into (3.39) and solve for the flow rate of water in stream 4:

$$\sigma = \frac{3.5 - \left(\frac{\sigma}{100-\sigma}\right) F_{W,4}}{100 - F_{W,4} - \left(\frac{\sigma}{100-\sigma}\right) F_{W,4}}, \tag{3.42}$$

$$F_{W,4} = -1.01\sigma + 101.05 - \frac{3.535}{\sigma}. \tag{3.43}$$

Finally, we derive an expression for the salt content of stream 4, which we will define as ε (epsilon), the Greek symbol commonly used for small quantities:

$$salt\ content\ of\ stream\ 4 \equiv \varepsilon = \frac{F_{S,4}}{F_{W,4} + F_{S,4}}. \tag{3.44}$$

Use Eq. (3.41) to substitute for $F_{S,4}$ and obtain

$$\varepsilon = \frac{\left(\frac{\sigma}{100-\sigma}\right) F_{W,4}}{F_{W,4} + \left(\frac{\sigma}{100-\sigma}\right) F_{W,4}} = \frac{\sigma}{100}. \tag{3.45}$$

Of course, the preceding derivation is not the only way to obtain these operating equations. You are encouraged to try your own strategy.

Again, we must check the operating equations, (3.43) and (3.45). Equation (3.45) is as we reasoned in our qualitative analysis of the desalinator; because the ice retains 1 wt% brine, the salt content of the melted ice is 0.01 times that of the brine. Equation (3.43) is more difficult to check because the water flow rate was less obvious. We reasoned that the water flow rate in stream 4 should be zero when the brine concentration equaled that of the seawater ($\sigma = 0.035$). When $\sigma = 0.035$ is substituted into Eq. (3.43), one calculates $F_{W,4} = 0$. Check. We also reasoned that the water flow rate increases as σ increases. When $\sigma = 0.04$ is substituted into Eq. (3.43), one calculates $F_{W,4} = 13$ kg/min. We don't know if this is correct, but at least it is positive. Check.

The maximum salt content for potable water is $\varepsilon = 0.0005$ wt fraction. From Eq. (3.45), this corresponds to $\sigma = 0.05$ wt fraction salt in the brine (stream 3). When $\sigma = 0.05$ is substituted into Eq. (3.43), one calculates $F_{W,4} = 30$ kg/min. Is this the maximum output of potable water? What is the maximum output, without regard for salt content? To find this we can take the derivative of the expression for $F_{W,4}$, Eqn. (3.43), with respect to σ, and then find the value of σ at which $dF_{W,4}/d\sigma = 0$:

$$\frac{dF_{W,4}}{d\sigma} = -1.01 + \frac{3.535}{\sigma^2}, \tag{3.46}$$

$$\frac{dF_{W,4}}{d\sigma} = 0 \quad \text{at} \quad \sigma = 1.87. \tag{3.47}$$

Since the maximum flow rate does not lie between $\sigma = 0.035$ and 0.05, we are guaranteed that the maximum output of potable water is at $\sigma = 0.05$, for which $F_{W,4} = 30$ kg/min.

The operating equations, (3.43) and (3.45), are useful for predicting the performance of this desalinator design. Depending on the salt content desired for the output stream, one can calculate the flow rate of potable water. Clearly it is more efficient to use these operating equations than it is to recalculate a mass balance for each new operating condition.

We can derive operating equations in terms of other parameters as well. Perhaps we wish to know the flow rate of potable water $F_{W,5}$ ($= F_{W,4}$) in terms of the wt fraction of salt in the potable water, ε. Combing Eqs. (3.43) and (3.45) we arrive at

$$F_{W,5} = -1.01\varepsilon + 101.05 - \frac{0.03535}{\varepsilon}. \tag{3.48}$$

Our mathematical analysis confirms our qualitative analysis: The process is limited by the amount of brine retained by the ice in stream 4. What is the benefit of reducing the amount of retained brine? Again we could apply mass balances and calculate the output of potable water for various levels of retained brine. However, it is more effective to define a variable for the weight fraction of retained brine. Let's use β, because "beta" is the Greek symbol for "b" and "b" is a good mnemonic for "brine":

weight fraction of retained brine $\equiv \beta = $ *retained brine*$/($*ice* $+$ *retained brine*$).$

$$\tag{3.49}$$

We repeat the preceding derivation, but replace "0.01" in Eq. (3.34) with β and obtain Eqs. (3.50) and (3.51), the operating equations in terms of the salt content of the waste brine, σ, and the weight fraction of brine retained by the ice, β:

salt content of stream 5 $= \sigma\beta,$ $\tag{3.50}$

$$F_{W,4} = \frac{(100\sigma - 3.5)(1 - \sigma\beta)}{\sigma(1 - \beta)}. \tag{3.51}$$

We can check Eq. (3.50) by comparing to two previous equations. Inserting $\beta = 0.01$ yields Eq. (3.43), the result when 1 wt% brine was retained by the ice. Check.

Inserting $\beta = 0$ yields Eq. (3.26), the result when nothing was retained by the water stream, which was the case in the evaporator. Check.

Let's use operating equations (3.50) and (3.51) to study the effect of the retained brine. Is the output rate sensitive to β? Is the purity sensitive to β? That is, is it worthwhile to reduce the amount of brine retained by the ice? The previous analysis revealed that the maximum production of potable water is obtained at the maximum allowable salt content, 0.0005 wt fraction salt. Substituting into Eq. (3.50) yields

$$\sigma\beta = 0.0005 \quad \text{or} \quad \sigma = \frac{1}{2000\beta}. \tag{3.52}$$

Because the salt content of the waste brine must be at least that of seawater and at most that of pure salt, σ is limited to the range $0.035 \le \sigma \le 1.0$. From the limits on σ and Eq. (3.52), the limits on the weight fraction of retained brine are $0.0005 \le \beta \le 0.0143$, assuming that the potable water has 0.05 wt% salt. This tells us that the process will not function if the ice retains more than 1.4 wt% brine, $\beta \ge 0.014$. This also tells us that for $\beta \le 0.0005$, the salt content of the potable water is less than 0.05 wt%.

Substituting Eq. (3.52) into Eq. (3.50) gives the operating equation for a desalinator that produces potable water with 0.05 wt% salt, the maximum allowable concentration:

$$F_{W,4} = \frac{1999(0.05 - 3.5\beta)}{1 - \beta} \quad \text{for} \quad 0.0005 \le \beta \le 0.014, \tag{3.53}$$

$$F_{W,4} \approx 100 - 7000\beta. \tag{3.54}$$

At $\beta = 0.01$, the original specification on the freezer–desalinator, Eq. (3.53) yields $F_{W,4} = 30$ kg/min, as we calculated before. At $\beta = 0.0005$, Eq. (3.53) yields $F_{W,4} = 96.5$ kg/min – all the water exits stream 5 (but with 0.05 wt% salt). Although the mass balance permits $\beta = 0.0005$, it would be impractical to freeze seawater to produce solid ice and solid salt, and then separate the solids. Some water must leave with the brine. What if we reduce the retained brine by a factor of 2, to $\beta = 0.005$? Equation (3.53) predicts we will double our output to $F_{W,4} = 65$ kg/min. Clearly, it is worth investigating designs to reduce retained brine.

3.5   An Even Better Desalinator

We now switch from analysis to design. Our analysis has indicated where to focus our design efforts – in reducing the amount of brine retained by the ice. What are some options? Perhaps a centrifuge? Or perhaps the ice chute could be a conveyor with a brine-absorbing belt? But what is the *real* problem?

The *real* problem is to reduce the amount of *salt* retained. Rather than try to produce liquid-free ice from a ice/brine slush, we might try to reduce the amount of salt in the retained liquid. We don't mind if the ice is wet, we just don't want the wetting liquid to contain salt. We propose that a better problem to solve is "reduce the amount of salt in the liquid retained by the ice." By redefining the problem we allow for other solutions. How about adding a unit in which warm air is blown onto the ice? The air melts some of the ice, dilutes the retained brine, and liquid drips from the ice until only *diluted* brine is retained. A mass balance on this process shows that melting just 5% of the ice will lower the brine concentration by a factor of six. And this does not reduce the flow rate of potable water by 5%. Why? When 1 wt% brine was retained the brine concentration in the ice/brine slush could not exceed 5 wt%. By melting just 5% of the ice we can form an ice/brine mixture in which the brine is a factor of six saltier! Higher salinity of the waste brine means more ice. (Remember the analysis of the second desalinator design?)

How else could we reduce the amount of salt in the retained liquid? A good source of ideas is to study other processes. How did we remove the liquid propanol retained by the solid heptane in Chapter 2? We washed the solid heptane with liquid heptane. So how do we remove liquid brine retained by the solid water? We wash solid water (*i.e.*, ice) with water. Where do we get the water? From melting the ice.

The desalinator is modified to recycle some water back to the separator to wash the ice, as shown in Figure 3.12. With an efficiently designed skimmer, the recycled wash stream, 6, is only 5% of stream 5. The product stream, 7, contains a negligible amount of salt.[5] Given that the brine leaving the skimmer is 7.0 wt% salt, calculate the flow rate of stream 7.

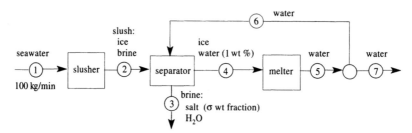

Figure 3.12. The freezer–desalinator with a water recycle to wash the brine from the ice in the ice brine skimmer separator.

The recycle stream severely complicates the analysis of a process. The trick here is to note that the details of the recycle stream and the other internal streams do not

---

[5] Note that when the process is *started*, the water product in stream 7 will be salty: too salty to be potable, but less salty than seawater. As the process runs, the water product becomes less salty, eventually containing negligible salt. How long must the process run before pure water is produced? Clearly, this is not a *steady-state* system during start-up. Rather it is a *transient* process, which we will consider in Chapter 6.

need to be calculated. If one draws system borders around the entire process, as in Figure 3.13,

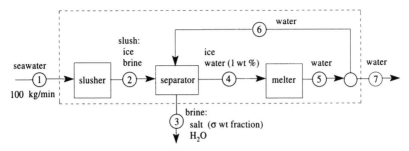

Figure 3.13. Mass balance around entire freezer–desalinator.

the process is reduced effectively to that shown in Figure 3.14, which is identical to the desalinator in Figure 3.6, except for the different stream labels. The operating

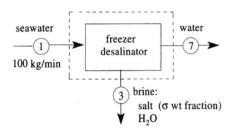

Figure 3.14. Effective result of a mass balance around entire freezer–desalinator.

equation for the process in Figure 3.14 is Eq. (3.26). Substituting $\sigma = 0.07$ into Eq. (3.26) yields $F_{W,7} = 50.$ kg/min.

3.6 Design Tools from Mathematical Modeling

Let's summarize our analysis of the desalinator so far. We started with the conservation of mass and derived operating equations for the process. We then used the operating equations as *design tools* to predict the performance of the desalinator. Creating design tools is the essence of chemical engineering design.

The chemical industry consists of thousands of processes, each composed of many units. It would be impractical for you to study the operating equations for every process, or every unit you might encounter during your career. It would also be unwise. What would you do when you encountered a new unit? But you don't need to memorize operating equations. Instead, you need only learn how to *create* tools to design and operate chemical engineering processes. Operating equations are examples of design tools.

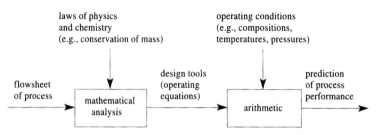

Figure 3.15. Schematic of the process to create and apply mathematical design tools.

We will introduce in this text two other methods to create design tools. These methods will also be summarized with figures similar to 3.15. The inputs to the "units" in Figure 3.15 will be the same. But the tools will be different and the manner in which the tools are used will be different. For example, in the next chapter we will create graphical methods rather than mathematical expressions. And we will use geometry, rather than arithmetic, to predict performance.

3.7 The Freezer Desalinator – Analysis of the Energy Flow

The chief reason that we explored desalination by freezing was its potential to use less energy than evaporation. Because evaporation requires seven times as much energy per kg as does melting, this seems qualitatively correct.[6] To confirm the qualitative analysis we need to analyze the energy requirements of the freezer–desalinator quantitatively.

Analysis of energy flow is inherently more complicated than analysis of mass flow. Mass is a tangible quantity. Energy is elusive. In fact, energy defies definition. One can quantify mass, for example, by the force an object exerts in a gravitational field or the acceleration of an object owing to an imposed force. Energy is more difficult to quantify. The problem is that energy can appear in many incarnations. To measure an object's internal energy, we must measure its thermal energy (what is its temperature?), determine its phase (is the material solid, liquid, or gas?), and determine its chemical energy (*e.g.*, is it the object coal, or sand, or whatever?); to measure an object's potential energy, we must measure its height; and so on for kinetic energy (its velocity), pressure energy (its pressure), etc.

Why are there so many different forms of energy? The many forms of energy evolved because of faith in the concept of the conservation of energy. When Leibnitz first proposed the principle in 1693 he included only two forms of energy: potential and kinetic. This was fine for analyzing a pendulum. But conservation of energy failed, for example, if one used a falling weight to drive a propeller that stirred and heated

---

[6] Of course, we could use heat exchangers to recover some of the energy used for evaporation or freezing. But we would still lose a fraction of the energy. The energy not recovered in evaporation would be greater than the energy not recovered in freezing.

a fluid. Rather than discard the conservation of energy, the principle was repaired by adding an additional term to account for energy in the form of heat. And thus each time conservation of energy appeared to fail, the definition of energy was expanded to include new terms and restore the principle. The last modification was to include the energy inherent in mass, $E = mc^2$.

The schematic of the freezer–desalinator in Figure 3.12 considers only the mass flow. We need to consider the energy flow as well. We will again assume the desalinator operates at steady state. To begin, we determine the amount of energy in each stream. We first indicate on each stream information sufficient to determine the energy. For the present, we will assume that the chief form of energy is heat. We have no chemical changes. We will assume that all streams are at the same height and pressure. Therefore, we add temperatures to the flowsheet, as shown in Figure 3.16.

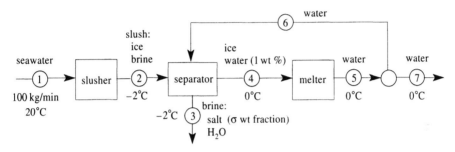

Figure 3.16. The freezer–desalinator with stream temperatures.

Clearly the energy is not balanced in the slusher. Streams 1 and 2 have the same mass flow, so mass is conserved. But the temperature in stream 1 is higher than in stream 2 and a portion of stream 2 has been converted to ice; more energy enters the slusher than leaves the slusher. We need to provide an outlet for the energy given up by cooling and freezing the seawater. We will indicate this with a squiggly arrow in Figure 3.17, to emphasize that this stream carries only energy, no mass. Likewise the melter requires energy. Again we add a squiggly arrow. We also introduce the

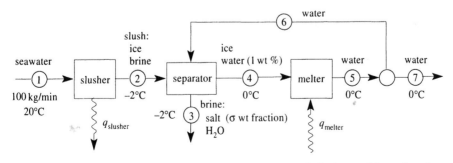

Figure 3.17. The freezer–desalinator with heat flow out of slusher and heat flow into melter.

convention q_i to represent the rate of energy flow associated with stream i. The dimensions of q are energy/time; in the mks system the units are kJ/sec.

Let's calculate rates of heat flow out of the slusher and into the melter. We begin with the conservation of energy for a system at *steady state*, which is given by Eq. (3.9):

rate in = rate out. (3.9)

To apply the conservation of energy we must define a system. We draw borders around the slusher (Figure 3.18).

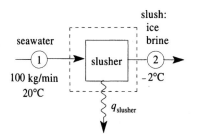

Figure 3.18. System borders for energy balance on the slusher.

Applying the conservation of energy to this system yields

$$q_1 = q_2 + q_{slusher}, \tag{3.55}$$

$$q_{slusher} = q_1 - q_2. \tag{3.56}$$

How do we compute the energy of streams 1 and 2? We must convert temperatures to energy. From thermodynamics we learn that the change in internal energy is linearly proportional to the change in temperature. The proportionality constant is the product of the two quantities: the mass and the heat capacity, the amount of energy needed to raise the temperature of 1 kg 1°C. Thus from thermodynamics we use the relation

change in internal energy =

change in [(mass) × *(heat capacity)* × *(temperature)*], (3.57)

$$\Delta U = \Delta[MC_PT]. \tag{3.58}$$

The heat capacity at constant pressure, C_P, varies from substance to substance. The heat capacity also varies with temperature, but the change is usually small. For the small temperature range in the desalinator, we will assume that the heat capacity is constant, which yields

$$\Delta U = MC_P(\Delta T). \tag{3.59}$$

Equation (3.59) illustrates another complexity of energy balances, compared to mass balances. Mass balances deal with absolute quantities; 100 kg/min enters the slusher and 100 kg/min leaves the slusher. However, it is usually not possible (nor convenient) to state an energy balance in absolute terms; for example it is not possible to state "100 kJ/min enters with the seawater and 50 kJ/min leaves with the seawater." Rather, we must consider *changes* in energy. Energy balances deal with relative quantities: the energy difference between the seawater entering and the seawater leaving is 50 kJ/min. When we calculate Eq. (3.56), we will not evaluate the individual energies, q_1 and q_2, associated with each stream. Rather we will calculate the change in energy, $q_1 - q_2$.

We need to convert a quantity of energy, ΔU, to a flow rate of energy, energy per time. We thus divide each side of Eq. (3.60) by Δt, an increment of time, to get

$$\frac{\Delta U}{\Delta t} = \frac{MC_P(\Delta T)}{\Delta t} = \frac{M}{\Delta t}C_P(\Delta T), \tag{3.60}$$

$$q = FC_P(\Delta T). \tag{3.61}$$

Equation (3.61) will allow us to calculate the energy change associated with cooling stream 1 to stream 2 because we know F (mass flow rate) and ΔT (temperature) and we can find C_P in a handbook.

But there is another term. We must also consider the energy change associated with forming ice from seawater. Again thermodynamics provides the necessary information: The energy change upon freezing is proportional to the amount frozen. The proportionality constant, which varies with composition, is the heat of fusion, ΔH_{fusion}. The flow rate of heat q associated with freezing a mass flow rate F is, analogous to Eqs. (3.60) and (3.61),

$$q = F(\Delta H_{fusion}). \tag{3.62}$$

Equation (3.62) again illustrates that energy balances involve relative energies and not absolute energies. The heat of fusion is the difference between the enthalpies of the liquid and the solid phases:

$$\Delta H_{fusion} = H_{liquid} - H_{solid}. \tag{3.63}$$

We will not calculate the absolute energies of ice and salt water. Rather we will calculate the energy change upon freezing.

Preparing for the energy balance has given us a different perspective on the slusher. Although the unit is trivial with regard to a mass balance, it is complicated with regard to an energy balance. That is, the slusher first cools all the seawater to $-2°C$ from $20°C$; it then freezes a portion of the seawater. The slusher seems to violate the "one unit – one operation" guideline of Chapter 2. The slusher is actually a cooler, and then becomes a splitter; a portion from the splitter enters a freezer, after which the frozen portion is recombined with the other stream from the splitter. An equivalent flowsheet for the slusher is shown in Figure 3.19. (We ignore the details of how the salt is excluded from the ice.)

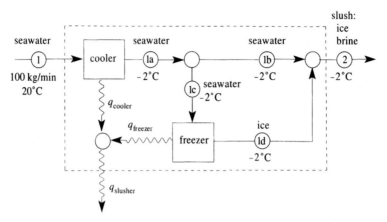

Figure 3.19. An equivalent cooler–freezer system for analysis of the slusher.

We now apply the equations from thermodynamics – Eqs. (3.62) and (3.63) – to write energy balances on the cooler and freezer of our imaginary slusher. The heat flow out of the cooler can be calculated with the energy balance:

$$q_{cooler} = q_1 - q_{1a}. \tag{3.64}$$

To apply Eq. (3.62) we use $F_{T,1} = F_{T,1a}$ and again assume the heat capacity is constant between $-2°C$ and $20°C$:

$$q_{cooler} = F_{T,1}C_{P,1}(T_1 - T_{1a}). \tag{3.65}$$

We assume further that the heat capacity of seawater equals that of water, $C_P = 4.18$ kJ/(kg · °C). We may now calculate the heat flow out of the cooler:

$$q_{cooler} = \left(\frac{100 \text{ kg}}{\text{min}}\right)\left(\frac{4.18 \text{ kJ}}{(\text{kg})(°\text{C})}\right)(20 - (-2)) \tag{3.66}$$

$$= 9.2 \times 10^3 \text{ kJ/min.} \tag{3.67}$$

Let's check this answer. A numerical answer has two parts: a sign and a magnitude. Let's begin with the sign. Our answer is positive, which says that heat leaves the cooler. This is correct. What about the magnitude, 9.2×10^3 kJ/min? We sympathize that quantities of "kJ/min" are probably not familiar to a first-year engineer. The answer is probably within a few orders of magnitude; 9.2×10^3 is not as suspicious as 9.2×10^{16} or 9.2×10^{-16} for example. One approach is to translate this energy flow rate to a common phenomenon. A hot tub contains about 5,000 kg of water. A mass flow rate of 100 kg/min would fill the hot tub in 50 minutes. A quick calculation shows that 9.2×10^3 kJ/min would heat the water from $10°C$ ($50°F$) to $30°C$ ($86°F$) in about 45 minutes. This seems reasonable. That is, if the hot tub warmed in a few seconds, the energy flow rate would seem too high. If the hot tub warmed in a few days, the energy flow rate would seem too low.

Another way to check an energy flow rate is to convert to energy per mass, or kJ/kg. Water flowing through the cooler at 100 kg/min releases heat at a rate of 9.2×10^3

kJ/min, so the energy per mass is 92 kJ/kg. Now compare this to some rules of thumb:

> 2 kJ/kg will increase the temperature 1°C.
> 20 kJ/kg will melt a nonpolar, molecular solid (butane, cyclohexane) at its melting point.
> 100 kJ/kg will melt a polar, molecular solid (water, ammonia) at its melting point.
> 400 kJ/kg will evaporate a nonpolar, molecular liquid at its boiling point.
> 1,000 kJ/kg will evaporate a polar, molecular liquid at its boiling point.

0 to 10,000 kJ/kg will be released (or absorbed) in chemical reaction.

Being rules of thumb, these numbers are rough estimates. If your answer is within a factor of 3 larger or smaller, your answer agrees with the rule of thumb. The rule of thumb for temperature change predicts 40 kJ/kg (one significant figure) for our cooler. Our answer of 92 kJ/kg is consistent.·

Returning to the desalinator, the heat flow out of the freezer can be calculated with the energy balance, the thermodynamic relation in Eq. (3.62), and the trivial mass balance, $F_{T,1c} = F_{T,1d}$:

$$q_{freezer} = F_{T,1c}(\textit{internal energy of water}) - F_{T,1d}(\textit{internal energy of ice}), \quad (3.68)$$

$$= F_{T,1c}(\textit{internal energy of water}) - (\textit{internal energy of ice}), \quad (3.69)$$

$$= F_{T,1c}\Delta H_{fusion,water}. \quad (3.70)$$

From a handbook, we find that the heat of fusion of water is 334 kJ/kg. We need the internal flow rates of our imaginary slusher. Let ϕ be the fraction of stream 1 that goes to the freezer via stream 1a (ϕ is Greek for "f", a nice mnemonic for frozen). Operating at the conditions described above, $\phi = 0.53$. (You are encouraged to verify that $\phi = 0.53$ with a mass balance.) So we substitute $F_{T,1c} = \phi F_{T,1a}$ into Eq. (3.70) to get

$$q_{freezer} = \phi F_{T,1a}\Delta H_{fusion,water} = 0.53 \left(\frac{100 \text{ kg}}{\text{min}}\right)\left(\frac{334 \text{ kJ}}{\text{kg}}\right) = 1.8 \times 10^4 \text{ kJ/min}.$$
$$(3.71)$$

Again, one should check one's answer. The sign is correct – energy leaves the freezer. The mass basis for the energy is 334 kJ/kg, which is consistent with the rule of thumb for freezing a polar molecular liquid.

Finally, an energy balance on the fictitious energy combiner yields

$$q_{slusher} = q_{cooler} + q_{freezer} = 9.2 \times 10^3 + 1.8 \times 10^4 = 2.7 \times 10^4 \text{ kJ/min}. \quad (3.72)$$

As emphasized earlier, mass and energy balances require careful accounting. Creating the fictitious slusher in Figure 3.19 clarifies the accounting. We recommend that you divide any complex unit into discrete energy operations until you are well experienced in applying energy balances.

A similar analysis of the melter yields $q_{melter} = 1.8 \times 10^4$ kJ/min. We recommend that you analyze the melter and verify that you can apply an energy balance correctly.

Although the energy requirements for the freezer–desalinator are less than that of an evaporator–desalinator, the requirements are still substantial. Because the raw material (seawater) is essentially free, the cost of energy is essentially the cost of operating the desalinator. Reducing the energy required by a factor of two reduces the cost of operating the desalinator by a factor of two. Let's explore some designs to reduce the energy required.

The slusher discards energy and the melter requires energy. Perhaps we could combine these two energy flows to transfer the energy from the freezer to the melter. That is, we could use the warm seawater to melt the ice and the melting ice would transform the seawater to slush. However, because $q_{slusher} > q_{melter}$ we need something more to absorb energy from the seawater. One possibility is the cold brine discarded at $-2°C$ via stream 3. Although the mass discarded in stream 3 has no value, the energy in stream 3 has value. Let's add a heat exchanger before the slusher to reclaim this value.

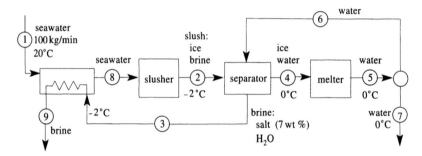

Figure 3.20. The freezer–desalinator with a precooler.

What temperature might one expect for the seawater in stream 8? First estimate the upper and lower limits. We expect stream 8 will be cooler than stream 1 so the upper limit is 20°C. Everyday experience tells us that the lower limit is determined by the temperature of the brine, $-2°C$. To refine these estimates we need to consider how the heat exchanger functions. As the symbol for the heat exchanger suggests, the brine flows through a pipe and the seawater flows past the pipe on the outside. The brine and seawater are isolated from each other, but energy is transferred through the pipe walls. We have two options for the flow in the pipe: *cocurrent* with the seawater flow or *countercurrent* to the seawater flow.

3.8 Cocurrent and Countercurrent Devices

Let's analyze qualitatively the differences between cocurrent and countercurrent flow in the heat exchanger. It is often useful to first consider ideal systems (see Figure 3.21).

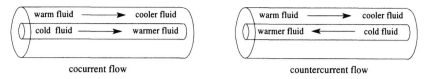

Figure 3.21. Cocurrent and countercurrent flow in a heat exchanger. In this case, the fluid being warmed moves inside the small pipe and the fluid being cooled moves outside the small pipe but inside the shell of the heat exchanger.

Assume that the fluids inside and outside the pipe have the same composition and the same flow rate. The only difference lies in the temperatures of the fluids. Which flow pattern – cocurrent or countercurrent – is capable of cooling the warm fluid the most? We can estimate graphically the temperatures of the fluids along the pipe for each configuration. We begin by plotting the inlet temperatures (Figure 3.22).

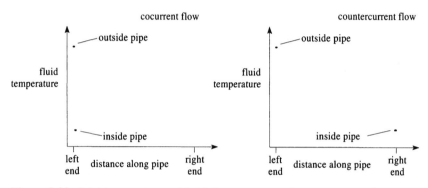

Figure 3.22. Inlet temperatures of fluids in cocurrent and countercurrent flows.

We now sketch lines from the inlet temperatures to the exit temperatures. The cold fluid should warm and the hot fluid should cool. And the lines should not cross. (Why?) And if the heat capacity is constant, because the flow rates are equal the temperature drop in the outside fluid must equal the temperature rise in the inside fluid. (Why?) One arrives at the sketches shown in Figure 3.23.

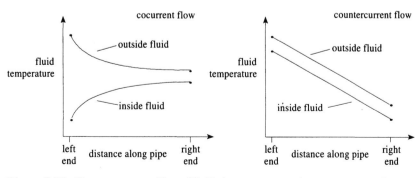

Figure 3.23. Temperature profiles of fluids in cocurrent and countercurrent flows.

Qualitatively, the countercurrent flow cools the outside fluid the most. We will see this theme again when we consider mass transfer between two streams. Chemical engineers can calculate the temperature along the pipes in a heat exchanger with the material learned in courses on fluid mechanics and heat transfer. And thus a chemical engineer can design a heat exchanger to meet specifications. We will not concern ourselves with those details here. For now we just arrange the piping for countercurrent flow.

3.9 Energy Balance in a Heat Exchanger

Assume the temperature of the brine warms to $14°C$ from $-2°C$ in the precooler. What is the exit temperature of the seawater? To apply the conservation of energy we again start by defining the system borders, as shown in Figure 3.24.

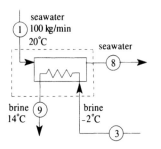

Figure 3.24. The precooler.

We write an energy balance on the precooler:

$$q_1 + q_3 = q_8 + q_9. \tag{3.73}$$

When applying an energy balance to a heat exchanger it is often useful to rearrange the equation from "rate in" on the left and "rate out" on the right to "fluid being cooled" on the left and "fluid being warmed" on the right:

$$q_1 - q_8 = q_9 - q_3. \tag{3.74}$$

Equation (3.74) shows that the energy lost by the seawater $(q_1 - q_8)$ equals the energy gained by the brine $(q_9 - q_3)$. To calculate the energy change for each stream we use the thermodynamic relation between energy change and temperature change, Eq. (3.61). For example, the energy gained by the brine is

$$q_{brine} = q_9 - q_3 = F_{brine} C_{P, brine} (\Delta T)_{brine}. \tag{3.75}$$

We need the mass flow rates for brine and seawater. The mass balance on the precooler is trivial because the brine and seawater flows are separated; seawater enters via stream

1 and leaves via stream 8, so $F_{T,1} = F_{T,8} = 100$ kg/min. Likewise, brine enters via stream 3 and leaves via stream 9, so $F_{T,3} = F_{T,9} = 50$ kg/min, from the previous analysis. Again we will assume that the heat capacity is independent of the salt contents here (3.5 wt% in streams 1 and 8, 7 wt% in streams 3 and 9) and is equal to the heat capacity of water. Equation (3.74) is transformed as follows:

$$F_{T,1}C_{P,\text{water}}(T_1 - T_8) = F_{T,3}C_{P,\text{water}}(T_9 - T_3), \tag{3.76}$$

$$T_1 - T_8 = \frac{F_{T,3}}{F_{T,1}}(T_9 - T_3), \tag{3.77}$$

$$T_8 = T_1 - \frac{F_{T,3}}{F_{T,1}}(T_9 - T_3), \tag{3.78}$$

$$T_8 = 20°\text{C} - \frac{50 \text{ kg/min}}{100 \text{ kg/min}}(14°\text{C} - (-2°\text{C})) = 12°\text{C}. \tag{3.79}$$

This calculation illustrates two characteristics of energy balances. First, energy is calculated as a *relative* amount, not as an *absolute* amount. We calculated how much the energy of the brine *changed* from stream 3 to stream 9. We did not calculate the energy of stream 3 or stream 9 individually. Second, energy differences are calculated indirectly, through changes in measurable quantities such as temperature. In other applications one might use changes in velocity, or pressure, or height, for example. Recognizing these two characteristics is key to solving energy balances. You should approach an energy balance asking "How can I rearrange these terms into differences between terms?" and "How can I recast an energy variable in terms of a measurable parameter?"

Let's return to the desalinator. Adding the precooler lowers the temperature of the seawater to 12°C and subsequently lowers the demand on the slusher; q_{cooler} in Eq. (3.65) is reduced to 6.7×10^3 kJ/min from 9.2×10^3 kJ/min and thus the entire demand is reduced to $q_{\text{slusher}} = 2.4 \times 10^4$ kJ/min. Recall that the energy absorbed by the melter is $q_{\text{melter}} = 1.8 \times 10^4$ kJ/min. Thus in principle one could add another heat exchanger to transfer energy from the cooled seawater to the ice, which would cool the seawater to $-2°$C and freeze some ice (although not as much ice as is required) and melt the ice to water at 0°C. Okay? No, this is not okay. Although this second heat exchanger would satisfy the first law of thermodynamics (an energy balance) it would violate the second law of thermodynamics. Heat does not spontaneously flow from a cold object (seawater at $-2°$C) to a warmer object (ice at 0°C), no more than mass spontaneously flows uphill. We must pump the heat from cold to warm just as we must pump fluids uphill.

The restrictions that physical laws place on creativity is a common frustration in design. Richard Feynman (1918–1988), the second-most famous physicist of the twentieth century, commented:

The whole question of imagination in science is often misunderstood by people in other disciplines. They try to test our imagination in the following way. They say "Here is a picture of some people in a situation. What do you imagine will happen next?" When we say, "I can't imagine," they may think we have a weak imagination. They overlook the fact that whatever we are allowed to imagine in science must be consistent with everything else we know. . . . We can't allow ourselves to seriously imagine things which are obviously in contradiction to the known laws of nature. And so our imagination is quite a difficult game. One has to have the imagination to think of something that has never been seen before, never been heard of before. At the same time the thoughts are restricted in a straitjacket, so to speak, limited by the conditions that come from our knowledge of the way nature really is. The problem of creating something which is new, but which is consistent with everything which has been seen before, is one of extreme difficulty.

3.10 Heat Pumps

Heat pumps are similar to fluid pumps. Consider this analogy: Figure 3.25 shows a scheme for moving fluid from a low container to a high container. The fluid drains from the low container to a bucket at a lower level; the bucket is raised and then drained into the high container.

Figure 3.26 shows an analogous scheme for moving heat from a cool fluid to a warm fluid. The heat drains spontaneously from the cool object to a colder object; the object is warmed and then the heat drains to the warm fluid.

Transferring another load of fluid is trivial; just lower the bucket. How does one cool the hot object to a cold object to transfer more heat? The solution will be obvious as we design the heat pump for the desalinator.

3.11 Refrigeration Cycle in the Desalinator

Let's design a system to remove heat from the seawater at $-2°C$ to form ice. We need a refrigerating fluid colder than $-2°C$. And the temperature of the refrigerant cannot rise above $-2°C$ as it absorbs heat from the seawater. Rather than absorb heat by *warming* the refrigerant, a more effective method is to absorb heat by *vaporizing* the refrigerant. We need a refrigerant that transforms from a liquid to a vapor at a temperature less than $-2°C$. The refrigerant has many other requirements imposed on it as well; it must be inexpensive, have a high heat of vaporization, be nonlethal, and be innocuous (so a leak is not a threat to one's health – refrigerators once used H_2S!). Chlorofluorocarbons were ideal refrigerants until it was proposed that they do not satisfy another requirement; refrigerants should be harmless to the environment.

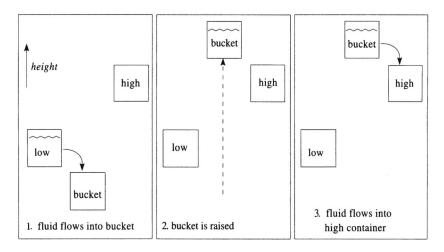

Figure 3.25. A method to transport mass from a lower container to a higher container.

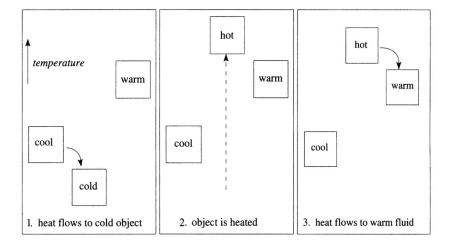

Figure 3.26. A method to transport heat from a cool fluid to a warm fluid.

So we consult a list of refrigerants in a handbook. Let's use a "natural" chemical as our refrigerant – ammonia.[7]

Let's arbitrarily specify that the ammonia vaporizes at $-22°C$. (We could have chosen any temperature less than $-2°C$.) However, ammonia boils at $-33°C$ at 1 atm. How do we cause ammonia to boil instead at $-22°C$? We increase the pressure of the ammonia. We consult a handbook and find that the saturation pressure of ammonia at $-22°C$ is 25 psi. In the handbook we also read that the heat of vaporization at $-22°C$ and 25 psi is $\Delta H_{vap} = 574.9$ Btu/lb. (Refrigeration data are usually in English units.)

---

[7] Note that ammonia is pungent and corrosive, and its condensed vapor causes edema of the respiratory tract as well as other problems. But at least ammonia is natural.

We convert to mks units:

$$\left(\frac{574.9\,\text{Btu}}{1\,\text{lb}}\right)\left(\frac{1\,\text{lb}}{0.4536\,\text{kg}}\right)\left(\frac{1.054\,\text{kJ}}{1\,\text{Btu}}\right) = 1{,}336\,\text{kJ/kg}. \tag{3.80}$$

Note that in pure mks units the pressure should also be converted to Pascals (Pa):

$$25\,\text{psi}\left(\frac{6895\,\text{Pa}}{1\,\text{psi}}\right) = 1.7 \times 10^5\,\text{Pa}. \tag{3.81}$$

But 1.7×10^5 is an awkward number. We'll use the more convenient units of atm instead; 25 psi = 1.7 atm. The design for the slusher is given in Figure 3.27.

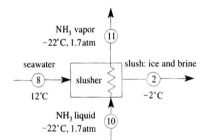

Figure 3.27. The slusher with ammonia refrigerant.

What flow rate of ammonia is needed to cool and partially freeze the seawater? An energy balance on the slusher yields

$$q_8 + q_{10} = q_2 + q_{11}. \tag{3.82}$$

Let's rearrange the equation from "rate of energy in" on the left, "rate of energy out" on the right to "rate of energy change in seawater" on the left, "rate of energy change in ammonia" on the right:

$$q_8 - q_2 = q_{11} - q_{10}. \tag{3.83}$$

The energy released by the seawater was calculated above, $q_8 - q_2 = 2.4 \times 10^4$ kJ/min. The right side of Eq. (3.83) is the energy absorbed by ammonia evaporating at $-22°$C and 1.7 atm:

$$q_8 - q_2 = F_{T,11}(\text{energy of NH}_3 \text{ vapor}) - F_{T,10}(\text{energy of NH}_3 \text{ liquid}), \tag{3.84}$$

$$q_8 - q_2 = F_{T,11}\Delta H_{\text{vap,NH}_3 \text{ at 25 psi}}, \tag{3.85}$$

$$F_{T,11} = \frac{q_8 - q_2}{\Delta H_{\text{vap,NH}_3 \text{ at 25 psi}}} = \frac{2.4 \times 10^4\,\text{kJ/min}}{1{,}336\,\text{kJ/kg}} = 18\,\text{kg/min}. \tag{3.86}$$

Likewise we pass ammonia through the melter. We condense ammonia vapor to form liquid ammonia, which releases heat. We must choose a temperature greater than $0°$C for the ammonia stream. We arbitrarily choose $32°$C. A handbook reveals that the saturation pressure for ammonia at $32°$C is 12.2 atm and $\Delta H_{\text{vap}} = 488.5$ Btu/lb, which converts via Eq. (3.80) to 1,135 kJ/kg. The design for the melter is given in Figure 3.28.

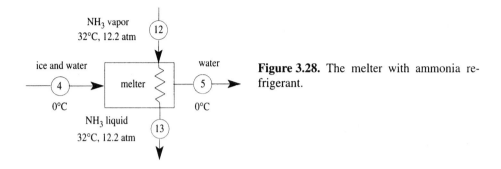

Figure 3.28. The melter with ammonia refrigerant.

What flow rate of ammonia is needed to melt the ice? An energy balance similar to that performed on the slusher yields $F_{T,12} = 15$ kg/min. We recommend that you attempt this analysis. Recall that $F_{T,4} = 52.6$ kg/min and stream 4 is 99 wt% ice and 1 wt% water at 0°C.

We need to close the ammonia loop. We connect the output from the slusher (stream 11) at 1.7 atm to the input to the melter (stream 12) at 12.2 atm by adding a compressor. We connect the output from the melter (stream 13) at 12.2 atm to the input to the slusher (stream 10) at 1.7 atm by adding a turbine or perhaps just an expansion valve. The ammonia loop is a typical refrigeration cycle, as found in refrigerators or air conditioners. We must supply energy to drive the compressor. It is possible that some of the energy could be recovered from the turbine.

The complete desalinator is shown in Figure 3.29. Is this acceptable? Almost. What detail remains? The flow rate of ammonia through the slusher is 18 kg/min and the flow

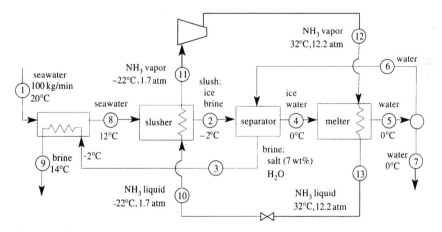

Figure 3.29. The melter with ammonia refrigerant.

rate of ammonia through the melter is 15 kg/min. A mass balance on the compressor (or the expansion valve) would reveal the problem. At what rate should ammonia flow through the refrigeration cycle? If ammonia flows at 15 kg/min, less energy is removed from the slusher and the rate of formation of ice is insufficient. If ammonia flows at 18 kg/min, excess heat is supplied to the melter. What becomes of the excess heat? It warms the melted ice. We calculate that the temperature of stream 5 would be 15°C. Again, we recommend that you attempt this calculation independently. (See Exercise 3.37 at the end of this chapter.)

Suppose we don't want to warm stream 5 to 15°C. Perhaps the temperature of the recycle stream 6 must be 0°C to minimize melting ice in the separator. Or perhaps we wish to feed the output stream 7 to a drinking fountain – it would be nice to have an ice-cold drink. What are the design options? Given an ammonia flow rate of 18 kg/min, we need to add less energy to the melter. As you will learn in physical chemistry, the heat of vaporization decreases as the pressure increases. Hence you could increase the pressure in stream 12 to lower the heat of vaporization by a factor of 15/18. Alternatively, we could flow ammonia at 15 kg/min and increase the heat of vaporization of ammonia in the slusher by 18/15.

3.12 Refrigeration Cycles without Refrigerants

The freezer desalinator in Figure 3.29 is moderately complex. It grew incrementally from the modest beginning in Figure 3.7. Most designs have humble beginnings and grow as they are analyzed or after they are constructed. The process of desalination by freezing has been taken yet another step forward. In our design, heat is pumped from the slusher to the melter by evaporating a refrigerant (ammonia) in the slusher and condensing the refrigerant in the melter. A conceptual breakthrough improved the design of the desalination process – the process fluid (H_2O) can also be used as the refrigerant. This avoids problems inherent with a foreign refrigerant. The slusher and melter are less expensive; no internal piping is needed because it is not necessary to keep the refrigerant isolated. And because the hot and cold fluids are intimately mixed, heat is transferred more efficiently.

The evolved desalinator is shown in Figure 3.30. Cooled seawater passes through a throttle valve to the slusher at low pressure. The temperature and pressure in the slusher are the triple point for seawater: the conditions at which ice, brine, and steam coexist. Heat is removed from the slusher by vaporizing water. The water vapor is compressed and heat is delivered to the melter by condensing water vapor. This is an exquisite design and yet it is a logical evolution from the design in Figure 3.29. A detailed analysis of desalination by evaporating at the triple point of seawater is presented in the textbook by Rudd, Powers, and Siirola (Chapter 7, "Fresh Water by Freezing"). Desalinators with the elegant design shown in Figure 3.29 have been used. It is important to recognize, however, that the optimum desalinator design depends on regional factors, including local costs of energy and real estate and the intended use for the desalinated water.

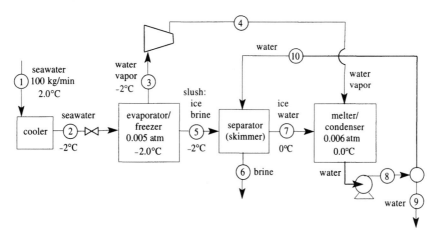

Figure 3.30. A freezer/evaporator desalinator with H_2O as the refrigerant.

A conceptual breakthrough in one process can be adapted to other processes. If the freezer desalinator can be modified to operate without a refrigeration loop and a coolant, perhaps an air conditioner can be modified as well.

An air conditioner has two functions: to remove heat from air and to remove water vapor from air. Cooling alone is insufficient. Air with high humidity is not comfortable, even if the air is cool. A flow diagram for a conventional air conditioner is shown in Exercise 3.16. Consider an unconventional air conditioner, shown in Figure 3.31. Water vapor is removed by adsorption onto a hydrophilic solid, called a *desiccant*.

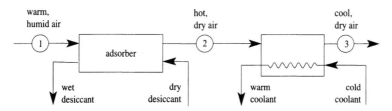

Figure 3.31. An air conditioner.

The air warms in the adsorber because water vapor releases heat when it adsorbs. Of course, the desiccant would be recycled, for example, by heating to drive off the water. The coolant would be recycled through a refrigeration loop, such as the loop we developed for the desalination process.

Most air conditioner refrigerants were chlorofluorocarbons (CFCs) until it was discovered that fugitive CFCs catalyze the destruction of ozone in the upper atmosphere. One solution is to replace CFCs with less-harmful substances. Another approach is to eliminate the need for a coolant by using the process fluid (air) to pump the heat.

To replace the CFC coolant with air we need to provide cold air to the heat exchanger in Figure 3.31. We could use a refrigeration cycle to cool the air – compress the air, cool the hot air, and then expand the air. But compressors are expensive. Instead, the air can be cooled by evaporating water, as described in Exercise 3.30. The air conditioner shown in Figure 3.31a uses air as a coolant (cold air is provided by the evaporator) and uses air to dry the desiccant (hot air is provided by the methane burner). The

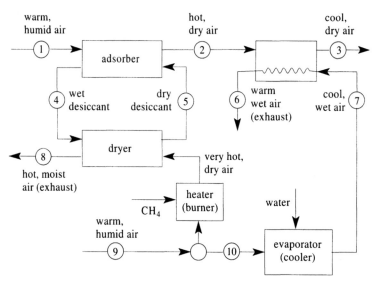

Figure 3.31a. An air conditioner that uses air as the refrigerant.

warm humid air in stream 9 is drawn from outside. The warm humid air in stream 1 may be outside air or air drawn from the air-conditioned space. Streams 6 and 8 are expelled to the outside.

The air conditioner in Figure 3.31a has limited application. The process will not function if the humidity in the air is too high. The air must be able to take on some water vapor in the evaporator.

Engineering the desiccant loop is challenging because it is not easy to circulate a solid. Two designs for the desiccant loop have been reported.[8] The Albers Corporation redefined the problem of pumping a solid by developing a liquid desiccant. ICC Technologies and LaRoche Chemicals each have designs in which the desiccant is a solid wheel on an axis parallel to the adsorber and the dryer. As the wheel turns, the desiccant moves between the two units.

3.13 Mathematical Modeling – Universal and Constrained Laws

Mathematical models allow one to select the better design from various alternatives or to predict the performance of a given process as a function of operating conditions (concentrations, temperatures, etc.). As eloquently stated by Isaac Newton

---

[8] "Desiccant Air-Conditioning System Uses No Refrigerants or Compressors," *Chemical and Engineering News*, November 15, 1993, pp. 15–16.

(1642–1727), mathematical models "subordinate the phenomena of nature to the laws of mathematics." A design's physical description is translated to mathematics using the laws of physics and chemistry. There are two types of laws: universal and constrained. Examples of universal laws are the conservation laws ("Energy is conserved") and Newton's laws of motion (Newton's second law – "A change in motion is proportional to the impressed force"). Universal laws, as the name implies, are valid everywhere. Constrained laws are valid only in restricted domains. As such, constrained laws are not really "laws," although the terminology is common. A constrained law is more appropriately called a *constitutive equation*. An example of a constitutive equation is the ideal gas "law," $PV = nRT$, which is valid only at high temperature and low pressure (see Figure 3.32). The adjectives "high" and "low" depend on the identity of the gas, as we shall see in Chapter 5.

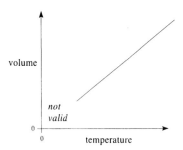

Figure 3.32. Volume of one mole of gas at 1 atm as a function of temperature in the region, the ideal gas law is valid. The slope of the straight line is R, the gas constant.

Another example of a constitutive equation is Stokes's "law": For a sphere of radius r moving at velocity v through a fluid of viscosity η, the resistance force is $6\pi\eta rv$. Stokes's law is valid only at low velocity for small spheres, where "low" and "small" are relative terms that depend on the viscosity and density of the fluid. Yet another example is Henry's law: Given a mixture of two liquids A and B, the partial pressure of B above the mixture is proportional to the mole fraction of B in the mixture. Henry's law is valid only for ideal solutions in which the mole fraction of B is low.

A mathematical model based on a universal law will be valid universally. However, a mathematical model based on a constrained law must be constrained. A common error is to forget the constraints acquired when using a constitutive equation to translate a physical description to mathematics.

3.14 Process Economics

We designed our desalinator to minimize energy requirements. But was that the true goal? Should we incorporate any modification that minimizes energy consumption? Why not? Because our true goal is to minimize cost. We focused on energy because energy is the chief operating cost. However, we must also consider the cost of the equipment. We introduce here mathematical models for calculating the cost of a design.

Profit is often the key consideration in designing a chemical process. Process economics is the formalism for calculating the finances of making and selling a product. Process economics also applies to the finances of providing a service. The concept of process economics is straightforward. The accounting, however, can be complicated. However, with this complication comes the opportunity for creativity. The economist John Maynard Keynes (1883–1946) remarked, "The avoidance of taxes is the only intellectual pursuit that still carries any reward."

We start by proposing the principle of Conservation of Asset. What is an asset? Like energy, assets can exist in many forms. An obvious form of asset is cash. But cash can be converted to other assets such as equipment. Is $1,000 cash equivalent to a piece of equipment worth $1,000? No, just as 1 kJ of thermal energy is not equivalent to 1 kJ of electrical energy. Electrical energy is a more usable form and converting thermal energy to electrical energy will cost you a fraction of the energy. Similarly, cash is a more liquid asset and converting equipment to cash may cost you a fraction of your asset. A liability (a debt) is a negative asset. A liability of $1,000 is equal to −$1,000 in cash.

By analogy with the principles of Conservation of Mass and Conservation of Energy we start with the general equation

$$\frac{d(assets)}{dt} = \text{rate of assets in} - \text{rate of assets out.} \tag{3.87}$$

As with previous conservation laws, we must be careful to define a system. The system might be your chemical company, as shown in Figure 3.33.

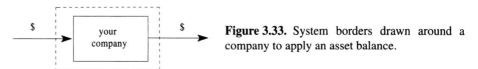

Figure 3.33. System borders drawn around a company to apply an asset balance.

The goal of every company is to have a positive flow of assets. Every company wants assets (such as cash) to accumulate. Like an energy balance, an asset balance is complicated by the many forms that assets can take. Writing an asset balance for a chemical process is never the simple task of counting piles of money or subtracting the rate that money flows out from the rate that money flows in. Just as energy is measured indirectly through parameters such as temperature and pressure, assets are measured indirectly through parameters such as equipment and inventory. For energy balances we appealed to thermodynamics for equations to convert parameters such as temperature and pressure into the currency of energy. For asset balances we appeal to accounting and tax codes for relations to convert equipment and inventory into units of assets. And just as there are proportionality constants for converting temperature to kJ, there are proportionality constants for converting inventory to dollars. Note that a proportionality "constant" in an asset calculation is rarely constant over time.

An asset balance, Eq. (3.87), is usually written in the terminology of Eq. (3.88).

$$\text{profit (or loss)} = \text{revenue} - \text{operating costs.} \tag{3.88}$$

This is where process economics can get confusing. As we shall see, operating costs include items such as paychecks to employees (assets flowing out) and the depreciation of equipment (assets consumed). More on the latter will be discussed later. The dimensions of Eq. (3.88) are money/time, typically expressed in dollars/year. Revenue in the chemical industry is usually derived from the sale of chemical commodities. Given a nonvarying price for the chemicals and the production rate, the revenue may be calculated with

$$revenue = \left(\frac{product\ sold\ (kg)}{year}\right)\left(\frac{price\ (\$)}{kg}\right), \tag{3.89}$$

which has units of $/year. Of course, revenue can also be derived from services, such as remediation of a toxic waste site or licensing a patent.

Let's consider a simplified example. Assume you inherit a desalinator. You produce 100,000 gal/day, which you sell at $12. per 1,000 gal. Your electric bill to run the desalinator is $13,500/month. The cost of labor to run the desalinator is $10.50/hour. The cost of space at the local pier is $2,400 per month. In this idealized example we will neglect all other costs, such as maintenance. We will also assume the desalinator is operating 24 hours a day, 52 weeks/year. We now calculate the profit (or loss). For the revenue we have

$$revenue = \left(\frac{100,000\ gal}{day}\right)\left(\frac{365\ days}{year}\right)\left(\frac{12.\$}{1,000\ gal}\right) = 438,000\ \$/year. \tag{3.90}$$

For the operating costs we get

$$operating\ costs = electricity + rent + labor \tag{3.91}$$

$$= \left(\frac{13,500\ \$}{month}\right)\left(\frac{12\ months}{year}\right) + \left(\frac{2,400\ \$}{month}\right)\left(\frac{12\ months}{year}\right)$$

$$+ \left(\frac{3\ shifts}{day}\right)\left(\frac{9\ hours}{shift}\right)\left(\frac{10.50\ \$}{hour}\right)\left(\frac{7\ days}{week}\right)\left(\frac{52\ weeks}{year}\right) \tag{3.92}$$

$$= 162,000\ \$/year + 28,800\ \$/year + 103,194\ \$/year \tag{3.93}$$

$$= 293,994\ \$/year. \tag{3.94}$$

And finally from the definition of profit in Eq. (3.89) we calculate

$$profit\ (or\ loss) = 438,000\ \$/year - 293,994\ \$/year = 144,006\ \$/year. \tag{3.95}$$

Let's make the example more realistic. Let's assume you first need to purchase the desalinator, which costs $440,000. Expenses for equipment are capital costs, not operating costs. As such, the expense of purchasing the desalinator is not included in Eq. (3.88) and thus does not affect the profit. But when we bought the desalinator, assets ($440,000 cash) left the system. Why is a capital cost not included in an asset balance? This doesn't seem correct. Did we neglect something when the Conservation of Asset, Eq. (3.87), was converted to economic terminology in Eq. (3.88)? Specifically, should not the term "rate of assets out" in Eq. (3.87) convert to "operating costs *plus capital costs*" in Eq. (3.88)?

The key is that capital expenses are not a net flow of assets out of your company. Rather, capital expenses are a *conversion* of cash into equipment. Although cash has left the system (your company), equipment of equal value has entered the system, as shown in Figure 3.34.

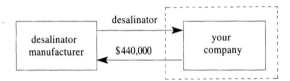

Figure 3.34. Asset flows during your desalinator purchase.

The net asset balance is zero. Thus, the asset balance, Eq. (3.88), is actually

$$profit\ (or\ loss) = rate\ of\ assets\ in - rate\ of\ assets\ out \tag{3.88}$$

$$= (revenue + equipment\ purchased)$$
$$- (operating\ costs + capital\ costs), \tag{3.96}$$

and because

$$equipment\ purchased = capital\ costs. \tag{3.97}$$

Equation (3.96) reduces to Eq. (3.88),

$$profit\ (or\ loss) = revenue - operating\ costs. \tag{3.88}$$

The equality in Eq. (3.97) is not to be taken literally. A piece of equipment worth $1,000 is not the same as $1,000 in cash. If you want to buy something, cash is a better form for one's assets. If you want to produce something, equipment is a better form for one's assets. Equation (3.97) states only that the asset value of the equipment is equal to the capital cost, in units of $/year.

But wait. One could draw an asset flowsheet similar to Figure 3.34 for anything your company purchases. Does this mean that every purchase should be excluded from the operating cost? Consider the ammonia synthesis plant in Chapter 2, which had the overall reaction $4CH_4 + 5O_2 + 2N_2 \rightarrow 4NH_3 + 4CO_2 + 2H_2O$. The ammonia synthesis plant buys methane from a local utility, as shown in Figure 3.35. The similarity between Figures 3.34 and 3.35 suggests that the utility bill for methane should not be included as an operating cost, for the same reason that

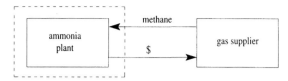

Figure 3.35. Asset flows during your methane purchase.

the payment for the desalinator was not an operating cost. In contrast to the desalinator, the methane does not remain as an asset within the ammonia plant; it is consumed by chemical reaction. The flowsheet should show that the methane is consumed when it enters the system, perhaps by extending the arrow to a "sink" (as opposed to a source). This "sink" is the equivalent of a black hole within the system borders.

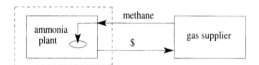

Figure 3.36. Asset flows during operation of your ammonia process.

We must modify Eq. (3.87) to account for assets that leave a system without crossing the system borders. We need to add another term: *rate of consumption*. Similarly, we must add a term to account for the ammonia asset that is created within the system border. The second new term is *rate of formation*. The Conservation of Assets is modified as follows:

$$\frac{d(assets)}{dt} = rate\ of\ assets\ in - rate\ of\ assets\ out$$
$$+ rate\ of\ formation - rate\ of\ consumption. \tag{3.98}$$

Equation (3.98) reverts to Eq. (3.88) if we redefine revenues as

$$revenue = rate\ of\ assets\ in + rate\ of\ formation \tag{3.99}$$

and we redefine operating costs as

$$operating\ costs = rate\ of\ assets\ out + rate\ of\ consumption. \tag{3.100}$$

For reactants, *rate of assets in = rate of consumption* and there is a net decrease in assets, caused by the flow of dollars out of the system to pay for the reactants. For capital expenses, such as a piece of equipment, the asset is not consumed, at least not immediately. This is the distinction between items purchased as capital costs and items purchased as operating costs.

But the asset value of a capital item decreases with time. After many years of use, the desalinator will have little value other than as scrap metal. As such, even capital assets are eventually consumed. The rate of consumption of capital assets is called *depreciation*. As defined in Eq. (3.100), depreciation is an operating cost. The definition for operating costs is therefore

$$operating\ costs = expenses + depreciation. \tag{3.101}$$

Depreciation includes only the rate of consumption of capital assets. We do not include the consumption of a noncapital item, such as a reactant, because its rate of consumption is canceled by its rate of flow into the system.

How does one calculate the decrease in asset value of a capital item, such as a piece of equipment? Depreciation is seldom calculated by determining the actual worth of the equipment. Depreciation is usually determined by tax laws, because the rate of depreciation affects the profit and profit is taxed. Thus, it is possible to have equipment that is still useful but has been depreciated to zero value. It is also possible to have useless equipment that still has value in the eyes of the Internal Revenue Service (IRS).

There are several algorithms for calculating depreciation. The simplest is the straight-line method, given by

$$depreciation = \frac{capital\ costs}{lifetime\ of\ equipment}. \tag{3.102}$$

The lifetime of equipment is dictated by accounting conventions and tax code, not necessarily the equipment's productive life. A typical lifetime is ten years.

Now that we know how to account for the capital items, let's return to the second example of calculating the profit (or loss) of buying and operating a desalinator. We will assume the same conditions as in the first example and add depreciation, assuming a straight-line 10-year rule:

$$depreciation = \frac{440,000\ \$}{10\ years} = 44,000\ \$/year. \tag{3.103}$$

Applying Eqs. (3.89) and (3.102), we get

$$profit\ (or\ loss) = revenue - operating\ costs \tag{3.88}$$
$$= revenue - (expenses + depreciation), \tag{3.104}$$

and substituting from Eqs. (3.91), (3.95), and (3.104), we have

$$profit\ (or\ loss) = 438,000\ \$/year - (293,994\ \$/year + 44,000\ \$/year) \tag{3.105}$$
$$= 100,006\ \$/year. \tag{3.106}$$

Another financial parameter important to company executives and investors is the *Return on Investment*, or ROI, defined by

$$return\ on\ investment = \frac{profit}{capital\ cost}. \tag{3.107}$$

Calculating the return on investment for our desalinator is trivial:

$$return \ on \ investment = \frac{100,006 \ \$/year}{440,000 \ \$} = 23 \ \%/year. \tag{3.108}$$

Test your ability to analyze process economics. Consider an alternative scheme that reduces the cost of labor but increases the capital cost. Assume that one employee can operate three desalinators. Also assume that the space you are renting on the pier can accommodate three desalinators. Calculate the profit and return on investment of purchasing three desalinators and operating for one 9-hour shift per day, 365 days per year. Although the worker is there for 9 hours a day, the desalinators are running for only 8 hours a day – it takes time to start up and shut down each day. We calculate a profit of 80,802 $/year and a substantially reduced return on investment of 6%.

As illustrated by our analysis of the desalinator, the profitability of a process is a balance between capital costs and operating costs. Capital costs are the expenses of building. The chief capital costs, in approximate order of appearance when building a chemical process, are the costs to

· acquire a site (rent or buy);
· develop a site (or adapt an existing site);
· add services – plumbing and heating, ventilation, lighting and electrical, etc.;
· purchase chemical processing equipment – major units are reactors, compressors, mixers, separators, and heat exchangers; and
· install processing equipment – piping, wiring, and controls.

Operating costs are the expenses of manufacturing. The chief operating costs, also known as plant costs or manufacturing costs, are the costs of

· materials – reactants, fuels;
· labor;
· maintenance and repair;
· plant overhead – administration, custodial, accounting, etc.;
· distribution costs – labor, advertising, market research, etc.;
· fixed costs (fees paid even if nothing is produced) – rent, interest, property taxes, insurance; and
· depreciation.

In addition to calculating the expenses of a complete process, one can also calculate the economic impact of modifying a process. Consider the precooler we added to the desalinator, shown in Figures 3.24 and 3.29. The precooler saves energy and thus lowers operating cost, but it requires an additional capital cost. Is a precooler justified economically? Let's apply mathematical modeling to compare the two options: no precooler and precooler.

Option 1: No precooler for the seawater. Let's take this option as our basis.

---

additional capital cost: 0
additional operating cost: 0
result: no change in profit

Option 2: Add a precooler. We must calculate the energy saved and convert to the units of the asset balance, $/year. The energy saved is the energy we would have expended cooling the seawater to 12°C from 20°C. The energy removed from the seawater is

$$q_{\text{seawater}} = F_{\text{seawater}} C_{P,\text{seawater}} (\Delta T)_{\text{seawater}} \tag{3.109}$$

$$= \left(\frac{100 \text{ kg}}{\text{min}}\right) \left(\frac{4.18 \text{ kJ/kg}}{(\text{kg})(°\text{C})}\right) (20°\text{C} - 12°\text{C})$$

$$\times \left(\frac{60 \text{ min}}{1 \text{ hr}}\right) \left(\frac{24 \text{ hr}}{1 \text{ day}}\right) \left(\frac{365 \text{ days}}{1 \text{ year}}\right) \tag{3.110}$$

$$= 1.8 \times 10^9 \text{ kJ/year.} \tag{3.111}$$

We must now convert to units of $/year. What is the operating cost to remove 1 kJ of energy by refrigeration? The text by Peters and Timmerhaus, *Plant Design and Economics for Chemical Engineers*, estimates the cost of refrigeration using ammonia in the range of 34°F to be $2.00/ton-day of refrigeration in 1990 (Table 5, p. 815). Converting to mks units and 1998 dollars, we estimate it costs $1 per 10^5 kJ of energy removed by refrigeration. Does this seem reasonable? It is difficult to tell with these units. Let's estimate how much energy is needed to remove 1 kJ by refrigeration. The same table also estimates that electricity costs 0.1 $/kilowatt-hour, which is 0.1 $/3,600 kJ. This yields

$$\left(\frac{1\$}{10^5 \text{ kJ removed by refrigeration}}\right) \left(\frac{3,600 \text{ kJ electricity}}{0.1 \$}\right)$$

$$= \frac{0.4 \text{ kJ electricity}}{1 \text{ kJ removed by refrigeration}}, \tag{3.112}$$

which seems reasonable. We can now calculate the annual cost of removing 1.8×10^9 kJ by refrigeration:

$$\left(\frac{1.8 \times 10^9 \text{ kJ refrigeration}}{\text{year}}\right) \left(\frac{1 \$}{10^5 \text{ kJ refrigeration}}\right) = \$18,000/\text{ year.} \tag{3.113}$$

The precooler costs $30,000 and is guaranteed to last three years. The result for option 2 is

---

additional capital cost: $30,000
additional operating cost: −$18,000/year

We use these data to calculate the profit. There is no change in revenue – the desalinator produces 100,000 gal/day with or without the precooler. The electricity bill decreases by 18,000 $/year. We convince the IRS that because the precooler lasts only three years we should be allowed to use a lifetime of three years. (It is usually better to depreciate equipment as quickly as possible, which decreases the profits on paper and thus decreases taxes.) Thus

$$depreciation = \frac{30,000 \ \$}{3 \ \text{years}} = 10,000 \ \$/\text{year}, \tag{3.114}$$

and the change is profit is

$$additional \ profit \ (\text{or} \ loss) = additional \ revenues - additional \ operating \ cost \tag{3.88}$$

$$= 0 - (-18,000 \ \$/\text{year} + 10,000 \ \$/\text{year}) \tag{3.115}$$

$$= \$8,000 \ \$/\text{year}. \tag{3.116}$$

So we decide to install the precooler.

Now test your skills with another example. After three years, when you are preparing to replace the precooler, a sales representative of an equipment supplier presents a third option. She can deliver a precooler that costs only $22,000, but because it is made of cheaper carbon steel it lasts only two years. Should you change to the cheaper precooler?

Comparing various economic options is often aided by graphing. Options 1 and 2 for the precooler are shown in the Figure 3.37. The cost of option 1 is assumed to be the energy needed to cool the seawater to 12°C from 20°C. The total cost increases linearly with time. The cost of option 2 begins with $30,000 to purchase the precooler; it then increases linearly for 3 years, the life of the precooler. The slope of the curve for option 2 is less than that of option 1. In fact, the difference in slopes is $18,000/year. Option 2 first crosses the curve for option 1 at 1.7 years, the first break-even point. Option 1 is briefly superior again at 3 years, when the precooler must be replaced, but thereafter option 2 remains the better choice.

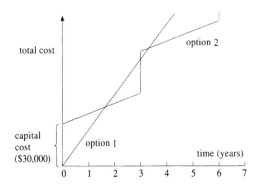

Figure 3.37. Economic comparison of precooler to no precooler.

Consider another example. You decide to insulate the pipes on your desalinator, as shown in Figure 3.38.

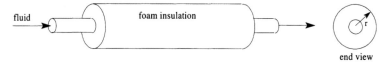

Figure 3.38. An insulated pipe.

What thickness of insulation will optimize your savings after one year? One learns in the chemical engineering course on heat and mass transfer that the heat loss is approximately proportional to $1/r^2$, where r is the outer radius of the insulated pipe. (This approximation is not true in general but it serves here.) Thus the cost of the heat loss is also proportional to $1/r^2$. Figure 3.39 shows the cost of heat loss (which we wish to minimize) as a function of insulation thickness.

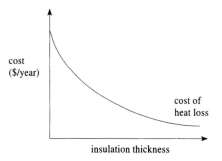

Figure 3.39. Cost of heat loss as a function of thickness.

The operating cost of insulating the steam pipes is the cost of the labor to install the insulation plus the depreciation cost of the insulating material. Assume that the material cost is proportional to the amount of insulation, which is proportional to r^2. Also assume that the cost of installing the insulation is independent of the insulation thickness; the labor to install 3" insulation costs as much as the labor to install 6" insulation. We add the cost of insulating the pipes in Figure 3.40.

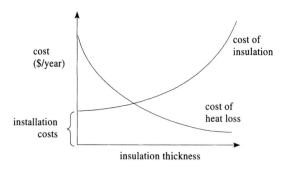

Figure 3.40. Cost of pipe insulation as a function of thickness.

The total cost is thus the sum of these two curves. The optimal cost after one year is the minimum in the total cost curve. How would you adjust the curves in Figure 3.41

to optimize your savings after two years instead of after one year? Would you install thicker or thinner insulation?

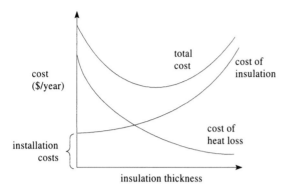

Figure 3.41. Total cost of pipe insulation as a function of thickness.

The capital costs of the pipe insulation follow a trend common to most equipment costs in chemical processes. Capital cost is not linearly proportional to capacity; a unit with twice the capacity does not cost twice as much. In general,

cost of equipment to produce N kg/year

$$= \left(\frac{N}{10^6}\right)^{\alpha} (\text{cost of equipment to produce} 10^6 \text{ kg/year}). \qquad (3.117)$$

A table of exponents for various types of equipment can be found in the textbook *Plant Design and Economics for Chemical Engineers* by Peters and Timmerhaus. Typically, $\alpha = 0.4$ to 1.0.

Consider one last example. Equipment with a capacity to produce 10^5 kg/year is purchased for $\$10^6$. The operating costs are a fixed cost of $\$150,000$/year plus a variable cost of $\$2$/kg. At what price must the product be sold to yield a 20% return on investment? Substituting into the expression for return on investment, Eq. (3.107), we have

$$0.20 = \frac{profit}{10^6}. \qquad (3.118)$$

We need to calculate the profit. From Eq. (3.88), assuming the equipment is operated at full capacity, and using a 10-year straight-line depreciation, we have

$$profit = revenue - operating\ cost \qquad (3.88)$$

$$= \left(\frac{10^5\ \text{kg}}{\text{year}}\right)\left(\frac{price}{\text{kg}}\right)$$

$$- \left[\$1.5 \times 10^5 + \left(\frac{10^5\ \text{kg}}{\text{year}}\right)\left(\frac{\$2}{\text{kg}}\right) + \$10^6\left(\frac{1\ \text{year}}{10\ \text{years}}\right)\right] \qquad (3.119)$$

$$= 10^5 \times price - [\$1.5 \times 10^5 + \$2 \times 10^5 + \$1 \times 10^5] \qquad (3.120)$$

$$= 10^5(price - 4.5). \qquad (3.121)$$

Substitute Eq. (3.122) into Eq. (3.119) to get

$$0.20 = \frac{10^5(price - 4.5)}{10^6} \tag{3.122}$$

$$price = 6.5 \text{ \$/kg.} \tag{3.123}$$

3.15 Summary

Complex designs begin as simple designs. To choose between design alternatives, or to the assess the impact of a design change, one must analyze and evaluate. One method of analysis is mathematical modeling.

Mathematical modeling is the translation of a physical description to mathematical expressions. One translates by applying relations from physics and chemistry. Some relations are universal laws, such as the conservation of energy. Some relations are constrained laws (constitutive equations), such as the ideal gas "law." The behavior of a physical process is governed by the laws of physics and chemistry. The behavior of a model is governed by the rules of mathematics.

In each of the examples in this chapter, we followed three key steps. First we studied the process to gain a qualitative understanding. What trends are expected? For example, how will the purity of the potable water depend on the amount of brine retained by the ice? What behavior is expected at extremes, such as no water in the brine stream? Second, we applied mathematical modeling to calculate a quantity or to yield operating equations. Third, we checked the result. Each step is important.

Mathematical modeling can be used to derive operating equations. These can predict the performance of a process given the operating conditions. For example, Eq. (3.43) predicts the flow rate of potable water given the salt content of the waste brine. And Eq. (3.121) predicts the profit given the price of the commodity produced. These equations comprise *design tools* for their respective processes.

REFERENCES

Felder, R. M., and Rousseau, R. W. 1986. *Elementary Principles of Chemical Processes*, 2nd ed., Wiley, New York.

Himmelblau, D. M. 1996. *Basic Principles and Calculations in Chemical Engineering*, 6th ed., Prentice Hall, Upper Saddle River, NJ.

Luyben, W. L., and Wenzel, L. A. 1988. *Chemical Process Analysis: Mass and Energy Balances*, Prentice Hall, Upper Saddle River, NJ.

Peters, M. S., and Timmerhaus, K. D. 1991. *Plant Design and Economics for Chemical Engineers*, 4th ed., Wiley, New York.

Reklaitis, G. V. 1983. *Introduction to Material and Energy Balances*, Wiley, New York.

Rudd, D. R., Powers, G. J., and Siirola, J. J. 1973. *Process Synthesis*, Prentice Hall, Upper Saddle River, NJ.

EXERCISES

Mass Balances

3.1 Calculate the flow rate (in kg/min) and composition of the effluent from the mixer shown below.

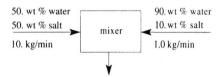

3.2 The unit shown below coats metal parts with plastic. A solution containing the plastic is applied to the metal. The solvent evaporates and leaves a plastic coating on the metal.

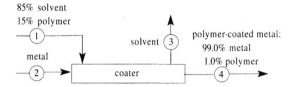

The metal parts enter the coater at a rate of 76 kg/min. Calculate the flow rates (in kg/min) of streams 1, 3, and 4.

3.3 Consider two options for cleaning a paint brush. In each option the brush contains 5.0 mL of paint and you have 0.30 L of solvent. The brush retains 10. mL of dirty solvent after cleaning. Dirty solvent is defined as solvent that contains paint.

Option 1: The paint brush is washed in a container containing 0.30 L of solvent. The brush is withdrawn and the solvent on the brush evaporates.

Option 2: The solvent is divided into three containers in 0.10 L aliquots. The brush is washed in the first container, withdrawn and washed in the second container, and then withdrawn and washed in the third container. The solvent retained by the brush does not evaporate before the brush is washed in the second and third containers. Finally, the brush is withdrawn from the third container and the solvent evaporates.

Calculate the amount of paint that remains on a brush with options 1 and 2.

3.4

(A) Calculate the flow rate (in kg/min) *and* composition of stream 3 for the evaporator shown below.

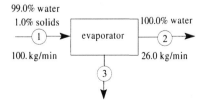

(B) Four evaporators are arranged in a cascade. Calculate the flow rate (in kg/min) *and* composition of stream 9 for the process shown below.

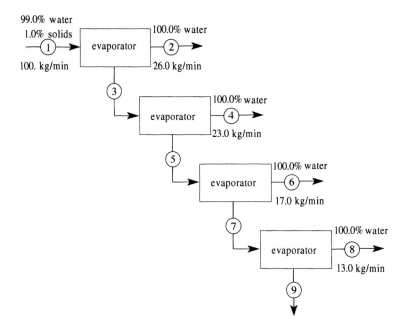

(This exercise appeared on an exam. It was estimated that it could be completed in 15 minutes.)

3.5 An artificial kidney, diagrammed below, removes wastes – chiefly urea, creatine, uric acid, and phosphate ions – from blood. Blood passes through tubes permeable to the wastes, and the wastes are absorbed by a dialyzing fluid. In this exercise we consider only the removal of urea. The flow rate of blood through the artificial kidney is 0.200 kg/min. The concentration of urea is 2.15 g urea/kg blood in the input stream and 1.60 g urea/kg blood in the output stream.

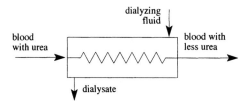

(A) Given that the flow rate of dialyzing fluid is 1.00 kg/min, calculate the concentration of urea in the dialysate (in g urea/kg dialysate).

(B) How long does it take to remove 5 g of urea from the blood (in minutes)?

(Adapted from Felder, R. M., and Rousseau, R. W. 1986. *Elementary Principles of Chemical Processes*, 2nd ed., Wiley, New York, exercise 22, p. 157. Reproduced by permission.)

3.6 The absorber diagrammed below removes benzene from contaminated air.

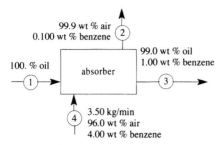

99.9 wt % air
0.100 wt % benzene ②

99.0 wt % oil
1.00 wt % benzene

100. % oil

absorber

①

③

3.50 kg/min
④ 96.0 wt % air
4.00 wt % benzene

Calculate the flow rates of streams 1 and 2 (in kg/min).

(This exercise appeared on an exam. It was estimated that it could be completed in 20 minutes.)

3.7 The separator shown below produces pure acetone from a mixture of acetone and toluene.

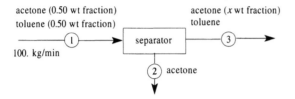

acetone (0.50 wt fraction)
toluene (0.50 wt fraction)

acetone (x wt fraction)
toluene

①

separator

③

100. kg/min

② acetone

(A) Derive an equation for the flow rate of stream 2 as a function of x.

(B) Check your equation at $x = 0.50$. Does your predicted flow rate for stream 2 seem correct?

(C) Check your equation at a different value of x. Choose a value of x that yields an obvious answer for the flow rate of stream 2.

(This exercise appeared on an exam. It was estimated that it could be completed in 15 minutes.)

3.8 Orange juice concentrate can be prepared in one step as shown below.

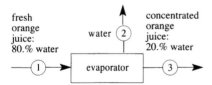

fresh
orange
juice:
80.% water

water ②

concentrated
orange
juice:
20.% water

①

evaporator

③

(A) The flow rate of stream 1 is 377 kg/min. Calculate the flow rate of stream 2 from the evaporator.

The above process compromises the taste of fresh orange juice. The taste is improved markedly by blending a small amount of fresh orange juice with the output from the evaporator. To improve the taste, the process is modified such that the fresh orange juice stream is split; one fraction goes to the evaporator and the other fraction bypasses the evaporator and is mixed with the concentrate. The evaporator is modified to remove a larger fraction of the water entering the evaporator.

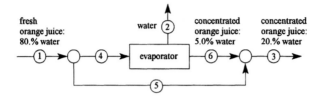

(B) The flow rate of stream 1 is 377 kg/min. Calculate the flow rate of water from the evaporator in the modified process.

(C) Calculate the flow rates of streams 4 and 5 that will produce an orange juice concentrate (stream 3) in which 5.0% of the nonwater component has bypassed the evaporator.

Note: Parts (B) and (C) can be simplified by judicious choices of system borders.

(From a suggestion by G. "Chip" Bettle, Cornell '65, '66.)

3.9 The steady-state process shown below extracts oil from soybeans.

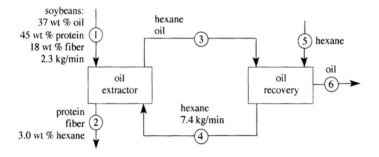

(A) Calculate the flow rate and composition of stream 2.

(B) Calculate the flow rate of stream 6.

(C) Calculate the flow rate of stream 5.

(Adapted from Reklaitis, G. V. 1983. *Introduction to Material and Energy Balances*, Wiley, New York, exercise 2.29, p. 99, and Rudd, D. F., Powers, G. J., and Siirola, J. J. 1973. *Process Synthesis*, Prentice Hall, Upper Saddle River, NJ, problem 7, p. 94.)

3.10 Air can be dried by bubbling through sulfuric acid. The absorber below passes air countercurrent to a sulfuric acid solution. 99.0 wt% sulfuric acid is added to the recycled sulfuric acid solution to maintain a concentration of 94.0 wt% sulfuric acid into the absorber.

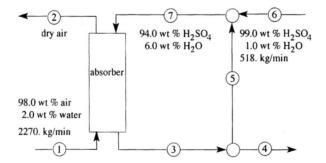

(A) Calculate the flow rate *and* composition of the effluent, stream 4.

(B) Calculate the flow rate *and* composition of the recycle, stream 5.

(Adapted from Luyben, W. L., and Wenzel, L. A. 1988. *Chemical Process Analysis: Mass and Energy Balances*, Prentice Hall, Upper Saddle River, NJ, exercise 4-9, p. 107. This exercise appeared on an exam. It was estimated that it could be completed in 25 minutes.)

3.11 Shown below is a portion of a process that produces styrene from benzene and ethylene (see Exercise 2.12). Benzene is recycled to an alkylation reactor and the by-product ethylbenzene is recycled to a dehydrogenation reactor. Stream 2 is 28.0% of stream 1. Also, 97.0% of the ethylbenzene in stream 3 leaves distillation column 2 via stream 4. Calculate the flow rate and composition of stream 5.

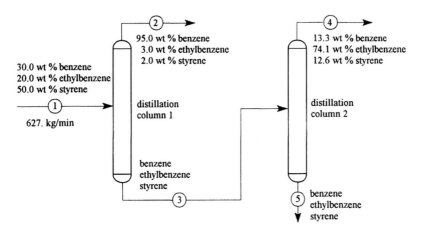

3.12 The process shown below removes water from a suspension of solid, insoluble particles of P.

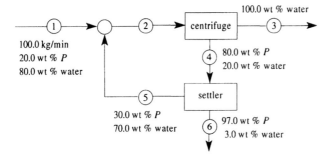

(A) Calculate the total flow rate of stream 6.

(B) Calculate the flow rate of the recycle, stream 5.

(Adapted from Himmelblau, D. M. 1996. *Basic Principles and Calculations in Chemical Engineering*, 6th ed. Prentice Hall, Upper Saddle River, NJ, pp. 210–212. This exercise appeared on an exam. It was estimated that it could be completed in 15 minutes.)

3.13 The synthesis of nitric acid from ammonia produces a solution of 60% HNO_3/40% H_2O. Concentrated nitric acid (99%) cannot be obtained by distillation owing to an azeotrope at

68% HNO_3. However, adding $Mg(NO_3)_2$ to dilute nitric acid allows one to distill a solution with greater than 68% HNO_3, which can then be distilled to produce concentrated nitric acid, as in the process diagrammed below.

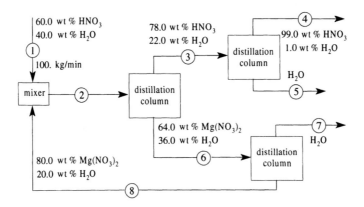

(A) Calculate the flow rate of the product, stream 4, in kg/min.

(B) Calculate the flow rate of the recycle, stream 8, in kg/min.

(C) The dilute nitric acid contains an impurity, $FeNO_3$. To prevent the impurity from accumulating in the process, the recycle stream is purged and fresh $Mg(NO_3)_2$ is added. 5.0 wt% of stream 8 is purged via stream 9. This is shown in the flowsheet below. Calculate the composition *and* flow rate of stream 10.

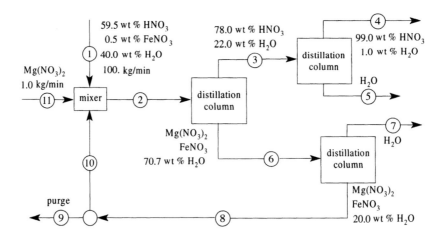

(Parts (A) and (B) of this exercise appeared on an exam. It was estimated that they could be completed in 30 minutes. Adapted from Luyben, W. L., and Wenzel, L. A. 1988. *Chemical Process Analysis: Mass and Energy Balances*, Prentice Hall, Upper Saddle River, NJ, exercise 4-28, p. 117.)

3.14 Instant coffee (the soluble portion of ground roasted coffee) is produced by the process shown in the simplified flowsheet on the following page.

Calculate the rate of production of dried soluble coffee (stream 9). Note that the ratio of water to soluble components is the same in streams 3, 4, 5, 6, and 7.

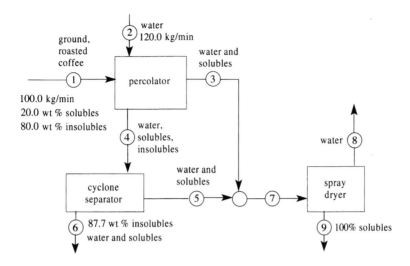

(Adapted from Reklaitis, G. V. 1983. *Introduction to Material and Energy Balances*, Wiley, New York, pp. 98–9, exercise 2.27; from Felder, R. M., and Rousseau, R. W., 1986. *Elementary Principles of Chemical Processes*, 2nd ed., Wiley, New York, pp. 161–2, problem 32; and Rudd, D. F., Powers, G. J., and Siirola, J. J. 1973. *Process Synthesis*, Prentice Hall, Upper Saddle River, NJ, pp. 69–70.)

3.15 Air at 95°F can hold a maximum of 3.52 wt% water, defined as 100% humidity at 95°F. On a hot humid day the temperature is 95°F and the air contains 3.20 wt% water (90.9% humidity). The air is cooled to 68°F. Because air at 68°F can hold a maximum of 1.44 wt% water (100% humidity at 68°F), water condenses from the air.

Hot, humid air (95°F, 90.9% humidity) flows into a cooler at a rate 154.0 kg/min. Calculate the flow rates of the two streams leaving the cooler: (1) air at 68°F and 100% humidity and (2) the water condensed from the air.

(This exercise appeared on an exam. It was estimated that it could be completed in 10 minutes.)

3.16 Air with 100% humidity is not pleasant, even at 68°F. Air with 50% humidity at 68°F can be produced by cooling some air below 68°F and then mixing with some hot humid air at 95°F. This process is shown in the schematic below.

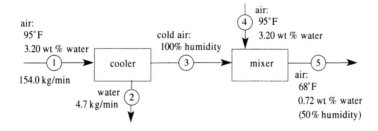

Calculate the flow rate of stream 4, the hot, humid air into the mixer.

(This exercise appeared on an exam. It was estimated that it could be completed in 15 minutes.)

3.17 The chemical Q reacts to form Z. Unreacted Q is separated from Z and recycled to the reactor. The feed contains an impurity, P, which is inert and is purged from the system via stream 7. The splitter purges 5.0% of stream 5. Note that a mass balance on Q must account for the Q that reacts to form Z. Likewise a mass balance on Z must account for the Z formed from Q.

(A) Which stream has the highest flow rate of Q?

(B) Calculate the flow rate of product stream 4, in kg/min.

(C) Calculate the composition of purge stream 7.

(D) Calculate the flow rate *and* composition of stream 2.

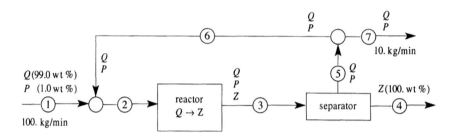

(This exercise appeared on an exam. It was estimated that it could be completed in 30 minutes.)

3.18 The moisture content of a solid is reduced to 8.0% from 30.0% in the unit diagrammed below. Some of the hot, wet air leaving the dryer is recycled, such that the air entering the dryer contains 1.0% water.

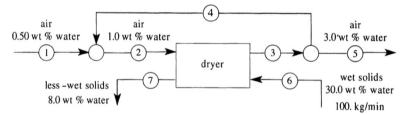

(A) Calculate the flow rate (in kg/min) of stream 7.

(B) Calculate the flow rate (in kg/min) of the recycle, stream 4.

(Adapted from Hougen, O. A., and Watson, K. M. 1943. *Chemical Process Principles*, Wiley, New York, p. 183; Henley, E. J., and Rosen, E. M. 1969. *Material and Energy Balance Computations*, Wiley, New York, pp. 480–1; and Luyben, W. L., and Wenzel, L. A. 1988. *Chemical Process Analysis: Mass and Energy Balances*, Prentice Hall, Upper Saddle River, NJ, pp. 98–100. This exercise appeared on an exam. It was estimated that it could be completed in 35 minutes.)

3.19 The process shown on the following page removes benzene from air by adsorption onto the surfaces of porous solids called *zeolites*. The adsorption unit shown below has a countercurrent flow of zeolites, and 1.00 kg zeolite can adsorb 0.10 kg benzene. Stream 6 purges 2% of the zeolite in stream 5.

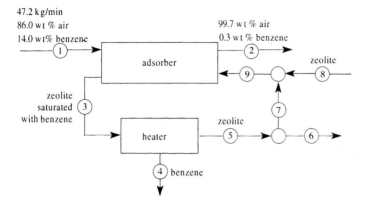

Calculate the flow rates of streams 4 and 8.

(This exercise appeared on an exam. It was estimated that it could be completed in 20 minutes.)

3.20 A valuable enzyme E is obtained from a biochemical process as a dilute water solution. Because E is heat-sensitive, it is purified by the three-step process shown below. Purge streams have been omitted.

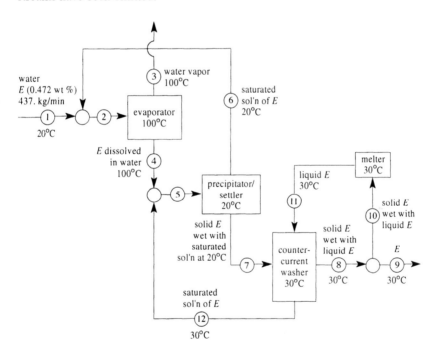

Temperature	Wt% E in a saturated solution
20°C	5.71
30°C	7.23
100°C	31.0

(A) Calculate the flow rate of stream 3, the water vapor leaving the evaporator.

(B) The total flow rate of recycle stream 6 is 35.0 kg/min. Calculate the flow rate and composition of stream 4, the liquid leaving the evaporator.

(C) The flow rate of stream 7 is 7.06 kg/min. Calculate the flow rate and composition of recycle stream 12.

(D) Stream 7 contains solid E and E in solution. Calculate the flow rate of solid E in stream 7.

(This exercise appeared on an exam. It was estimated that the exercise could be completed in 35 minutes.)

3.21 We wish to react acetylene ($HC \equiv CH$) and hydrogen chloride (HCl) to produce vinylchloride ($CH_2 = CHCl$), which is used to manufacture polyvinylchloride (PVC):

$$C_2H_2 + HCl \rightarrow C_2H_3Cl \quad \text{(desired reaction)}.$$

We wish to minimize the subsequent reaction of vinylchloride to dichloroethane:

$$C_2H_3Cl + HCl \rightarrow C_2H_4Cl_2 \quad \text{(undesired reaction)}.$$

Consider the three process schemes diagrammed below. All compositions are mol percent.

Scheme I.

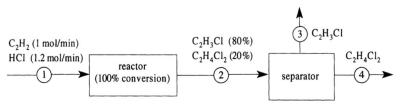

Scheme II.

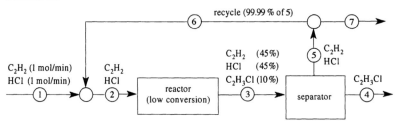

Scheme III.

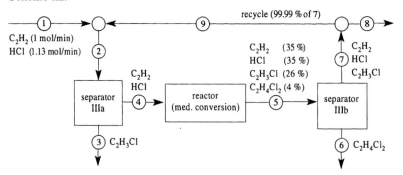

Some physical properties at 1 atm

	C_2H_2	HCl	C_2H_3Cl	$C_2H_4Cl_2$
molecular wt	26	36.5	62.5	99
melting point (°C)	-81^1	-115	-154	-97
boiling point (°C)	-84^2	-85	-13	57

[1] At 1.2 atm. [2] Sublimation of solid.

The following questions can be answered by comparing the qualitative differences in the three schemes. You need not calculate the flow rates. **Briefly** explain your qualitative analysis.

(A) Which scheme has the highest rate of production of desired product, C_2H_3Cl? **Explain.**

(B) Which scheme has the highest rate of production of undesired by-product, $C_2H_4Cl_2$? **Explain.**

(C) In which scheme is the reactor the largest, as measured by the total mol/min through the reactor? **Explain.**

(D) Which separator is the largest, as measured by the total mol/min through the separator? **Explain.**

(E) What are the approximate average temperatures in separators IIIa and IIIb? **Explain.**

(F) Improve the design of Scheme III. Your improved scheme must use the reactor with medium conversion. Focus on minimizing the size(s) of the separator(s). Ignore heat exchangers.

(This exercise appeared on an exam. It was estimated that it could be completed in 25 minutes.)

3.22 The process to desalinate water is improved vastly by recycling some of the melted water to wash the ice in the skimmer, as shown in the flowsheet below. With an efficient skimmer, the recycled wash stream, 6, is only 5% of stream 5. The product stream, 7, contains a negligible amount of salt. Given that the brine leaving the skimmer is 7.0 wt% salt, calculate the flow rate of stream 7.

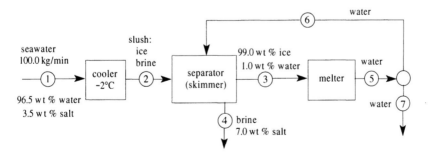

3.23 The flash drum shown on the following page separates a mixture of methanol and ethanol into a vapor stream and a liquid stream.

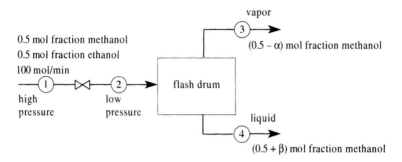

Calculate the flow rate of stream 3 (in mol/min) in terms of α and β. If you derived this correctly you proved the lever rule (see Chapter 4). You are not permitted to invoke the lever rule to calculate the flow rate of stream 3.

(This exercise appeared on an exam. It was estimated that it could be completed in 15 minutes.)

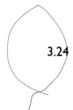

3.24 In this exercise the oxidation of glucose and the elimination of oxidation products are modeled. The chemical reaction is

$$C_6H_{12}O_6 + xO_2 \rightarrow yCO_2 + zH_2O.$$

(A) Balance the stoichiometry in the above chemical equation (determine x, y, and z).

(B) Consider the following flowsheet.

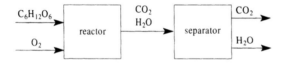

Given that the flow rate of glucose is 0.10 kg/day, calculate the stoichiometric flow rate of oxygen (in kg/day).

(C) Calculate the flow rate (kg/day) and composition (wt% CO_2 and wt% H_2O) in the stream leaving the reactor.

(D) Calculate the flow rates of the two streams leaving the separator.

3.25 The process diagrammed below proposes to produce hydrogen bromate by the reaction

$$3Ca(BrO_3)_2 + 2H_3PO_4 \rightarrow Ca_3(PO_4)_2 + 6HBrO_3.$$

Calculate the rate of production of $HBrO_3$. That is, calculate the flow rate of stream 7.

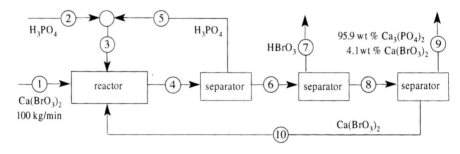

Hint: Calculate the composition of a stream in terms of cations (Ca^{2+} or H^+) and anions (BrO_3^- or PO_4^{3-}). For example, stream 1 is 13.5 wt% Ca and 86.5 wt% BrO_3.

(Adapted from Reklaitis, G. V. 1983. *Introduction to Material and Energy Balances*, Wiley, New York, exercise 3.13, p. 168. This exercise appeared on an exam. It was estimated that it could be completed in 30 minutes.)

3.26 Methane (CH_4) is burned in air (21.0 mol% O_2, 79.0 mol% N_2) to yield CO_2 and water.

(A) Write a balanced chemical equation for this reaction.

(B) The inputs to the burner are 4.70 kg/min CH_4 and 92.0 kg/min air. Assume that the CH_4 is completely consumed in the burner. Calculate the flow rate and composition (by wt) of the exhaust from the burner.

3.27 Coal can be transformed to methanol (CH_3OH) by the process shown below. The partial oxidation reactor converts coal into CO_2, CO, and H_2 in molar ratio of 1:3:5.

$$coal + O_2 + 3H_2O \rightarrow CO_2 + 3CO + 5H_2.$$

CO and CO_2 are converted to methanol by the following reactions:

$$CO_2 + 3H_2 \rightarrow CH_3OH + H_2O,$$

$$CO + 2H_2 \rightarrow CH_3OH.$$

To convert all the CO and CO_2 leaving the partial oxidation reactor, we need 3 mol of H_2 to react with the CO_2 and $3 \times 2 = 6$ mol of H_2 to react with the CO, for a total of $3 + 6 = 9$ mol of H_2. Because only 5 mol of H_2 leave the partial oxidation reactor, some of the CO_2 and CO are removed before the mixture is fed to the methanol reactor.

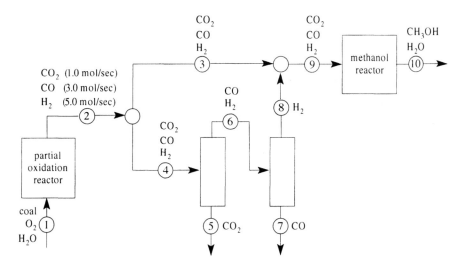

(A) What flow rate in stream 4 (mol/sec) will deliver a stoichiometric mixture (CO_2: $H_2 = 1{:}3$ and $CO{:}H_2 = 1{:}2$) into the methanol reactor?

(B) The process to convert coal into methanol is modified slightly. The second distillation column is changed to remove some, but not all, of the CO. The flow rates in streams 3 and 4 are equal.

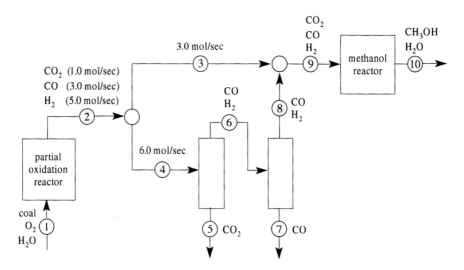

What composition of stream 8 will deliver a stoichiometric mixture (CO_2:H_2 = 1:3 and CO:H_2 = 1:2) into the methanol reactor?

(Adapted from Luyben, W. L., and Wenzel, L. A. 1988. *Chemical Process Analysis: Mass and Energy Balances,* Prentice Hall, Upper Saddle River, NJ, exercise 4-32, p. 119. Part (A) appeared on an exam. It was estimated that it could be completed in 20 minutes.)

Mass Balances on Spreadsheets

Spreadsheets are adept at mass balances. The mass balance exercises in this chapter may be solved with spreadsheets. Indeed, most commercial software to model chemical processes are embellished spreadsheets. We offer here a tutorial in two examples: a simple separation and a process with a recycle. The tutorial assumes a basic knowledge of spreadsheets, such as how to use a mouse to select a cell and how to copy and paste.

Spreadsheet Example 1   The process below separates grain extract (50 kg/min, 97 wt% water and 3 wt% ethanol) into a product stream and a waste stream. The product stream must recover 99% of the ethanol in the grain extract. In addition, the product stream must be 50 wt% ethanol.

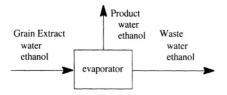

(A) Calculate the total flow rates of the product and waste streams. Also calculate the flow rates of water and ethanol in the product and waste streams.

(B) Repeat the calculation in (A), but with a grain extract composed of 98.5 wt% water, 1.5 wt% ethanol.

(C) Repeat the calculation in (A), but with the relaxed criterion that only 95% of the ethanol is recovered.

Solution to Spreadsheet Example 1 A spreadsheet solution to this exercise is shown on the following page. The steps below will lead you through the process of creating this spreadsheet. The steps will refer to locations on the spreadsheet, known as cells, by their column (an uppercase letter) and their row (a number).

1. Enter the title. Type "Example 1" in cell A1. For those new to spreadsheets, first select cell A1 with the mouse. This should highlight the border around the cell. Then type "Example 1" and hit *enter*.
2. Enter the list of process specifications. Type "Process Specifications" in cell A3. Do not be concerned that the text extends beyond cell A3. Because there is nothing in the adjacent cell, the text appears in full. Enter the other specification descriptions in cells A4 through A7, and enter their values in cells C4 through C7 as shown in the example.
3. Draw the flowsheet for the process. Most spreadsheet software allows one to open a drawing palette or drawing toolbar to create the rectangles, circles, and arrows you will need. In Microsoft Excel®, pull down the *Options* menu to list *Toolbars*. Select the *Drawing* toolbar. Practice drawing various objects, moving the objects, resizing objects, and reformatting patterns such as line widths and arrowhead sizes.

   After experimenting, create a flowsheet with the general features shown in the example spreadsheet. Place an evaporator in column G with two stream arrows on the border between rows 12 and 13 and a stream arrow extending upward in column G.
4. Label the input stream. Type "Grain Extract" in cell D8, list the components in cells D9 and D10, and type "Total" in cell D11.
5. Calculate the input flow rate and input composition. Do not copy the example; do not type "48.5," "1.5," and "50" into cells E9, E10, and E11. These cells show numbers but actually contain formulas. The formulas do not appear in the example spreadsheet printed here, only the values calculated by the formulas.

   Let's begin with what we know. The total flow rate is one of our specifications. Type the simple formula "= C4" into cell E11. The number "50" should appear in cell E11 after you hit *enter*. The flow rate of ethanol is not specified explicitly, but we can write a simple formula to calculate it. The fraction of ethanol in the feed is one of the process specifications; it appears in cell C5. Type the formula "= E11* C5" into cell E10. What is the water flow rate? Again, it is not specified but is trivial to calculate. From a mass balance, water = total − ethanol. Translate this into a formula for cell E9; type "= E11 − E10."

   It is tempting to calculate the flow rates for water and ethanol in one's head, and then type in the values. Don't. The power of spreadsheets lies in referencing to key specifications. To illustrate, change the ethanol fraction (cell C5) to 0.015. Because you typed formulas, the numbers in cells E9 and E10 change. Similarly, try changing the total flow rate of grain extract in cell C4. Now reset all specifications back to the given values.

   As veterans of spreadsheets will attest, creating formulas is expedited by selecting cells rather than typing cell addresses. For example, the formula for cell E10, the ethanol flow rate, can be created by first typing "=" into cell E10. This tells the spreadsheet you are entering a formula. Then select cell E11 with the mouse. The characters "E11" will appear in cell E10. Type the symbol for multiplication, "*." Now select cell C5 with the mouse. This completes the formula. Hit *enter*.

	A	B	C	D	E	F	G	H	I	J	K
1	Example 1										
2											
3	Process Specifications										
4	flow rate of grain extract		50								
5	fraction of ethanol in feed		0.03								
6	fraction of ethanol recovered		0.99								
7	fraction of ethanol in product		0.5								
8				Grain Extract	(kg/min)			Product	(kg/min)	Waste	(kg/min)
9				water	48.5			water	1.485	water	47.015
10				ethanol	1.5			ethanol	1.485	ethanol	0.015
11				Total	50			Total	2.97	Total	47.03
12							Evaporator				
13											
14	Example 2										
15											
16	Process Specifications										
17	flow rate of feed		100								
18	fraction of grain in feed		0.2								
19	solution/grain ratio in slurry		0.1								
20	fraction of slurry purged		0.05								
21							Reactor				
22	Feed	(kg/min)		Reactor Feed	(kg/min)			Effluent			
23	grain	20.		grain	69.57		grain	52.17			
24	water	80.		water	84.51		water	93.21			
25				ethanol	0.44	yeast	ethanol	9.14			
26	Total	100.		Total	154.52	reactor	Total	154.52			
27						efficiency=					
28						0.25	solution composition		Filter	Grain Extract	(kg/min)
29							water	0.91		water	88.46
30							ethanol	0.09		ethanol	8.67
31										Total	97.13
32											
33											
34											
35				Recycle	(kg/min)			Purge	(kg/min)	Slurry	(kg/min)
36				grain	49.57			grain	2.61	grain	52.17
37				water	4.51			water	0.24	water	4.75
38				ethanol	0.44			ethanol	0.02	ethanol	0.47
39				Total	54.52			Total	2.87	Total	57.39

6. Calculate the product stream. The fraction of ethanol recovered is one of the process specifications. For cell I4, type the formula "= C6 * E10." The fraction of ethanol in the product, a process specification we listed in cell C7, can be used to calculate the total flow rate. Into cell I5 type the formula "= I4/C7." Finally, the water flow rate is obtained as before, by the formula "water = total − ethanol." Type "= I5 − I4" into cell I3.

   Because the formula for water in the product stream is the same as the formula for water in the input stream, we could have copied the formula from cell E9 into cell I3. Try this. First, delete the formula in cell I3. Select cell E9 and *copy*. Now select cell I3 and *paste*. Note that the formula pasted into cell I3 is not "= E11 − E10" as it appeared in cell E9, but is just what we wanted, "= I5 − I4." Unless one specifies otherwise, formulas use relative references, not absolute references. The formula for cell E9 is actually "subtract the number in the cell below from the number two cells below."

7. Calculate the waste stream. There are many ways to accomplish this. If you followed steps 1 through 6 you can probably do this step on your own. Try it.

   Compare your formulas to the formulas we devised. The total waste flow obtains from a total mass balance on the entire process, "waste = input − product." For cell J11, we used the formula "= E11 − I5." Likewise the ethanol flow rate obtains from an ethanol mass balance on the entire process. For cell J10, we used the formula "= E10 − I4." And we entered the formula by pasting, rather than typing. We selected cell J11 and copied, and then selected cell J10 and pasted. Likewise for cell J9.

   Clearly, there are many ways to calculate the waste stream. If you get the correct values, your process is correct. Indeed, there are many ways to model this process.

Your spreadsheet should now resemble the example given. Learn how to change the formats for the numbers to show only the significant figures, or perhaps one or two more than significant. Adjust the column widths to fit more of your spreadsheet onto the monitor screen.

Now complete parts (B) and (C) of Example 1. Change only the numbers in the process specifications: the numbers in column C. You need not change any formulas in your spreadsheet. These relations are correct, regardless of the process specifications. Parts (B) and (C) are intended to illustrate the utility of a spreadsheet for exploring the effects of changing process specifications.

Example 1 was a simple mass balance. You probably could have worked it with pencil and paper in less time than it took to create the spreadsheet. The advantage of spreadsheets lies in modeling more complicated processes, such as Example 2.

Spreadsheet Example 2   The grain extract in Example 1 is produced by the process on the following page. Ethanol is synthesized by yeast in the reactor. In this idealistic process, the yeast converts 2 kg of grain into 1 kg of ethanol and 1 kg of water. A perfectly efficient yeast reactor (efficiency = 1) would convert all the grain entering the reactor. A reactor with efficiency = 0.5 would convert half the grain entering the reactor, and so on. Here are some additional specifications:

· The feed is 100. kg/min, 20 wt% grain, and 80 wt% water.
· The reactor efficiency is 0.25.
· The grain in the slurry retains 1 kg of ethanol/water solution for every 10 kg of grain.
· 5 wt% of the slurry is purged.

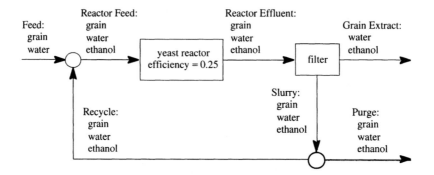

(A) Calculate the flow rates of all streams. Also calculate the flow rates of all components in each stream.

(B) Repeat the calculation in (A), but with a reactor efficiency of 0.10.

(C) Repeat the calculation in (A), but with a purge fraction of 0.01.

(D) Repeat the calculation in (A), but with a feed composed of 10 wt% grain and 90 wt% water.

Solution to Spreadsheet Example 2 We recommend that you prohibit the spreadsheet from iterating on circular references until you are ready. Otherwise, you may encounter frustrating error messages as you create your spreadsheet. For Excel® users, pull down the *Options* menu and open the *Calculations* window. There should be a box labeled *Iterations*. Be sure the box does not contain an "x."

1. As in Example 1, type the title and list the process specifications in column A. Type the values for the specifications in column C.
2. Draw a flowsheet for the process and label the units. This is another activity that benefits from liberal copying and pasting. We chose not to include the reactor efficiency in the list of process specifications (although it would be appropriate to do so). Rather we placed this specification in cell F28, inside the rectangle denoting the reactor.
3. Enter labels for all streams. Your experience in Example 1 showed you it is wise to maintain the same order of components in each stream. That is, the relative position of "grain," "water," and "ethanol" should be the same in all streams. Again, copying and pasting will save time and reduce drudgery.
4. Calculate the input stream in column B. Cells B24, B25, and B26 should all contain formulas in terms of cells C17 and C18. Do not simply copy the numbers from the example flowsheet. The correct numbers will appear in your spreadsheet after you have entered formulas.
5. Calculate the reactor feed flow rates in cells E23 through E26. The reactor feed is the input feed plus the recycle. Because we do not yet know the recycle flow rates, we cannot yet know the flow rates of the reactor feed. But this is not a problem. We only need to know the formulas to calculate the flow rates in the reactor feed.

   The total flow rate of the reactor feed is the total input feed plus the total recycle. The formula for cell E26 is "= B26 + E39." Because the recycle flow rate is presently zero, the total reactor feed is presently 100, and not 154.52 as shown in the completed spreadsheet.

Enter similar formulas for the flow rates of grain, water, and ethanol in the reactor feed.

6. Calculate the reactor effluent flow rates. Again we must devise formulas for each of the component flow rates. The formulas will be in terms of the reactor feed flow rates and the reactor efficiency. The grain in the effluent is the grain not converted to ethanol and water, because the reactor efficiency is less than 1. Thus cell H23 has the formula "= E23 − E23 * F28," or "= E23 * (1 − F28)." The ethanol flow rate is half the flow rate of the grain that reacted plus the ethanol that entered the reactor (from the recycle). Recall that when 2 kg of grain is consumed, 1 kg of ethanol is synthesized. The other 1 kg is water. So the formula for ethanol in the reactor effluent, cell H25, is "= E25 + E23 * F28/2." Similarly, the formula for water in the reactor effluent, cell H24, is "= E24 + E23 * F28/2."

The formula for the total flow rate is trivial; total in = total out. But because the component flow rates are susceptible to errors, it is better to use the total flow rate of the effluent to check the formulas for the components. Set the total equal to the sum of the components. For cell H26 enter "= H23 + H24 + H25." If the reactor effluent total does not equal the reactor feed total, one or more of your formulas in cells H23 through H25 are probably wrong.

Again, because the recycle stream is yet to be calculated, the reactor effluent numbers in your spreadsheet will be different from the example spreadsheet given here.

7. Calculate the slurry flow rates. At this point, the next step is to calculate either the grain extract or slurry streams. The slurry is easier to calculate first because all the grain leaves via the slurry. The formula for the grain in the slurry, cell K29, is thus "= H23."

Water and ethanol appear in the slurry because each kilogram of solid grain retains 0.1 kg of ethanol/water solution. We have found it is easier to first calculate the composition of the solution. This appears in cells H29 and H30. The fraction of water in the solution, cell H29, is thus "= H24/(H24 + H25)." The fraction of ethanol, cell H30, is "= 1 − H29."

The ratio of ethanol/water solution to grain in the slurry is the process specification in cell C19. Thus the formula for water in the slurry, cell K30, is "= K29 * C19 * H29." Similarly, the formula for ethanol in the slurry, cell K31, is "= K29 * C19 * H30."

The total slurry flow rate is the sum of the components. The formula of cell K32 is "= K29 + K30 + K31." This formula can be pasted from cell H26.

8. Calculate the grain extract flow rates. The grain extract is the reactor effluent minus the slurry. Enter formulas for cells K23, K24, and K25.

9. Calculate the purge stream. The purge is a fraction of the slurry, as given by the specification in cell C20. Enter formulas for cells K36, K37, K38, and K39.

10. Calculate the recycle stream. The recycle is the slurry stream minus the purge. Enter formulas for cells E36, E37, E38, and E39.

11. Iterate until values in the cells converge. You have now completed the circular references, so your spreadsheet software should not complain when you ask it to iterate.

For Excel® users, again pull down the *Options* menus and open the *Calculations* window. Select the box labeled *Iterations*.

The spreadsheet is useful for demonstrating the dynamics of starting a process. Set the number of iterations to a small number, something in the range 1 to 10. Then manually induce iterations and watch the numbers converge.

Complete parts (B), (C), and (D) by changing the specifications in column C. As a final challenge link the results of Example 2 to Example 1. That is, change the Example 1 specifications for "flow rate of grain extract" and "fraction of ethanol in feed" to use the values calculated by Example 2.

Use what you have learned in this tutorial to repeat some of the mass balance exercises you solved by pencil and paper for homework assignments or in calculation sessions.

(The two spreadsheet examples presented here are derived from a tutorial developed by Mary Carmen Gascó, Cornell Chemical Engineering '97.)

Energy Balances

3.28   Water is used to cool a stream of cyclohexane (C_6H_{12}) in the heat exchanger shown below. The water is available at 10°C and the maximum discharge temperature is 30°C. What flow rate of cyclohexane can be accommodated with this unit? Assume that the pressure in all streams is 1 atm. Tables of thermodynamic properties appear at the end of this chapter.

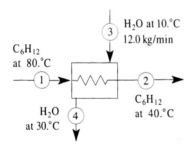

3.29   Steam is used to heat a stream of cyclohexane (C_6H_{12}) in the heat exchanger shown below. What flow rate of cyclohexane can be accommodated with this unit? Assume that the pressure in all streams is 1 atm. Tables of thermodynamic properties appear at the end of this chapter.

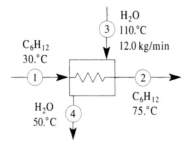

3.30   A unit is designed to lower the temperature of hot, dry air by evaporating water. There is concern that the cooled air (stream 2) contains so much water that it is uncomfortable, even though it is cool.

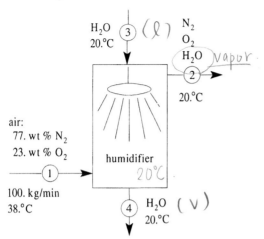

(A) Calculate the weight fraction of water in stream 2. You may assume that the temperature inside the humidifier is 20°C. Thus the water evaporates at 20°C. All the H_2O in stream 2 is vapor. All the H_2O in streams 3 and 4 is liquid. Tables of thermodynamic parameters appear at the end of this chapter.

(B) Air at 20°C can hold a maximum of 1.44 wt% H_2O, defined as 100% humidity. What is the maximum air temperature in stream 1 that can be cooled to 20°C?

(This exercise appeared on an exam. It was estimated that it could be completed in 10 minutes.)

3.31 Cornell University pipes chilled water to buildings for air conditioning and humidity control. Presently the water is chilled by electric-powered refrigerators, which use chlorofluorocarbons (CFCs) as the refrigerant. Cornell proposes to replace the refrigerators and their offending CFCs with lake-source cooling. In the proposed scheme, water would be drawn from deep in Cayuga Lake where the temperature is a constant 40°F. The cold lake water would be used to cool the chilled water to 45°F from 60°F. The lake water would then be returned to Cayuga Lake at 55°F.

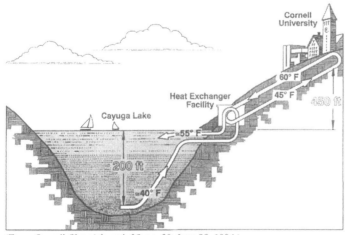

Schematic diagram for proposed lake source cooling system.

(From *Cornell Chronicle*, vol. 25, no. 39, June 30, 1994.)

(A) Estimate the flow rate of chilled water (in gal/hr) needed to cool 2×10^5 m^3 of air per minute to 68°F from 90°F.

(B) Repeat the calculation in (A), but include the effect of condensing water from the air. Assume air at 90°F and 90% humidity (3.02 wt% water) is cooled to air at 68°F and 50.% humidity (0.72 wt% water). For this estimate, you may assume that the water vapor condenses at an average temperature of 68°F and that the heat capacity of air is independent of humidity.

(C) Use the flow rate of chilled water calculated in (B) to calculate the flow rate of water from Cayuga Lake, in gal/hr.

3.32 Superheated steam is used to warm a stream of styrene ($C_6H_5CH{=}CH_2$) in the heat exchanger diagrammed below. Calculate the temperature of the styrene leaving the unit. All streams are at a pressure of 1 atm. Tables of thermodynamic properties appear at the end of this chapter.

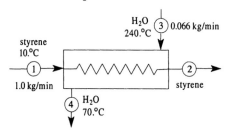

3.33 A stream of *formula X* is to be gently warmed to 80°C from 30°C. Because *formula X* is heat-sensitive, the heating source must be exactly 100°C. The temperature of the heating source is maintained by a steam/water mixture at 1 atm.

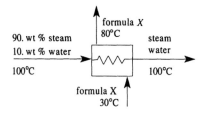

However, steam is available only at 360°C and 1 atm. The process below cools the hot steam before using it to warm *formula X*.

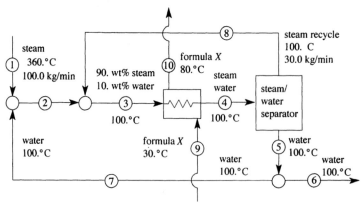

(A) Calculate the flow rate of stream 6.

(B) What flow rate of *formula X* (stream 9) can be accommodated by this design?

(C) What is the flow rate of the water recycle, stream 7?

A table of thermodynamic properties appears at the end of this chapter.

(This exercise appeared on an exam. It was estimated that it could be completed in 30 minutes.)

3.34 Water enters a boiler at 50.°C and exits as two streams: steam at 100.°C and water at 100.°C. The pressure in the boiler is 1 atm. Given that the heater supplies 4.5×10^6 joules/min, calculate the flow rates (in kg/min) of the two exit streams 2 and 3. Tables of thermodynamic properties appear at the end of this chapter.

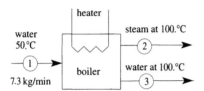

3.35 Calculate the temperature of the stream leaving the mixer below.

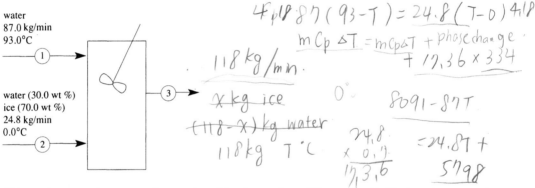

(This exercise appeared on an exam. It was estimated that the exercise could be completed in 15 minutes.)

3.36 The mixer shown below combines ice, water, and steam at 1 atm. For stream 4 leaving the mixer calculate (A) the flow rate (in kg/min), (B) the temperature (in°C), and (C) the composition (weight fraction of ice, water, and steam). Tables of thermodynamic properties appear at the end of this chapter.

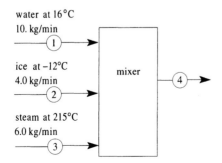

You might consider dividing the mixer into several equivalent units, such as a melter, a condenser, warmers, coolers, and combiners.

3.37 Complete the energy balance on the melter in the desalination process (shown below) to calculate the temperature of stream 5. Tables of thermodynamic properties appear at the end of this chapter.

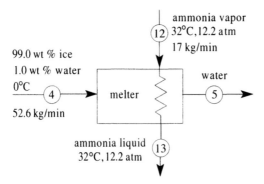

3.38 The "exquisite" desalinator diagrammed in Figure 3.30 uses water as the refrigerant. Consequently the heat transfer is improved, the cost of each unit is reduced, and there is no need for a foreign refrigerant.

As shown in the flowsheet in Figure 3.30, cooled seawater is admitted to the freezer at low pressure. The conditions in the freezer are the triple point for seawater; the temperature and pressure at which ice, brine, and steam coexist. Heat is removed from the freezer by vaporizing water. Water vapor is compressed and heat is delivered to the melter by the condensation of water vapor.

Recall that the sum of the flow rate of water vapor in stream 3 plus ice in stream 5 is 53 kg/min. Given an input rate of 100. kg/min of seawater, calculate the flow rate of water vapor in stream 3. You may neglect any energy flow into the freezer owing to changes in pressure. Tables of thermodynamic properties appear at the end of this chapter.

3.39 Ancient civilizations in hot, arid climates used earthenware pitchers to chill water. An earthenware pitcher was filled with water, the temperature of which was initially the same as the air. After a few hours, the water was much cooler than the air. Here is how it worked: The porous earthenware allows water to diffuse through the walls. When the water reaches the surface, it evaporates. The evaporation draws energy from the water in the pitcher, which cools the water. The lower temperature is maintained because the porous earthenware also has a low thermal conductivity.

Consider a pitcher containing 5.2 kg of water for which the rate of evaporation is 0.060 kg/hour and the rate of heat conduction into the pitcher is given by the equation

$$q_{conduction} = U(T_{outside} - T_{water}),$$

where $U = 6.0 \times 10^3$ joules hr^{-1} K^{-1} and $T_{outside} = 39°C$. Assume the water has been in this pitcher for a long time and its temperature is constant. Calculate the temperature of the water.

You may assume:

- The water evaporates at 39°C (and the heat of vaporization is the same as at 100°C).
- The rate of evaporation is independent of the amount of water in the pitcher.

(Adapted from Zubizarreta, J. I., and Pinto, G. 1995. *Chemical Engineering Education*, Spring, *Equation*, 29(2), p. 96.)

3.40 Heat escapes from my house by conduction through the walls and roof. The rate of heat loss, q_{loss} in kJ/hour, is proportional to the difference between the inside temperature and the outside temperature,

$$q_{loss} = k(T_{inside} - T_{outside}),$$

where $k = 740$ kJ/(°C hour). The volume of my house is approximately 870 m³ and the temperature inside is uniform. The "heat capacity" of the house is 3,300 kJ/°C. That is, 3,300 kJ raises the temperature of my house 1°C.

(A) At what rate must I heat my house (q_{heat} in kJ/hour) to maintain a steady temperature of 20°C (68°F) when the outside temperature is −5°C (23°F)?

(B) The heater in my house can supply heat at a maximum rate of 5.2×10^4 kJ/hour. I wish to keep the temperature in my home at a steady 20°C (68°F). What is the lowest outside temperature (in °C) that will still allow me to maintain an inside temperature of 20°C?

(C) Air leaks into (and leaks out of) my house at a rate of 0.10 m³/sec. What heating rate (in kJ/hour) do I need to maintain a steady temperature of 20°C inside if the outside temperature is −5°C? You may assume that the volumetric flow rate of 0.10 m³/sec is referenced to 20°C.

(D) The air that leaks into my drafty house is very dry. I maintain a comfortable humidity by placing pans of water on the heaters. To maintain 50% humidity the heater must evaporate 9.0 g water/m³ dry air that leaks in. Again assume air flows through my house at a rate of 0.10 m³/sec and the outside temperature is −5°C. Cold dry air leaks in and warm wet air leaks out. What heating rate do I need (in kJ/hour) to maintain a steady temperature of 20°C and a humidity of 50% inside my home? You may assume that the water comes out of the faucet at 20°C and evaporates at 20°C.

3.41 Chemical engineering students at the University of Michigan designed a coffee mug that keeps coffee at the ideal temperature of 72°C longer than ceramic mugs.

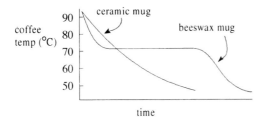

Here is how the mug works. When hot coffee is poured into the mug it melts a reservoir of beeswax sealed inside the mug. The temperature of the coffee remains at the melting point of beeswax, 72°C, until all the beeswax has solidified, after which, the temperature of the coffee falls below 72°C. The coffee mug holds 250 mL of coffee and contains 80. g of beeswax. You may assume that the properties of beeswax are the same as its main component, myricyl palmitate $(CH_3(CH_2)_{14}COO(CH_2)_{30}CH_3)$, and the properties of coffee are the same as its main component, water.

Coffee at what temperature will melt all the beeswax initially at a temperature 20.°C? (You may assume the coffee gives up heat only to the beeswax and that the heat capacity of the other components of the mug are negligible.)

Tables of thermodynamic properties appear at the end of this chapter.

(This exercise appeared on an exam. It was estimated that it could be completed in 20 minutes.)

Process Economics

3.42 An electric company, which burns coal to produce electricity, must reduce the emission of sulfur compounds from its smokestacks. The company is presently burning 17,000 tons of low-sulfur coal per year. Consider these three options:

(A) Continue to release sulfur at the present levels and pay a fine.

Fine: 2.7×10^6 $/year

(B) Install scrubbers to reduce sulfur emissions from the smokestacks to the level mandated by law.

Cost of scrubbers: 2.3×10^6
Cost of operating scrubbers: 1.8×10^5 $/year

(C) Install scrubbers to reduce sulfur emissions from the smokestacks to a level substantially below that mandated by law. This will allow the company to burn high-sulfur coal and still remain within legal limits for sulfur emissions.

Cost of scrubbers: 3.5×10^6
Cost of operating scrubbers: 2.1×10^5 $/year
Low-sulfur coal: $45/ton ($2.5 \times 10^7$ Btu/ton)
High-sulfur coal: $35/ton ($2.9 \times 10^7$ Btu/ton)

You may assume a straight-line depreciation on the scrubbers, with a lifetime of 10 years. Calculate the annual profit (in this case, annual loss) associated with each option.

3.43 A 1,000-megawatt power plant produces an average of 20. tons of SO_2 per hour. (Wattage is a measure of power. One watt equals one joule per second.) 90.% of the SO_2 must be removed from the smokestack emissions. The SO_2 may be removed by reaction with limestone $(CaCO_3)$ or by reaction with lime (CaO). 1.7 ton of limestone is required to remove one ton of SO_2. 1.0 ton of lime is required to remove one ton of SO_2.

Raw materials cost $15/ton limestone and $60./ton lime. The capital costs for the scrubbers depend on the capacity of the power plants, rated in kW. The limestone scrubber costs $215/kW and the lime scrubber costs $175/kW.

Which system, the limestone scrubber or the lime scrubber, gives the lower yearly cost? That is, which system gives the least-negative profit? Assume a 10-year straight-line depreciation.

(From Professor P. Harriott (Cornell University, 1993). This exercise appeared on an exam. It was estimated that it could be completed in 15 minutes.)

3.44 The electric bill of your small chemical company is $110,000/year. Your company operates only during the normal business day, when the cost of electricity is the highest. Consider this scheme to reduce the electric bill: Install rechargeable batteries will be charged from midnight to 6 am (when the cost of electricity is lower) and that will be discharged during the day to power the plant. The capital cost of this scheme (batteries plus installation) is $65,000. The batteries are expected to last 5 years. The maintenance and inspection of the batteries is expected to be $4,000/year. With these batteries, you predict your electric bill will drop to $80,000/year.

Calculate the return on investment for this scheme to install batteries. (You may assume a straight-line depreciation on the batteries.)

3.45 You form a small company to produce the specialty chemical Z. You purchase equipment that will produce 100 kg Z/year if operated 40 hours/week. The equipment has a capital cost of 1.5×10^5, which includes installation. The equipment has a lifetime of 5 years. The operating costs are:

fixed costs	$40,000./year	rent, taxes, etc.
payroll: you	$120,000./year	$100,000. salary plus $20,000. benefits (health insurance, retirement, etc.)
employees	$55,000./year/employee	$40,000. wages plus $15,000. benefits
variable costs	$400./(kg of Z produced)	

(A) Three employees are needed to operate the process. Calculate the minimum selling price for the chemical Z (in $/kg) that will yield a return on investment (ROI) of 20%.

(B) You increase production to 200 kg Z/year by adding a night shift of three additional employees. The salary of the night-shift employees is 25% higher than the day-shift employees. Calculate the minimum selling price for chemical Z that will yield a ROI of 20% if your company produces 200 kg Z/year using two work shifts.

(This exercise appeared on an exam. It was estimated that it could be completed in 25 minutes.)

3.46 The price of a chemical DPT is fixed at $235/(kg DPT). The cost of equipment to produce x kg of DPT per year is given by the formula

$$captial\ cost = \$700.x - \$0.10x^2.$$

This formula is valid for pricing equipment with capacity larger than 500 kg/year and smaller than 2,500 kg/year. The variable operating cost is $30./(kg of DPT) and the fixed operating cost is 1.0×10^5. Tax laws restrict your company to a straight-line depreciation calculated over 10 years.

(A) Calculate the profit (or loss) of producing 1,000. kg of DPT per year.

(B) Calculate the amount of DPT that must be produced per year to yield a 20.% ROI.

(This exercise appeared on an exam. It was estimated that it could be completed in 20 minutes.)

3.47 The lake-source cooling proposed by Cornell University in the *Cornell Chronicle* (June 30, 1994) is claimed to have economic benefits as well as environmental virtues. (See Exercise 3.31 for a description of the scheme.) Cornell has two options.

> *Option 1.* Replace present refrigerators with similar refrigerators at a cost of 20×10^6.
> *Option 2.* Replace present refrigerators with lake-source cooling at a cost of 50×10^6. However, the lake-source cooling would save 1.5×10^6 \$/year in electricity costs.

Assume a 10-year straight-line depreciation for both options. Do you agree with the utilities engineers' conclusion that lake-source cooling is economically advantageous?

3.48 Consider the two options for producing *B* from *A*.

Option I.

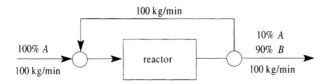

Option II.

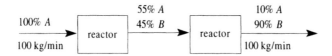

The capital cost of a reactor and the cost to operate a reactor are given by the charts below. Which option has the lower operating cost? Assume each process operates 24 hours per day, 7 days per week, 50 weeks per year. You may assume straight-line depreciation over 10 years.

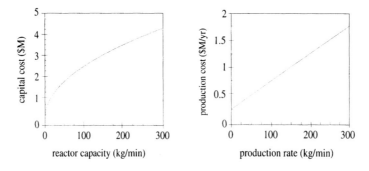

(This exercise appeared on an exam. It was estimated that it could be completed in 15 minutes.)

3.49 Consider the two schemes below for separating a mixture of pentane, hexane, and heptane.

Scheme I

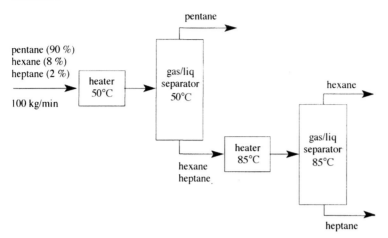

Scheme II

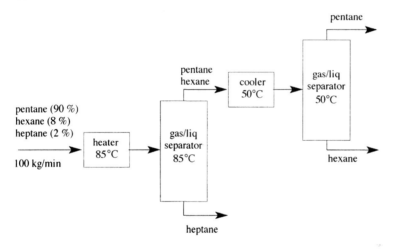

	Boiling pt (°C)
pentane	36
hexane	69
heptane	98

Which scheme is superior economically? Explain *succinctly*. State the basis for your choice. For example, is your choice based on the lowest depreciation? Assume the cost of a unit is directly proportional to its input capacity, in kg/min: A unit twice as large costs twice as much. Energy costs are directly proportional to the amount of energy consumed.

(This exercise appeared on an exam. It was estimated that the exercise could be completed in 20 minutes.)

3.50 The graphs on the following page give the capital costs and operating costs for the chemical produced by your company. Indicate the changes in these costs caused by the following three events.

(A) The property tax on your factory increases by 50%. Sketch the qualitative change in the cost curve(s) on either (or both) of the two graphs directly below.

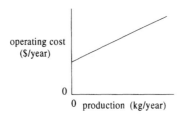 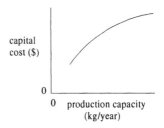

(B) The cost of chemical reactants increases by 50%. Sketch the qualitative change in the cost curve(s) on either (or both) of the two graphs directly below. **Do not** include the effect of part (A).

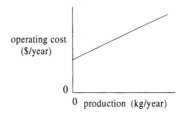 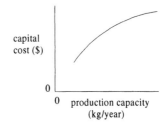

(C) The local government increases the tax on new equipment by 50%. The tax is paid when a manufacturer installs a new piece of equipment and the amount is independent of the cost of the equipment. Sketch the qualitative change in the cost curve(s) on either (or both) of the two graphs directly below. **Do not** include the effects of parts (A) and (B).

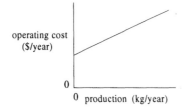 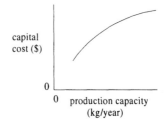

3.51 You buy equipment for $523,000 that has a capacity of 340. kg/week of *formula X*. The equipment operates 50. weeks per year; there are one-week maintenance periods every 6.0 months. *Formula X* sells for $28.7/kg. The production costs (reactants, labor, electricity, maintenance, repairs, etc.) are $11.4/kg. Rent, insurance, and miscellaneous fees are $62,500./year.

(A) Calculate the profit and return on investment (ROI). Assume a straight-line depreciation with a lifetime of 10 years.

(B) A key part on your equipment breaks once a year, on average. When you purchased the equipment for \$523,000 you also purchased a spare part for \$24,900 (total price = \$547,900.). It takes one week to replace the part, during which no *formula X* is produced. The additional costs for each repair are \$31,200 (\$6,300 for labor plus \$24,900 for a new spare part to replace the part in inventory.) Calculate the profit and ROI during a year in which the part is replaced once.

(C) Because you know the key part will fail every year or so, you decide to avoid any production interruptions by replacing the part during the biannual one-week maintenance periods. Because the equipment is disassembled for maintenance, there is no additional labor to replace the part. The cost of replacing the part is only the price of the part, \$24,900. Calculate the profit and ROI for replacing the part twice a year during the normal maintenance periods.

(D) You decide to replace the part only after it fails, but with a different strategy. When the equipment fails, the first two days of the repair are spent disassembling the equipment and removing the broken part. Instead of having a spare part on hand, you decide to rush-order the part when it fails. The rush order costs \$24,900 for the part plus \$1,000 for special rush delivery. Calculate the profit and ROI for this *just-in-time* strategy during a year when the part is replaced once.

(E) A different manufacturer also sells a process to produce 340. kg/week of *formula X*. But her equipment has been proved to never fail during its lifetime. This superior piece of equipment costs \$725,000. All other parameters are as stated in part (A). Which has the better ROI – (B), (C), (D), or (E)?

(Parts (A) and (B) appeared on an exam. It was estimated that they could be completed in 12 minutes.)

Engineering Calculations

3.52 If one travels at 55 miles per hour in a car with a fuel tank capacity of 30.4 liters and a fuel efficiency of 23 miles per gallon, how many minutes can one travel before the fuel tank is empty? Calculate the answer in one line, starting with 30.4 liters on the left and ending with your answer on the right.

3.53 On January 4, 1994 a storm deposited 18 inches of snow on the 200 scenic acres of the Cornell campus. An article in the *Cornell Chronicle*, January 13, 1994, claimed that the storm "dumped what seemed like tons of the stuff" on Cornell. Use the rule of thumb that 12 inches of snow melts to 1 inch of water to assess the *Chronicle's* claim. Did the *Chronicle* exaggerate or understate the snowfall on the Cornell campus?

3.54 Optical fibers are fine threads of silica drawn from a rod of quartz, called a *boule*. One end of the boule is heated to its melting point and a thread of silica 0.125 mm in diameter is drawn from the melt. By careful control of the boule temperature and the drawing tension, long fibers of uniform diameter can be drawn.

(A) Calculate the length of fiber (in km) that can be drawn from one boule 1.0 m in length and 2.5 cm diameter.

(B) The fiber is drawn at a rate of 10 m/s. Calculate the time required to draw the boule into fiber.

(C) The drawn fiber is wound onto a cylindrical spool of diameter 20 cm and height 10 cm. Estimate the diameter of the spool after all the fiber is wound onto the spool.

3.55 Optical fibers are silica threads with a protective polymer coating. Given that the silica fiber diameter is 0.125 mm and the coating is 0.050 mm thick, how many kilograms of polymer are needed to coat 26 km of fiber? (Density of polymer $= 1,740.$ kg/m^3.)

3.56 In the preparation of integrated circuits (chips) silicon wafers are coated with a photolithographic polymer. The wafers are 8 inches in diameter and the coating is 1,000 Å thick. Each day of a five-day work week 4,000 wafers are coated. How many liters of polymer are needed to coat wafers for a year?

Thermodynamic properties for compounds at 1 atm

	C_6H_{12}	Acetone	Myricyl palmitate	Styrene	Air	Formula X
molecular wt (g mol^{-1})	84	58	690	104	28.8	103
C_p (joules °C^{-1} mol^{-1})	157. (liq)	126. (liq)	1,100 (sol)	160. (liq)	42. (gas)	451. (liq)
	150. (gas)	74.9 (gas)	2,300 (liq)	217. (gas)	—	243. (gas)
boiling point (°C)	80.7	56.	—	145.2	—	103.
melting point (°C)	6.5	−95.	72	−30.6	—	21.
heat of vaporization (joules mol^{-1})	3.00×10^4	2.91×10^4	—	3.70×10^4	—	3.2×10^4
heat of melting (joules mol^{-1})	2.68×10^3	5.72×10^3	1.3×10^5	1.10×10^3	—	7.7×10^3

Thermodynamic properties for H$_2$O

	$P = 1$ atm	$P = 0.005$ atm
C_p (joules °C^{-1} mol^{-1})[1]	33. (solid)	33. (solid)
	75. (liq)	75. (liq)
	35. (vapor)	33. (vapor)
boiling point (°C)	100.0	−2.0
melting point (°C)	0.0	−2.0
heat of vaporization (joules mol^{-1})	4.1×10^4	4.5×10^4
heat of melting (joules mol^{-1})	6.0×10^3	6.8×10^3

Notes: [1]C_P varies with temperature. These values are valid to two significant figures in the range −50°C to 200°C.

Thermodynamic properties for ammonia

	$P = 1.7$ atm	$P = 12.2$ atm
C_p (joules °C^{-1} mol^{-1})	77 (liq)	82 (liq)
	24 (gas)	8 (gas)
boiling point (°C)	−22	32
heat of vaporization (joules mol^{-1})	2.27×10^4	1.93×10^4

4

Graphical Analysis

C OMPLEX DESIGNS evolve from simple designs. Each step in the evolution is guided by analysis and evaluation. One method of analysis, mathematical modeling, entails translating the physical or chemical description into equations. For example, in Chapter 3 we applied the conservation of mass, a universal law, to translate steady-state systems into the equation "rate in = rate out." For other systems one might apply the ideal gas law (a constrained law) and translate into the expression "$PV = nRT$."

But what if we wanted to model a system that operated at conditions outside the ideal gas law's range of validity? How might we model, for example, a gas at high pressure and low temperature? One could search for a constitutive equation that applies in these limits; the basic course in chemical engineering thermodynamics offers several constitutive equations for specific conditions or types of mixtures. However, to apply these equations one needs to know parameters specific to each gas or each mixture. Suppose these parameters have never been reported. What is one to do? One approach is to measure data at the conditions of interest. For example, one might measure the volume of one mole of the gas at temperatures and pressures relevant to your process. You could then extend the plot from the ideal region (Figure 3.32) to the plot shown in Figure 4.1. Although we may not understand the fundamental

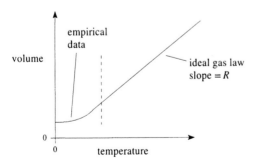

Figure 4.1. Volume of one mole of a gas at 1 atm as a function of temperature.

phenomena at low temperatures and/or high pressures, we can still use the data to model a process, for example, to pressurize the gas, and likewise for other systems in regions outside the validity of constrained laws. In this chapter we explore analysis based on empirical data. Whereas Chapter 3 involved mathematical manipulations of symbols, this chapter will involve graphical analysis and geometry.

For completeness we mention a third method of modeling, dimensional analysis. In Chapter 5 we will explore methods to scale constrained laws or empirical data to dynamically similar systems. For example, the data shown in Figure 4.1 for one particular gas can be used to predict the results for another gas. Whereas a plot such as shown in Figures 3.32 and 4.1 contain data for a specific gas, dimensional analysis allows one to apply these data to any gas, with appropriate scaling of the volume, temperature, and pressure.

In this chapter we study empirical methods of analysis in the context of process units for separating mixtures, just as we studied mathematical modeling in the context of a desalinator. The physical basis for most separations is equilibrium thermodynamics of multiphase mixtures. Universal laws, and even constrained laws, are sometimes not available. Instead, one's analysis is based on empirical data, which is often presented graphically.

The cost of producing a chemical commodity is often determined by the difficulty of purifying the commodity. Separating mixtures is often 50 to 90% of the capital cost of a chemical process. The separation units contribute a significant portion of the operating costs as well. The value of pharmaceuticals, precious elements, and rare isotopes is generally independent of the costs of reactants, or reactors, for example. The selling price of a chemical commodity is inversely proportional to the concentration of the commodity, as shown in Figure 4.2, which is known as the Sherwood plot. Remarkably, the correlation spans ten orders of magnitude!

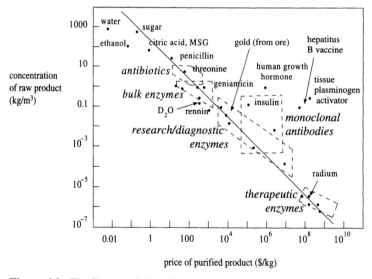

Figure 4.2. The Sherwood plot. [Adapted from a similar figure (Nystrom 1984) and discussed in an article by Lightfoot (1988). Additional data taken from texts by King (1971): Figure 1.16; Sherwood, Pigford, and Wilke (1975): Figure 1.2; and Blanch and Clark (1995): Figure 6.2.]

4.1 Air Pollution

Let's start with an important separations problem: How does one separate pollutants from an effluent to be discharged to the air? Note that this is a poor statement of the *real* problem. The real problem is: How does one reduce air pollution? This broader definition invites creative approaches such as "Don't create the pollutant in the first place." Chemical engineers examine processes and modify the chemistry to avoid pollutant by-products. Producing pollutants may also be avoided by changing solvents or catalysts. Such changes may increase the cost of purifying the product, but the overall cost – which includes pollution control – may be less.

Assume that we don't have the option of modifying the chemical process. Let's further assume that our pollutant is a volatile organic compound. Volatile organics are often used as solvents in chemical processes, although much of the air pollution by volatile organics comes from oil-based paints and fugitive fumes at automotive service stations. Common pollutants are benzene, xylenes, methanol, and acetone.

There are two ways to deal with pollutants in an effluent to be discharged into the air: Capture the pollutants or convert the pollutants. The four chief processes for reducing pollution by volatile organics are: reaction (such as incineration), condensation, adsorption, and absorption. In this chapter we will analyze condensation and absorption. For simplicity we will assume we have only one pollutant: benzene in air.

4.2 Separation by Condensation

A condenser removes a chemical component by condensing it from a vapor to a liquid or solid. To design a condenser that removes benzene from air we need to know the vapor pressure of benzene at a given temperature. The vapor pressure is the pressure exerted by a pure substance on its surroundings at a particular temperature. The substance may exist as any combination of solid, liquid, and vapor phases. How might we obtain such data – not just for benzene but for any substance or mixture? One should first search handbooks and the numerous databases of thermodynamic data. In the course on equilibrium thermodynamics, chemical engineering students study theories for predicting vapor pressures. Assume for the present that our exotic substance (okay – its only benzene, but play along) is not documented anywhere. Let's measure some data.

Our experiment must measure the temperature, pressure, and phase (solid and/or liquid and/or vapor) of benzene at equilibrium. So we seal some benzene in a cylinder fitted with a thermometer, a pressure gauge, and a viewing port. The top of the cylinder is a piston as shown in Figure 4.3.

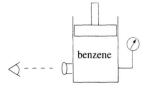

Figure 4.3. A device for measuring phase data. The object on the left is an eye, looking into the viewport.

We use this device to observe the phase(s) present at various temperatures, pressures, and volumes. We arbitrarily decide to systematically decrease the volume while holding the temperature constant, and we record the pressure and the phase(s) present. Table 4.1 contains the results of one such experiment.

Table 4.1. Phase data for 1 mol benzene at 80°C

Volume (m^3)	Pressure (atm)	Viewport shows	Phase(s)
10.14	0.27	transparent gas	vapor
3.67	0.77	transparent gas	vapor
2.92	0.99	transparent gas	vapor
2.897	1.00	mist	vapor/liquid
1.04	1.00	liquid below gas	vapor/liquid
0.49	1.00	liquid below gas	vapor/liquid
0.062	1.00	liquid below gas	vapor/liquid
0.0013	1.00	liquid below gas	vapor/liquid
0.0000887	1.00	liquid only	liquid
0.0000887	1.38	liquid only	liquid
0.0000885	96.57	liquid only	liquid

For safety, we dare not exceed 100 atm in our device. At this limit, the benzene is still entirely liquid. The experiment is repeated at different temperatures. At lower temperatures, the first evidence of liquid is detected at higher volumes; at 50°C, a minuscule liquid film forms at 8.134 m^3/mol and 0.357 atm. At even lower temperatures, solid benzene is detected in the liquid at high pressures. And at even lower temperatures, the first phase to condense is solid benzene.

How do we plot these data so they are useful for designing a condenser? We have *two* independent variables – volume and temperature – and two dependent variables – pressure and the phase.[1] The usual circumstance of one independent variable and one dependent variable can be accommodated on a standard x–y plot with a line representing the correlation. How do we show phase as a function of volume *and* temperature? Perhaps a three-dimensional plot? Three-dimensional plots are difficult to prepare and difficult to use. Another option is to prepare a phase diagram, which is in the general category of a "map."

---

[1] How do we know we don't need all three variables – temperature, pressure, and volume – to determine the phase of the system? Because Gibbs's Phase Rule, the zeroth law of thermodynamics, guarantees that you only need two for a pure substance.

4.3 Maps

A *map* is a convenient means of representing the state of a system as a function of two independent variables. The independent variables on a conventional map are "distance East" and "distance North" from a reference point, as shown in Figure 4.4.

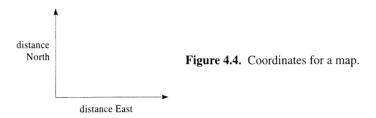

Figure 4.4. Coordinates for a map.

Figure 4.4 is not a graph. A graph, such as shown in Figure 4.5, has an independent variable, typically x, a dependent variable, typically y, and a line to represent $y = f(x)$.

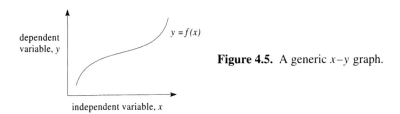

Figure 4.5. A generic x–y graph.

Given a value of x, one can read a value of y from the line on the graph. Clearly, no function "$y = f(x)$" exists for the map in Figure 4.4. If you tell us your distance East from a reference point, there is no function that tells us your distance North. Rather, you must tell us your distance East from a reference point *and* your distance North from a reference point. Then we could use a physical map to determine if you are in a lake, on a hill, or in a forest. If the map is a topographical map, such as shown in Figure 4.6, your East/North coordinates will tell us your elevation above sea level.

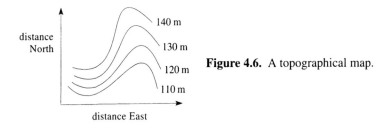

Figure 4.6. A topographical map.

Let's "map" the phase data we collected for benzene. Shown in Figure 4.7 are the data from one series of measurements at a given temperature, such as presented in Table 4.1. Our qualitative map uses logarithmic scales (see Appendix D) for both pressure and volume.

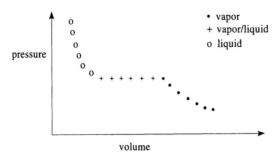

Figure 4.7. A map of the thermodynamic data in Table 4.1.

After we plot data from experiments at other temperatures, as shown in Figure 4.8,

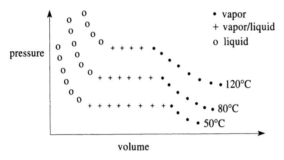

Figure 4.8. A map of the thermodynamic data for one mol benzene.

regions on the map become evident, as shown in Figure 4.9. At high volume and low pressure, benzene is entirely vapor. At low volume and high pressure, benzene is entirely liquid. In between is the region of "vapor/liquid."

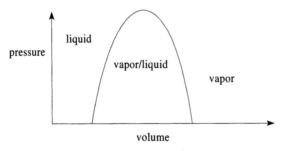

Figure 4.9. A pressure–volume phase diagram for one mol benzene.

Lines of constant temperature can be added to the map, as shown in Figure 4.10. These lines are analogous to the constant elevation lines on a topographic map. In the vapor region, a constant temperature line shows that the pressure increases as

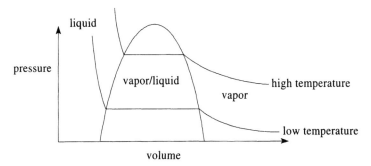

Figure 4.10. A pressure–volume phase diagram for benzene with isothermal contours.

the volume decreases. If the vapor is ideal, the relation is $P = nRT/V$. In the liquid region, a line of constant temperature is very steep. Why? Because a very large pressure increase is needed to decrease the volume of a liquid at constant temperature.

Figure 4.10 is useful for defining two terms: the dew point and the bubble point. If we follow an isotherm from the vapor region to smaller molar volumes, the dew point is the condition where the first iota of liquid forms. If we continue to reduce the volume, the bubble point is the condition at which the last iota of vapor remains. Note that the dew point and bubble point are not unique points as their names might imply. Rather the dew point can be anywhere on the border between vapor and vapor/liquid. Like-wise, the bubble point can be anywhere on the border between liquid and vapor/liquid.

The maps we prepared in Figures 4.9 and 4.10 are useful in modeling processes whose operating parameters are pressure and molar volume. But we want to vary the pressure and *temperature* to operate our condenser. So we also need a map with axes pressure and temperature. Let's get our bearings on this new map. Where do you expect the regions corresponding to vapor, liquid, and solid? Vapor should be at high temperature and low pressure. Solid should be at low temperature and high pressure. Liquid should lie in between. Qualitatively our diagram looks like Figure 4.11. (We continue to use a logarithmic scale for pressure but a standard scale for temperature.)

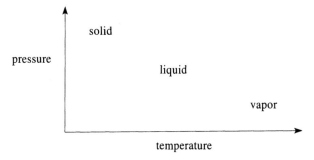

Figure 4.11. A qualitative pressure–temperature phase diagram.

When we plot our experimental data, we arrive at the phase diagram shown in Figure 4.12. In contrast to the phase diagram in Figure 4.10, the two-phase systems, such as vapor/liquid mixtures, correspond to a line, not a region, and likewise for the other two-phase systems, solid/vapor and solid/liquid. The three two-phase borders meet at the triple point, the unique condition at which solid, liquid, and vapor coexist. The triple point is usually at a temperature slightly colder than the melting point at 1 atm, but at a much lower pressure – typically less than 0.01 atm.

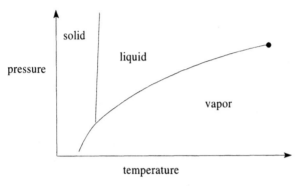

Figure 4.12. A pressure–temperature phase diagram.

The solid/liquid border extends upward indefinitely. However, the vapor/liquid border terminates at a condition called the *critical point*. Liquid and vapor are indistinguishable above the critical point. Vapor transforms into liquid without condensation. The vapor gets denser and denser until it is a liquid. Why? That's thermodynamics – we warned you in the first chapter that we would occasionally have to postpone "why?" until later in the curriculum. The critical point is at very high temperature and pressure for substances that are liquid at 25°C and 1 atm. For H_2O, the critical temperature is 374°C and the critical pressure is 218 atm. For *n*-hexane, the critical point is 234°C and 29 atm.

When using any map, it is important to begin by locating landmarks. On thermodynamic maps, the important landmarks are the borders between phases, the critical point, and the triple point.

4.4 Condensation of Benzene from Air

We now use the maps in Figures 4.10 and 4.12 to determine how to condense benzene from the vapor phase to the liquid phase. Let's first do this for pure benzene. Then we'll add the (minor) complication of a mixture with air.

First, we find the point on the map that corresponds to benzene vapor and label this on the map as shown in Figure 4.13. Because the critical point for benzene is at high temperature (289°C) and high pressure (48 atm or 37,000 torr) the vapor/liquid border extends off the map.

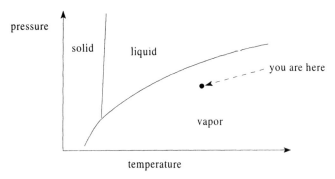

Figure 4.13. A pressure–temperature phase diagram for benzene.

What is our destination? To condense benzene, we wish to travel to the vapor/liquid border. Our map tells us that we need to lower the temperature and/or raise the pressure. Let's first lower the temperature at constant pressure. The process is diagrammed in Figure 4.14.

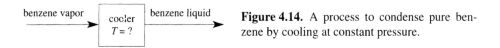

Figure 4.14. A process to condense pure benzene by cooling at constant pressure.

The path on the phase map is shown in Figure 4.15. Liquid benzene begins to condense when we reach the vapor/liquid border. Below $T_{condense}$ the pure benzene is all liquid. The map indicates the minimum temperature for our condenser. Any temperature below $T_{condense}$ will condense all the liquid (at the particular unspecified pressure).

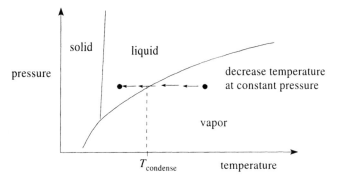

Figure 4.15. A path corresponding to decreasing the temperature at constant pressure.

Likewise we could condense the benzene by increasing the pressure at constant temperature, as in the process diagrammed in Figure 4.16 and as shown by the route in Figure 4.17. Again, liquid benzene begins to condense when we reach the vapor/liquid border. When the pressure is increased above $P_{condense}$ the pure benzene is entirely liquid.

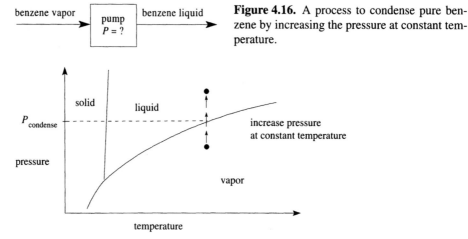

benzene vapor → pump P = ? → benzene liquid

Figure 4.16. A process to condense pure benzene by increasing the pressure at constant temperature.

Figure 4.17. A path corresponding to increasing the pressure at constant temperature.

We now consider benzene vapor in air. Air and benzene are very different. At 1 atm pure benzene condenses at 80°C, whereas air condenses at about −193°C. We will assume that benzene and air form an "ideal solution" so our data for the vapor pressure of benzene can be used for the partial pressure of benzene in air. We further assume that air is not affected by the presence of a small fraction of benzene.

Let's quantify our design. Assume that we have approximately 1.3 mol% benzene in air at 1 atm and 26°C. Because benzene and air behave ideally, the partial pressure of benzene is proportional to the mol fraction of benzene:

$$\frac{\text{partial pressure of benzene}}{\text{total pressure}} = \frac{\text{moles of benzene}}{\text{total moles}}, \tag{4.1}$$

$$\frac{P_{\text{benzene}}}{P_{\text{total}}} = y_{\text{benzene}}, \tag{4.2}$$

$$P_{\text{benzene}} = (0.013)(1.0\,\text{atm}) = 0.013\,\text{atm}\left(\frac{760\,\text{torr}}{1\,\text{atm}}\right) = 10\,\text{torr}. \tag{4.3}$$

We locate our starting point on the map, as shown in Figure 4.18.

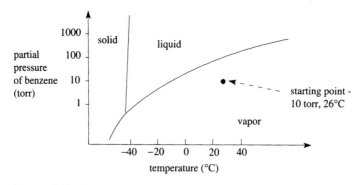

Figure 4.18. The starting point for 1.3 mol% benzene in air.

Whereas pure benzene could be condensed entirely to liquid, it is not possible to condense all the benzene from air by cooling. Even if we cool below the dew point for benzene at 10 torr, some benzene remains in the vapor phase. Consider water vapor in air. A temperature decrease may cause some water vapor to condense (as rain or snow), but not all the water vapor is condensed; the relative humidity does not decrease to zero below the dew point. Because some benzene vapor will remain in the air, our destination on the phase map will be on the vapor/liquid border. This is important to remember.

We first design a process to condense benzene by decreasing the temperature. Such a process is diagrammed in Figure 4.19. Let's determine the temperature decrease needed to condense 90% of the benzene vapor.

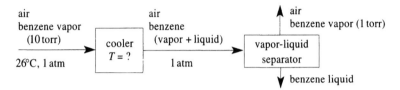

Figure 4.19. A process to condense benzene from air by cooling at constant total pressure.

We now use the pressure–temperature map in Figure 4.18 to determine the temperature of our destination. Our condenser maintains 1 atm total pressure, so we plot a path of constant pressure. We move horizontally on the map until we reach the two phase boundary between vapor and liquid, also known as the dew point. One finds in the *CRC Handbook of Chemistry and Physics* (59th edition, p. D-246) that the dew point of benzene at 10 torr is $-12°C$. Thus at 10 torr and $-12°C$ benzene begins to condense from the air and the partial pressure of benzene falls. The total pressure in the system, however, falls little because the system is predominately (98.7%) air.

How does the path continue after reaching the vapor/liquid border? The temperature continues to decrease, so the path is to the left. But recall that the system is a mixture of benzene liquid and benzene vapor. So we are constrained to move along the vapor/liquid border. As we move along the border, the partial pressure of benzene decreases. Benzene began to condense at a partial pressure of 10 torr. Therefore 90% will be condensed when we reach 1 torr. The *CRC Handbook* tells us that the dew point of benzene at 1 torr (the design goal) is $-37°C$. Thus we continue cooling the system until we reach $-37°C$, as shown in Figure 4.20. Our graphical analysis predicts the condenser should operate at $-37°C$.

The map in Figure 4.20 shows how the operating temperature affects the partial pressure of benzene in the effluent. This map is a *design tool* for our condenser analogous to the equations we developed as design tools in the previous chapter. More on graphical design tools will be presented later.

Let's use the graphical design tools to determine the pressure needed to condense the benzene from the air at constant temperature. The process in Figure 4.21 compresses the benzene–air mixture to condense 90% of the benzene. The liquid is separated from

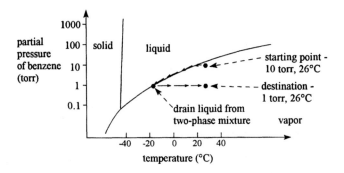

Figure 4.20. Condensing benzene from air by cooling at constant total pressure.

the vapor (at high pressure) and the air is then expanded to 1 atm. Note that again the benzene is a liquid–vapor mixture in the separator.

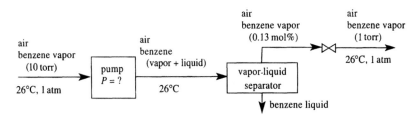

Figure 4.21. A process to condense benzene from air by increasing the pressure at constant temperature.

What is the operating pressure of the process? Again we start at 10 torr and 26°C. But now we proceed vertically (temperature is constant) until we reach the vapor/liquid border (see Figure 4.22). Again we consult the *CRC Handbook* to put numbers on our map. We find that at room temperature (26°C) the dew point of benzene is 100 torr.

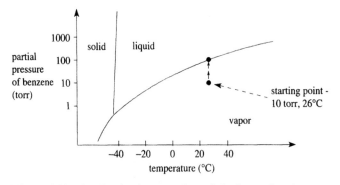

Figure 4.22. Condensing benzene from air by increasing the pressure at constant total temperature.

Thus to condense benzene we must increase its partial pressure to 100 torr, a factor of 10, which means we must increase the *total* pressure by a factor of 10 from 760 torr to 7,600 torr to condense the first iota of benzene. This is a key concept.

To condense more benzene, we must increase the pressure further. Recall that pure benzene condensed entirely to liquid when we increased the pressure above the dew point. We crossed into the liquid region on the phase map. With an air–benzene mixture we must remain on the vapor/liquid border. And with the further constraint of constant temperature, we are obliged to remain at the point we first contacted the vapor/liquid border.

How much higher must we raise the pressure to achieve the design specification of 0.13 mol% benzene in the air? Or phrased differently: What is the total pressure if the partial pressure of benzene is 100 torr and it comprises 0.13 mol%? We use Eq. (4.2),

$$\frac{P_{benzene}}{P_{total}} = y_{benzene}, \tag{4.2}$$

$$P_{total} = \frac{P_{benzene}}{y_{benzene}} = \frac{100 \text{ torr}}{0.0013} = 76{,}000 \text{ torr} = 100 \text{ atm}. \tag{4.4}$$

Let's compare the visual effectiveness of our graphical analyses in Figures 4.20 and 4.22. The path in Figure 4.20 clearly illustrates the extent of condensation. The farther we move downward along the vapor/liquid border, the more benzene condenses. In contrast, the graphical analysis in Figure 4.22 is obtuse. The system remains at the same point on the map regardless of the amount of benzene condensed. We need a map that better shows the progress of condensation at constant temperature.

The condensation in Figure 4.22 remains at one point because the two map coordinates – temperature and partial pressure – are constant as the benzene condenses. To move on a map, at least one coordinate must change. What is changing? Well, if we used a device such as shown in Figure 4.3 to increase the total pressure at constant temperature, we would depress the piston to decrease the volume. Volume is changing. Let's use a map with pressure–volume coordinates, as shown in Figure 4.10.

We begin in the vapor region of the map and follow a constant temperature path to the vapor/liquid border, as shown in Figure 4.23. As the volume decreases at constant

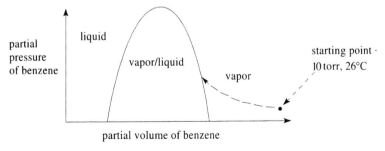

Figure 4.23. Increasing the pressure (by decreasing the volume) until the first drop of benzene condenses.

temperature, the pressure increases, as one would expect from the ideal gas law, $PV = nRT$. When the path reaches the border, the first drop of benzene condenses. This is the dew point. Again the total pressure increased by a factor of 10 to 7,600 torr and the partial pressure of benzene increased by a factor of 10, from 10 torr to 100 torr.

What happens when we increase the total pressure further by a factor of 2? The reflexive response is that the partial pressure of benzene increases by a factor of 2, to 200 Torr. But this is incorrect: The phase map in Figure 4.22 tells us that the partial pressure of benzene cannot exceed 100 torr at 26°C. Visualize the pressure increase in terms of the system volume. How did we increase the total pressure by a factor of 2? We decreased the total volume by a factor of 2 (assuming the ideal gas law applies and neglecting the volume of the benzene that condensed). What path on the map in Figure 4.23 corresponds to increasing the total pressure by a factor of 2? The partial pressure of benzene is constant at 100 torr (the path is horizontal) and the volume decreases by a factor of 2 (the path is to the left). We have moved into the two-phase region of vapor/liquid mixtures as shown in Figure 4.24. Recall that the volume scale

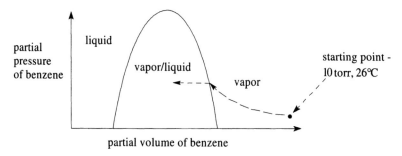

Figure 4.24. Increasing the total pressure to condense about half the benzene.

is logarithmic; a factor of two does not move one half the distance to the pressure axis. So the partial volume of the benzene vapor decreased by a factor of 2. But the partial pressure of benzene remained constant at 100 torr and the temperature remained constant at 26°C. How can this be if $PV = nRT$? If V changes, something else must change. In this case, n (the moles of benzene in the vapor phase) decreased by a factor of 2. How did this happen? Half of the benzene in the vapor phase condensed to the liquid phase. The extent of condensation at constant temperature is illustrated by the distance traveled across the vapor/liquid region.

Systems are generally described by several parameters. You must choose the map style appropriate to your needs. Consider a road map of Ithaca, New York, or Berkeley, California. Each map shows road locations in terms of east–west/north–south coordinates. Most road maps don't show elevations. Likewise, most temperature–pressure maps don't show volumes. Elevation changes can be very important when bicycling. Volume changes can be important when designing process equipment. Just as one soon learns that traveling west on Buffalo Avenue in Ithaca or west on Hearst Avenue in Berkeley entails a decrease in elevation, one should be aware that increasing the pressure at constant temperature entails a decrease in volume. Crossing the vapor/liquid border on a pressure–temperature map decreases the volume by a factor of about 1,000 – like driving off a cliff on a road map.

An alternative to the two-dimensional maps such as Figures 4.9 and 4.12 is a three-dimensional map. Figure 4.25 shows a map with three axes: temperature, pressure,

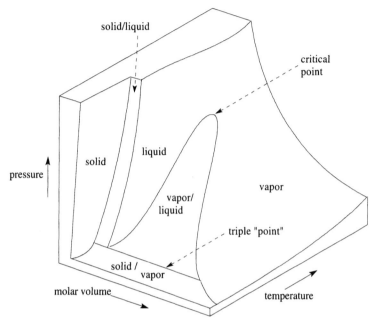

Figure 4.25. A three-dimensional phase map of a pure substance. (Adapted from Sandler, S. I. 1989. *Chemical and Engineering Thermodynamics*, Wiley, New York, p. 219, Figure 5.6.)

and volume. For a pure substance, the valid combinations of T, P, and V form a three-dimensional surface. If you project the solid object in Figure 4.25 onto the P–T plane, a two-dimensional map like Figure 4.12 results. Similarly, if you project the solid object onto the P–V plane, Figure 4.9 results. Actually, Figure 4.9 is only the upper portion of the map. A larger P–V phase map with lower pressures and lower volumes is shown in Figure 4.26. Two-phase mixtures, such as solid/vapor, appear as regions on a P–V

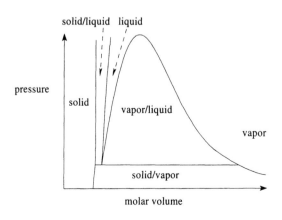

Figure 4.26. A pressure–volume phase diagram for a pure substance.

153

phase map. As with any new map, it is helpful to identify landmarks. You are advised to locate the critical point and the triple "point" on the phase map in Figure 4.26.

We can use the phase maps in Figures 4.12, 4.25, or 4.26 to evaluate any combination of pressure and temperature to operate our condenser. It is generally preferable to condense organic liquids by reducing the temperature. Moreover, condensation is often used as a pretreatment for adsorption: Lower temperatures favor adsorption of organics on adsorbents. Furthermore, lower temperatures help decrease the humidity of the air before it reaches the adsorber; water often competes with the organic for adsorption sites on the adsorbent. Condensation is effective for high concentrations of organic pollutants where the partial pressure of the pollutant can be substantially decreased by only modest changes in temperature. We arrive at the more general scheme for condensing organic compounds shown in Figure 4.27. The water removed by condensation will have traces of benzene and must be treated (with a charcoal adsorber, for example) before released.

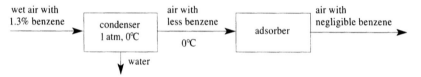

Figure 4.27. A general scheme for removing benzene from air.

4.5 Separation by Absorption

Another method to remove pollutants from air is to absorb the pollutants into a nonvolatile liquid such as oil. During absorption, pollutants move from the gas phase to the liquid phase. This movement is an example of *mass transfer*. To move between phases, the pollutants must cross the liquid–gas interface. Increasing the interfacial area increases the rate of mass transfer. An easy way to increase the interfacial area is to bubble the gas through the liquid. This concept is effective in delivering oxygen to water in fish tanks and it is effective in delivering benzene to oil.

Contaminated air is bubbled through oil and the oil absorbs benzene, as shown in Figure 4.28. For now we will ignore the rate of mass transfer from the gas phase to the liquid phase. We will assume that the bubble size and the time a bubble spends in the oil are such that mass transfer is complete when a bubble reaches the top of the absorber.

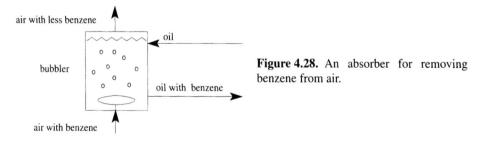

Figure 4.28. An absorber for removing benzene from air.

In other words, we will assume that the air/oil/benzene system reaches equilibrium in our absorber. But note – when mass transfer is complete there will still be benzene in the air. How much benzene? That is what we will calculate in this section.

First, we complete our absorber flowsheet. Of course, the oil should be recycled. Because the oil is nonvolatile, it is a simple matter to distill benzene vapor from the oil.

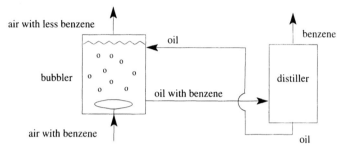

Figure 4.29. An absorber for removing benzene from air, with oil recycle.

And of course, one would want to consider adding a heat exchanger so the hot oil leaving the distiller could heat the oil/benzene mixture entering the distiller.

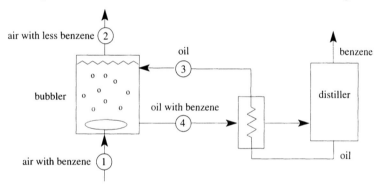

Figure 4.30. An absorber for removing benzene from air, with oil recycle and heat recovery.

We will assume that the bubbler is designed such that streams 2 and 4 are in thermodynamic equilibrium. Given a flow rate of polluted air, say 100 kg/min, what flow rate of oil is needed to reduce the benzene content of the air to an acceptable level? As always, it is illustrative to begin with a qualitative analysis. We start with extremes – assume the air leaving the bubbler contains *no* benzene. Thus all the benzene that enters with the polluted air leaves with the oil. Assume that the flow rate of benzene is 1 kg/min in both streams 1 and 4 and that the equilibrium concentration of benzene in the oil is 1 kg benzene/100 kg oil. The flow rate of oil is therefore 100 kg/min. If the oil could hold twice as much benzene, the flow rate of oil would be half as large, and so on. We need to know the oil's capacity for absorbing benzene.

As before, we search the literature for the data. Again, assume we find no data, so we perform some simple experiments. We fill a container with air and oil. We then inject a known quantity of benzene into the container and wait for the benzene to equilibrate between the two phases (Figure 4.31). We then measure the concentration of benzene in the air (or in the oil) and then calculate the other concentration using a mass balance. (You might verify that you could calculate this.)

Figure 4.31. An apparatus for measuring the absorption of benzene in an oil.

We inject more benzene into the container and repeat the process. Of course, the temperature is kept constant. We plot the data as shown in Figure 4.32, which looks very much like a standard x–y graph. That is, given a weight fraction of benzene in oil, the line indicates the weight fraction of benzene in air, for a system at equilibrium.

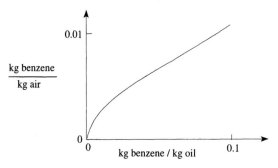

Figure 4.32. A map of the vapor–liquid two-phase system of air/benzene–oil/benzene.

But Figure 4.32 is actually a map, with three regions. The first region is the line – this represents systems at equilibrium. What is the region below the line? Obviously, it represents a system not at equilibrium. We have two choices for nonequilibrium: excess benzene in the oil or excess benzene in the air. Mark an arbitrary point in the region below the equilibrium line. How will a system at this point move toward equilibrium? It will move upward and to the left. How do these directions translate on this map? "Upward" means the system will increase the amount of benzene in the air. "To the left" means the system will decrease the amount of benzene in the oil. Thus the region below the line represents "excess benzene in the oil." Similarly, the region above the line represents "excess benzene in the air." These regions are labeled on the map in Figure 4.33.

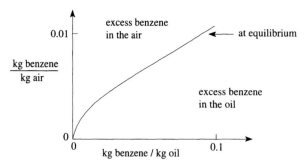

Figure 4.33. A map of the vapor–liquid two-phase system of air/benzene–oil/benzene.

Before we analyze the bubbler, let's start with a simpler system: a closed system of air, oil, and benzene. Assume there is initially no benzene in the oil. We begin by locating our starting point on the map in Figure 4.34. We choose an arbitrary point on the ordinate. Check this – is our starting point in the correct region of the map?

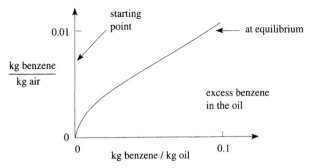

Figure 4.34. Starting point on the map for a closed system with benzene only in the air.

Yes – it is in the region "excess benzene in the air." What will be the concentrations of benzene in the air and in the oil at equilibrium? The system must move to the equilibrium line. But where on the equilibrium line? What else do we need to know?

Again, let's consider extremes. Consider a system with a lot of air and only a tiny drop of oil. The tiny drop of oil absorbs an even tinier amount of benzene. When this system comes to equilibrium only a infinitesimal amount of benzene has left the air. The concentration of benzene in the air remains essentially constant. The path on the map is shown in Figure 4.35.

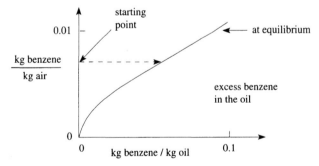

Figure 4.35. The path to equilibrium for lots of air and a tiny amount of oil.

Now consider the converse – a tiny amount of air and lots of oil. The oil has a large capacity for benzene. The benzene content of the air will fall practically to zero whereas the benzene content of the oil will barely increase because there is so much oil to dilute the benzene. This path, shown in Figure 4.36, follows the ordinate downward.

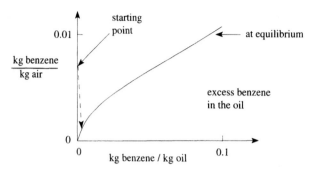

Figure 4.36. The path to equilibrium for a tiny amount of air and lots of oil.

Note what we have done to familiarize ourselves with the new map. We found our starting point and we explored some paths corresponding to extreme circumstances. This is a good practice to adopt.

Our quantitative analysis suggests that the relative size of the air and oil phases will determine the path to the equilibrium line. Lots of air corresponds to a horizontal path. Lots of oil corresponds to a vertical path downward. An intermediate ratio of air to oil would yield a path between these two extremes.

We now apply the conservation of mass to this closed system to calculate the slope of the path from the initial point to the equilibrium line. We need some nomenclature to represent quantities such as "kg benzene per kg air." Define M_X as the mass of component X. Conservation of mass requires that the amount of benzene in the system initially equals the amount of benzene in the system at any time later:

$$[(M_{benz})_{vapor} + (M_{benz})_{liquid}]_{initially} = [(M_{benz})_{vapor} + (M_{benz})_{liquid}]_{at\ equilbrium}.$$

$$(4.5)$$

We now multiply each term by carefully chosen ratios equal to 1 (such as M_{air}/M_{air}) and substitute $w_{benz/air}$ for the weight ratio of benzene to air and $w_{benz/oil}$ for the weight ratio of benzene to oil. This gives

$$[(M_B)_{vapor}]_{init}\left(\frac{M_{air}}{M_{air}}\right) + [(M_B)_{liq}]_{init}\left(\frac{M_{oil}}{M_{oil}}\right)$$

$$= [(M_B)_{vapor}]_{eq}\left(\frac{M_{air}}{M_{air}}\right) + [(M_B)_{liq}]_{eq}\left(\frac{M_{oil}}{M_{oil}}\right), \qquad (4.6)$$

$$(w_{benz/air})_{init} M_{air} + (w_{benz/oil})_{init} M_{oil}$$

$$= (w_{benz/air})_{eq} M_{air} + (w_{benz/oil})_{eq} M_{oil}, \qquad (4.7)$$

$$-[(w_{\text{benz/air}})_{\text{eq}} - (w_{\text{benz/air}})_{\text{init}}] \, M_{\text{air}} = [(w_{\text{benz/oil}})_{\text{eq}} - (w_{\text{benz/oil}})_{\text{init}}] M_{\text{oil}}, \quad (4.8)$$

$$\frac{(w_{\text{benz/air}})_{\text{eq}} - (w_{\text{benz/air}})_{\text{init}}}{(w_{\text{benz/oil}})_{\text{eq}} - (w_{\text{benz/oil}})_{\text{init}}} = -\frac{M_{\text{oil}}}{M_{\text{air}}} \quad (4.9)$$

We have assumed implicitly that no air dissolves in the oil and no oil evaporates into the air.

Let's plot the starting point and equilibrium point on the map. A line from the initial point to the equilibrium point is the hypotenuse of a right triangle whose sides are the terms in the ratio of Eq. (4.9), as shown in Figure 4.37. The numerator in Eq. (4.9) is the rise of our path to the equilibrium line. The denominator in Eq. (4.9) is the run

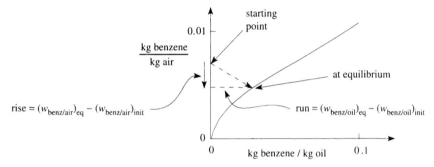

Figure 4.37. The path to equilibrium for M_{air} kg of air and M_{oil} kg of oil.

of the path. The ratio, rise/run, is the slope of the line from the initial conditions to the equilibrium conditions. The slope of the line is $-M_{\text{oil}}/M_{\text{air}}$, the ratio of oil to air. Let's check if this equation agrees with our qualitative analysis. In the limit of almost no oil, $M_{\text{oil}} \approx 0$, so $-M_{\text{oil}}/M_{\text{air}} \approx 0$, and a slope of 0 is a horizontal line. Check. In the limit of infinite oil, $M_{\text{oil}} \approx \infty$, so $-M_{\text{oil}}/M_{\text{air}} \approx -\infty$, and a slope of $-\infty$ is a vertical line directed downward. Check.

So we can predict graphically the equilibrium state of any nonequilibrium closed system of air, oil, and benzene. Note that Eq. (4.9) also applies to systems below the equilibrium line, systems with excess benzene in the oil.

Now, how about open systems at steady state, such as the bubbler? The derivation above can be repeated with F_X substituted for M_X, and the result is

$$\frac{(w_{\text{benz/air}})_{\text{out}} - (w_{\text{benz/air}})_{\text{in}}}{(w_{\text{benz/oil}})_{\text{out}} - (w_{\text{benz/oil}})_{\text{in}}} = -\frac{F_{\text{oil}}}{F_{\text{air}}}. \quad (4.10)$$

This equation assumes that the two streams leaving the bubbler are in equilibrium.

Let's apply our design tool to a more complicated absorber: the two-stage absorber shown in Figure 4.38. Assume the oil-to-air flow rate ratio is the same in each bubbler. Use the graphical techniques to determine the concentration of benzene in the air leaving the second bubbler.

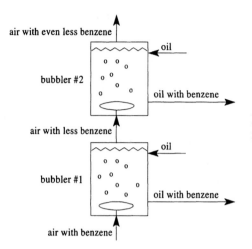

Figure 4.38. A two-stage absorber for removing benzene from air.

The first bubbler is trivial to analyze. It is the same problem we solved for the closed system. The solution is shown in Figure 4.39.

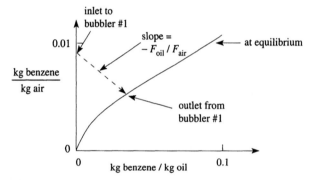

Figure 4.39. Graphical analysis of the first bubbler in a two-stage absorber.

But what is the initial point for the second bubbler? We need two coordinates: the concentration of benzene in the air entering bubbler #2 (which is given by the outlet from bubbler #1) and the concentration of benzene in the oil entering bubbler #2 (which is Zero). The inlet condition for bubbler #2 is obtained by moving horizontally from the equilibrium line to the ordinate. From here we move toward the equilibrium line at a slope $-F_{oil}/F_{air}$, as shown in Figure 4.40.

And how would one analyze a four-stage absorber designed like the process in Figure 4.38? Just add two more steps to the map in Figure 4.40. One can appreciate that this method is very efficient for exploring different designs or different flow rates. For example, given that the equilibrium line is concave upward as shown in Figure 4.40, and given that one has a limited flow of fresh oil to share between the bubblers, how would you adjust the relative flow rates to minimize the benzene in the outlet air? Increase the flow to bubbler #1 and decrease the flow to bubbler #2? Or vice versa?

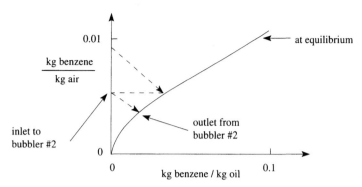

Figure 4.40. Graphical analysis of the second bubbler in a two-stage absorber.

As illustrated in Exercise 4.7, a multistage absorber removes more benzene from the air than does a single-stage absorber, given the same total flow of oil through both systems. However, an even better design exists. Can you guess what it is? Think of the countercurrent mechanism of the heat exchanger we analyzed in Chapter 3. Each stream, the hot and the cold, entered the heat exchanger at only one port and exited at only one port. Analogously, a better multistage absorber does not inject fresh oil at each stage. Rather the oil leaving the top stage is fed into the stage below, and so on until the bottom stage. Likewise, the air bubbles up through each stage in succession, as shown in Figure 4.41.

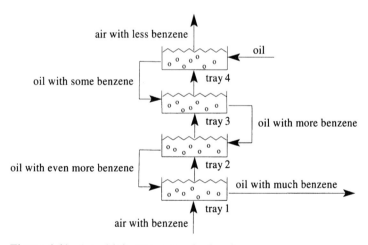

Figure 4.41. A multiple-stage cascade absorber.

How do we analyze the process in Figure 4.41? We could use the graphical method developed for multistage absorbers, such as in Figure 4.38. However, this method is awkward for analyzing *cascade* absorbers. To start at the bottom tray we need to know the concentration of benzene in the air entering the bottom tray (no problem) and the concentration of benzene in the oil entering the bottom tray (problem). We don't know the composition of the oil stream entering the bottom tray or the composition of any intermediate stream. We only know the compositions of the streams entering and leaving the overall process.

If we were to apply a mass balance to every stage in the cascade and then plot the results on our map, the pattern in Figure 4.42 would emerge. The inlet coordinates

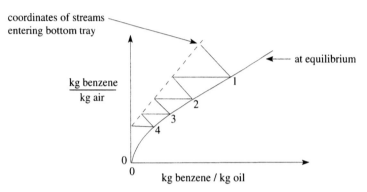

Figure 4.42. Analysis of a multiple-stage cascade absorber.

to the bottom tray give the upper point on the dashed line. The equilibrium condition leaving the bottom tray (tray 1) is the point on the equilibrium line labeled "1," and so on for trays 2, 3, and 4. The difficulty is that subsequent trays don't lie on the ordinate and thus cannot be plotted without a complete analysis of the cascade unit. However, as suggested by the dashed line, these points lie on a line, called the *operating line*. This will always be the case. Before you can ask "why?" let's derive an equation for an operating line.

To derive the equation for any line, we can use the coordinates of two points on the line. But we do not want to use the points in Figure 4.42, because we do not know those points. Instead, we want to plot points derived from what we know. Two known points are the (x, y) pair given by the outlet oil and the inlet air (bottom stage) and the pair given by the inlet oil and the outlet air (top stage). Let's derive an equation that contains these points. We start with a mass balance around the entire unit:

$$\text{rate of benzene in} = \text{rate of benzene out}, \tag{4.11}$$

$$\text{benzene in polluted air} = \text{benzene in oil effluent}$$
$$+ \text{ benzene in cleaned air}, \tag{4.12}$$

$$[(F_{\text{benz}})_{\text{vapor}}]_{\text{in}} = [(F_{\text{benz}})_{\text{liquid}}]_{\text{out}} + [(F_{\text{benz}})_{\text{vapor}}]_{\text{out}} \tag{4.13}$$

$$(w_{\text{benz/air}})_{\text{in}}(F_{\text{air}})_{\text{in}} = (w_{\text{benz/oil}})_{\text{out}}(F_{\text{oil}})_{\text{out}} + (w_{\text{benz/air}})_{\text{out}}(F_{\text{air}})_{\text{out}}. \tag{4.14}$$

As before we assume that air does not dissolve in oil and vice versa, so $F_{in} = F_{out}$ for air and oil. Equation (4.14) thus simplifies to

$$(w_{benz/air})_{in} = \left(\frac{F_{oil}}{F_{air}}\right)(w_{benz/oil})_{out} + (w_{benz/air})_{out}. \qquad (4.15)$$

If one repeats the mass balance on the top tray, one obtains a similar equation:

$$(w_{benz/air})_{into\ top\ tray} = \left(\frac{F_{oil}}{F_{air}}\right)(w_{benz/oil})_{out\ of\ top\ tray} + (w_{benz/air})_{out}. \qquad (4.16)$$

Indeed, if one performs a mass balance around a system that includes the top tray and the subsequent trays below down to tray n, one finds a similar relationship:

$$(w_{benz/air})_{into\ nth\ tray} = \left(\frac{F_{oil}}{F_{air}}\right)(w_{benz/oil})_{out\ of\ nth\ tray} + (w_{benz/air})_{out}. \qquad (4.17)$$

Equations (4.15), (4.16), and (4.17) all have the form of a straight line: $y = mx + b$, where m is the slope and b is the y intercept. The coordinates formed by the (x, y) pair ($w_{benz/oil}$ leaving a tray, $w_{benz/air}$ entering a tray) lie on a straight line. This line has slope F_{oil}/F_{air} and y intercept given by the weight ratio of benzene to air leaving the cascade absorber, $(w_{benz/air})_{out}$. This line is the operating line, although not the same operating line as shown in Figure 4.42.

Let's use the operating line to analyze a multistage cascade absorber. We begin in Figure 4.43 by drawing the operating line based on its slope, F_{oil}/F_{air}, and its intercept, $(w_{benz/air})_{out}$.

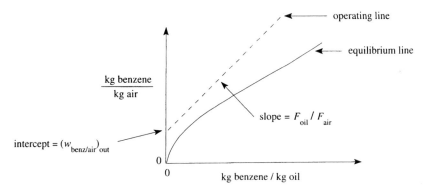

Figure 4.43. Analysis of a multiple-stage cascade absorber.

We now determine the compositions of the liquid and vapor phases in equilibrium in the top stage. This coordinate must lie on the equilibrium line. And the $w_{benz/air}$ coordinate is the same as the $w_{benz/air}$ coordinate of the stream leaving the top stage, y-axis the intercept of the operating line. Thus the equilibrium in the top stage is directly right of the y-axis intercept, as shown in Figure 4.44.

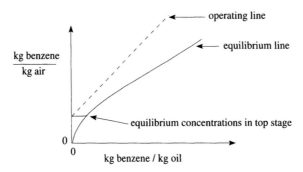

Figure 4.44. Analysis of a multiple-stage cascade absorber.

We now find the next point on the operating line. The x coordinate of this point is the concentration of benzene in the oil leaving the top stage, which is the x coordinate of the equilibrium point we just plotted. So the point we seek lies on the operating line directly above the point we just plotted on the equilibrium line. From here we draw a horizontal line to find the equilibrium condition in the second stage down, as shown in Figure 4.45.

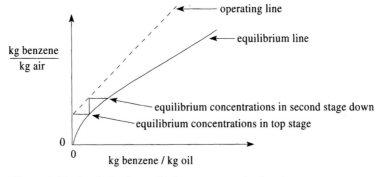

Figure 4.45. Analysis of a multiple-stage cascade absorber.

We continue to mark off the stages in our four-stage unit, which yields the analysis shown in Figure 4.46. The coordinates corresponding to the equilibrium stages are

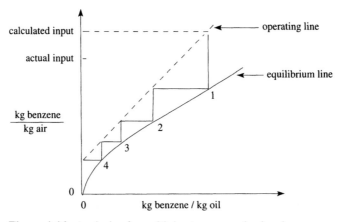

Figure 4.46. Analysis of a multiple-stage cascade absorber.

labeled 1 through 4. This method is much more convenient than the method shown in Figure 4.42 because one steps vertically and horizontally. There are no slopes to calculate after the operating line is drawn.

This analysis predicts that we can obtain the desired purity of air with an input composition labeled "*calculated input*" in Figure 4.46. But what if the concentration of benzene in the incoming air (labeled as "*actual input*" in Figure 4.46) is less polluted than "*calculated input*"? Could we make more efficient use of our absorber? You bet. We could change the slope of the operating line, by changing the flow rates of air and oil, until the stages we stepped off matched the actual concentration, as shown in Figure 4.47.

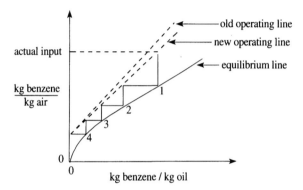

Figure 4.47. Analysis of a multiple-stage cascade absorber.

To match the actual inlet concentration, we decreased the slope of the operating line. So how should the operation of the absorber be changed? F_{oil}/F_{air} should be decreased. Since the flow rate of polluted air is presumably constant, this means we can use a lower flow rate of oil. Thus we will reduce our operating cost. What is the minimum flow rate of oil? We decrease the slope of the operating line until it intersects the equilibrium line at the composition of the inlet polluted air as shown in Figure 4.48.

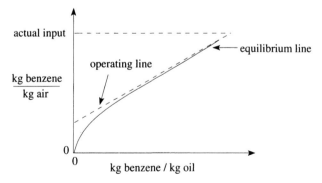

Figure 4.48. Analysis of a multiple-stage cascade absorber.

How many stages are needed to operate with this minimum flow rate of oil? If one steps off stages, one finds that an infinite number are needed to reach the intersection of the operating line and the equilibrium line. Thus we reduce operating cost, but we increase capital cost (to infinity).

The capital costs and operating costs are generally inversely related, as sketched in Figure 4.49. At either extreme, small decreases in either cost require large increases in the other cost. The curve line shown in Figure 4.49 is for well-designed absorbers. Although one cannot go below the line, a poor design could lie substantially above the curve. If the air does not bubble effectively through the oil, for example, one could design a column with high operating cost *and* high capital cost.

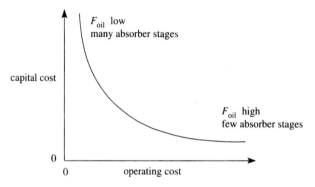

Figure 4.49. Analysis of a multiple-stage cascade absorber.

Figure 4.49 represents yet another type of design map. The curve represents efficiently designed absorbers for the particular operating conditions. A different design – perhaps a different type of perforated tray – would have a different curve. A cost map with many curves corresponding to many different designs would be useful in choosing the optimum.

4.6 Design Tools from Graphical Analysis

Recall the theme introduced in mathematical modeling. Chemical engineers create *design tools* to analyze processes. In Chapter 3 we developed equations to analyze processes. So far in this chapter we have developed a graphical tool to analyze multistage countercurrent absorbers. The method we followed to develop the graphical design tool is diagrammed in Figure 4.50.

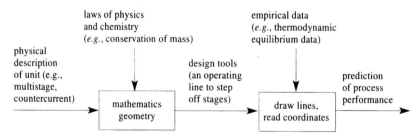

Figure 4.50. Schematic of the process to create and apply graphical design tools.

We will use this method to develop two more graphical tools – one for the analysis of flash drums and another for multistage countercurrent distillation columns. There are many others – one cannot possibly learn them all. Again the key is to know how to *create* graphical design tools.

4.7 Distillation – Flash Drums

The previous example (air polluted with benzene) involved a mixture of two components with very different boiling points. We now consider separating a mixture in which the components have comparable volatility – for example, a mixture of benzene (which boils at 80°C at 1 atm) and toluene (which boils at 111°C at 1 atm).

A flash drum is a simple unit to effect a crude separation (see Figure 4.51). A liquid mixture enters a container in which the pressure is low. The mixture "flashes"

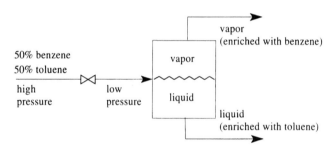

Figure 4.51. A low-pressure flash drum.

into two phases – vapor and liquid. One would expect that the vapor would be enriched with the more volatile component (*e.g.*, benzene) and the liquid would be enriched in the less volatile component (*e.g.*, toluene). Flash drums can also operate by increasing the temperature or by both decreasing the pressure and increasing the temperature.

Given an operating pressure in the flash drum, what are the compositions of the vapor and liquid streams? Or, at what pressure should we operate the flash drum to achieve a desired separation? What is the best separation we can expect from this simple unit?

Because benzene and toluene are common solvents one can find in the literature the thermodynamic data needed to model the flash drum. However, it is instructive to consider how one might measure the requisite data. We need a map of the phases (vapor, liquid, or two phase) for benzene/toluene mixtures. The coordinates of the map should be the composition of the mixture (*x* axis) and the pressure of the system (*y* axis). We dust off the device we used for benzene, diagrammed in Figure 4.3, and measure the phase of the system as a function of pressure, at constant temperature, for various benzene/toluene mixtures. We arbitrarily decide to operate our flash drum at 80°C, the boiling point of benzene at 1 atm.

As always, it is wise to start with an extreme – pure benzene. The data for pure benzene have already been measured and are reported in Table 4.1. We plot these data on our map in Figure 4.52. As with all new maps it is important to get one's bearings. Pure benzene is the right side of the map; pure toluene is the left side.

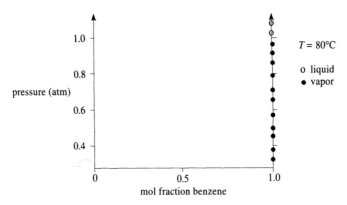

Figure 4.52. Phase map for benzene/toluene mixtures.

We perform the same experiment with the other extreme, pure toluene. Some of the data we measure are given in Table 4.2.

Table 4.2. Phase data for
1 mol toluene at 80°C

Pressure (atm)	Phase(s)
0.20	vapor
0.30	vapor
0.39	vapor
0.40	vapor/liquid
0.40	vapor/liquid
0.40	vapor/liquid
0.40	vapor/liquid
0.40	vapor/liquid
0.40	liquid
0.42	liquid
0.55	liquid

We add the data from Table 4.2 to our benzene/toluene phase map in Figure 4.53.

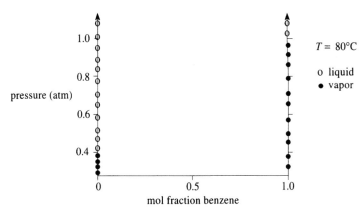

Figure 4.53. Phase map for benzene/toluene mixtures.

We proceed to a mixture of 50 mol% benzene/50 mol% toluene. But now we have additional data to measure – the compositions of the vapor and liquid phases. So we add an instrument to the cell in Figure 4.3 to measure compositions. We also record the amount of each phase.

Table 4.3. Phase data for a mixture of 0.5 mol benzene / 0.5 mol toluene at 80°C

Pressure (atm)	Phase(s)	Vapor phase		Liquid phase	
		Amount (mol)	Mol% benzene	Amount (mol)	Mol% benzene
0.31	vapor	1.0	0.5		
0.37	vapor	1.0	0.5		
0.45	vapor	1.0	0.5		
0.47	vapor/liquid	0.999	0.50	0.001	0.12
0.51	vapor/liquid	0.79	0.59	0.21	0.18
0.55	vapor/liquid	0.59	0.67	0.41	0.26
0.61	vapor/liquid	0.35	0.76	0.65	0.36
0.69	vapor/liquid	0.001	0.83	0.999	0.50
0.70	liquid			1.0	0.5
0.78	liquid			1.0	0.5
0.96	liquid			1.0	0.5

We plot these data on our phase map for benzene/toluene mixtures, in Figure 4.54. The dew point at 80°C for 50/50 benzene/toluene is 0.47 atm. The bubble point is at 0.69 atm. In contrast to pure substances, the dew point and bubble point of a mixture may occur at different pressures, at a given temperature.

We also run experiments with 25/75 benzene/toluene and 75/25 benzene toluene. These data are also added to our phase map (Figure 4.55).

The data begin to define the boundaries on our map. After experiments with many more compositions, the map in Figure 4.56 emerges. This map has a new type of

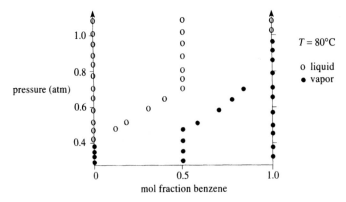

Figure 4.54. Phase map for benzene/toluene mixtures.

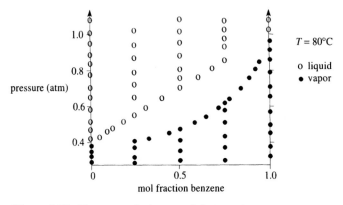

Figure 4.55. Phase map for benzene/toluene mixtures.

region – a forbidden zone,[2] indicated by the "keep out" label on our map. When we plotted our data, we never plotted a point in the forbidden zone.

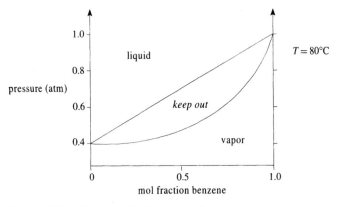

Figure 4.56. Phase map for benzene/toluene mixtures.

[2] We thought the term "forbidden zone" would appeal to aficionados of intergalactic encounters. If your background is the trench warfare of World War I, you may wish to label this area "No Man's Land." If your reference is one of the more recent conflicts on the Pacific rim, the term "DMZ" may be more appropriate.

If a mixture has a composition and pressure in the forbidden zone, the mixture is a two-phase system: liquid and vapor. The composition of the liquid phase is determined by moving horizontally left to the liquid border. The composition of the vapor phase is determined by moving horizontally right to the vapor border. This is illustrated in Figure 4.57.

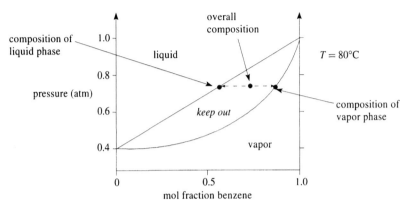

Figure 4.57. Determining the liquid and vapor compositions for a coordinate in the forbidden zone.

With this new map comes new terminology. The horizontal line from the liquid border to the vapor border is a *tie line*. The mol fraction of component *i* in the liquid phase is defined as x_i. The mol fraction of component *i* in the vapor phase is defined as y_i. It follows that

$$x_B \equiv \frac{\text{moles of benzene in the liquid phase}}{\text{total moles in the liquid phase}}, \tag{4.18}$$

$$y_B \equiv \frac{\text{moles of benzene in the vapor phase}}{\text{total moles in the vapor phase}}. \tag{4.19}$$

Tie lines can be used to determine the compositions *and* relative amounts of the vapor and liquid phases. Figure 4.58 shows a magnified view of the tie line in Figure 4.57.

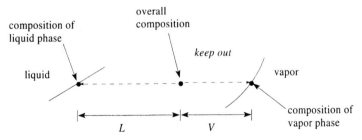

Figure 4.58. Determining the liquid and vapor compositions for a coordinate in the forbidden zone.

The length of the tie line segment to the liquid border is defined as L. The length of the tie line segment to the vapor border is defined as V. The total length of the tie line is T, so that $T = L + V$. The amounts of the liquid and vapor phases are

$$\text{mole fraction liquid} = \frac{\text{moles in the liquid phase}}{\text{total moles in the system}} = \frac{V}{T}, \tag{4.20}$$

$$\text{mole fraction vapor} = \frac{\text{moles in the vapor phase}}{\text{total moles in the system}} = \frac{L}{T}. \tag{4.21}$$

Let's apply Eqs. (4.20) and (4.21) to the data measured for the 50/50 benzene/toluene mixture. We started with all vapor and increased the pressure. At 0.47 atm a tiny drop of liquid condensed. A tie line at this pressure has $L \gg V$. From Eqs. (4.20) and (4.21) we calculate that the system is mostly vapor. Check. As we increased the pressure, more liquid formed and the tie line shifts to the right. Eventually $L \approx V$ and there are equal amounts (in moles) of liquid and vapor. Finally, at 0.69 atm, there is only a tiny amount of vapor left. And the tie line is such that $L \ll V$. From Eqs. (4.20) and (4.21) we calculate that the system is mostly liquid. Check.

Equations (4.20) and (4.21) are called the *lever rule*. The lever rule is easily verified by analyzing the system with a mass balance (see Exercise 3.23). Think of the tie line as a fulcrum. What mass is needed on the right side to balance the fulcrum shown in Figure 4.59?

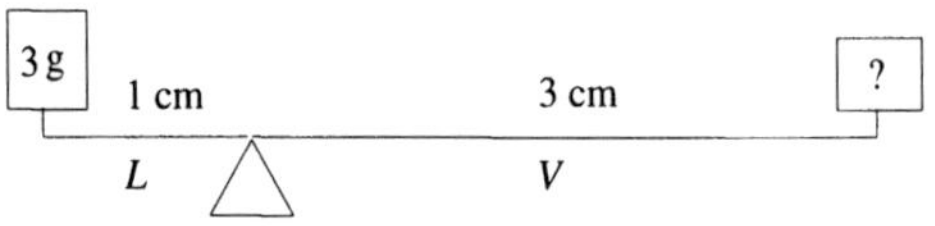

Figure 4.59. A lever.

If the fulcrum is balanced, it is not rotating. If it is not rotating, the torques caused by the masses at either end are balanced:

$$\text{torque on left side} = \text{torque on right side}, \tag{4.22}$$

$$(\text{moment arm on left})(\text{force on left}) = (\text{moment arm on right})(\text{force on right}). \tag{4.23}$$

The force on either end is $F = ma$, where a is the acceleration due to gravity, $g = 9.8 \text{ m/s}^2$:

$$L \times m_{\text{left}} \times g = V \times m_{\text{right}} \times g, \tag{4.24}$$

$$m_{\text{right}} = \frac{L}{V} m_{\text{left}} = \frac{1 \text{ cm}}{3 \text{ cm}} 3 \text{ gram} = 1 \text{ gram}. \tag{4.25}$$

The shorter arm on a balanced lever has the greater mass. The lever rule tells us that the shorter arm on a tie line has the greater number of moles.

We now have a geometric design tool to analyze a flash drum. For example, we can determine the pressure needed to produce a vapor with 75 mol% benzene, the relative flow rates, and the composition of the liquid stream (Figure 4.60).

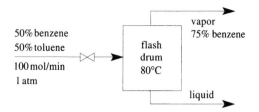

50% benzene
50% toluene

100 mol/min
1 atm

vapor
75% benzene

flash
drum
80°C

liquid

Figure 4.60. A low-pressure flash drum.

The graphical solution is shown in Figure 4.61. Because the vapor is 75 mol% benzene, we know the tie line (the solid horizontal line in the forbidden zone) must intersect the vapor border at $y_B = 0.75$. The point that the tie line intersects the liquid border gives the composition of the liquid phase: 0.38 mol% benzene. Extending the tie line to the ordinate axis gives the pressure: 0.62 atm. With a ruler we measure $T = 19$ mm, $L = 7$ mm, and $V = 12$ mm. From Eqs. (4.20) and (4.21) we calculate flow rates of $100 \times 7/19 = 37$ mol/min in the vapor and $100 \times 12/19 = 63$ mol/min in the liquid.

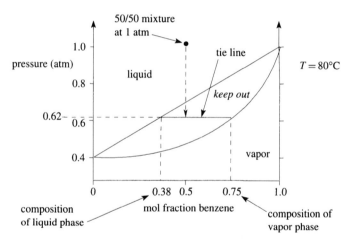

Figure 4.61. Graphical analysis of the low-pressure flash drum.

Let's check our graphical analysis with a mole balance on benzene. (Note that a mole balance is obtained from a mass balance with the conversion factor, for example, 1 mol benzene/78 g benzene.) We have

$$\text{molar flow rate of benzene in} \overset{?}{=} \text{molar flow rate of benzene out,} \tag{4.26}$$

$$(100\,\text{mol/min})(0.50) \overset{?}{=} (63\,\text{mol/min})(0.38) + (37\,\text{mol/min})(0.75), \tag{4.27}$$

$$50\,\text{mol/min} \approx 52\,\text{mol/min}. \tag{4.28}$$

Check.

What if we wanted a vapor stream with a higher benzene content? Our map tells us that the highest benzene content we can expect with this design is about 83 mol%, and at the maximum concentration, the flow rate of the liquid distillate is nearly zero.

How can we obtain a vapor stream with more than 83 mol% benzene? Recall how we improved the performance of the absorber. We added more stages. So we add another stage to the vapor stream, as shown in Figure 4.62. The second stage takes in a vapor and gives out a vapor stream and a liquid stream. We know that we will need to increase the pressure to condense some of the vapor, so we add a compressor before the second stage. But what is the pressure in the second stage?

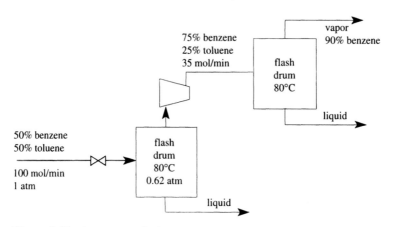

Figure 4.62. A two-stage flash system.

The analysis of this two-stage flash system is shown in Figure 4.63. We estimate that the second flash drum will operate at 0.78 atm and the flow rate of the vapor at 90 mol% benzene is 15 mol/min. Study the graphical analysis in Figure 4.63 and verify this analysis.

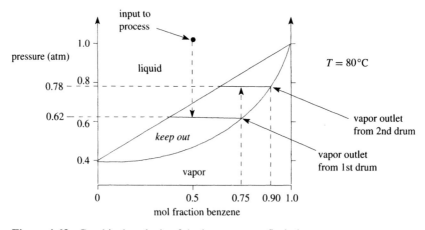

Figure 4.63. Graphical analysis of the low-pressure flash drum.

What is the composition of the liquid stream leaving the second flash drum? What do you propose we do with this liquid stream? Recycle it to the first drum? Good idea. In fact, assuming we also wanted to purify the toluene stream, we would add a third flash drum to the liquid output of the first flash drum. And what would we do with the vapor effluent of this third flash drum? That's right – recycle it to the first flash drum. If we continue in this manner we soon have a complicated process of flash drums that accepts 50/50 benzene/toluene and produces very pure benzene and toluene, with no other waste streams. Such a process is shown in Figure 4.64. (Compressors and expansion valves have been omitted.)

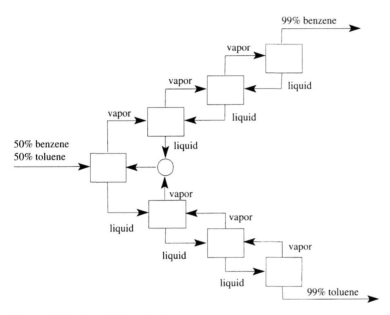

Figure 4.64. A multistage flash drum process.

It is impractical to analyze the multistage flash drum process with the graphical method we devised for a single flash drum. The problem is the recycle streams – it is awkward to determine the net composition of the flow into a particular drum. For example, how would you account for the liquid effluent from the second unit in Figure 4.62 being recycled to the first unit? Yet multistage distillation should be no more difficult to analyze than multistage absorption. What is the problem?

The problem is our map. We have the wrong coordinates for analyzing multistage countercurrent flow. The map that worked well for the absorber had "benzene in the liquid (oil)" on the abscissa and "benzene in the vapor (air)" on the ordinate. We need to transcribe our map in Figure 4.63 to a map of "mol fraction benzene in the liquid (x_B)" on the abscissa and "mol fraction benzene in the vapor (y_B)" on the ordinate.

Figure 4.65 shows three points transcribed from the plot on the left to the plot on the right. Point 1 corresponds to pure benzene. Point 2 corresponds to pure toluene. Point

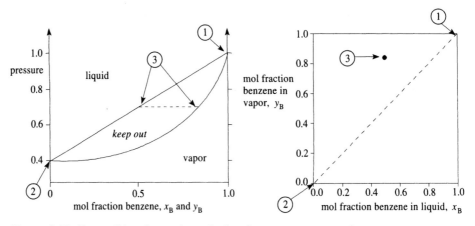

Figure 4.65. Transcribing thermodynamic data from one map to another.

3 is an arbitrary point that corresponds to $x_B = 0.50$ and $y_B = 0.83$. In this manner many pairs of points are transferred to the new plot. Eventually a line can be drawn through the data, as shown in Figure 4.66. (The dashed diagonal line is to guide one's eye – it does not represent anything physical . . . yet.)

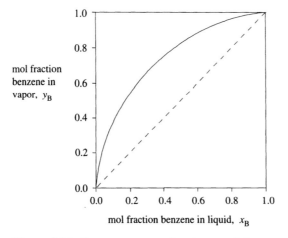

Figure 4.66. Our new map for benzene/toluene mixtures.

What is a good practice when one encounters a new map? Label the landmarks. Attempt to label some landmarks and regions on the map in Figure 4.66. Then inspect the labeled map in Figure 4.67. Generally, phase equilibria data are plotted so that the equilibrium line lies above the diagonal. That is, the more-volatile component (the component with the lower boiling point) is chosen as the basis for x and y.

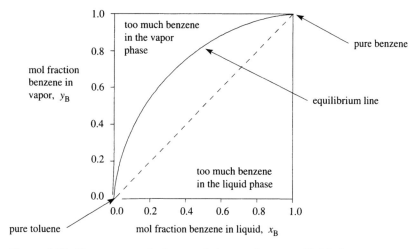

Figure 4.67. Our new map for benzene/toluene mixtures, with labels.

In Figure 4.68 are shown phase-equilibrium maps for benzene mixed with other chemicals (not toluene). What is the volatility of the second component, compared to benzene? Compared to toluene? (**Hint:** Use the maps in Figures 4.67 and 4.68 to determine the composition of the vapor above a 50/50 liquid mixture.) Which benzene mixture will be easier to separate by distillation?

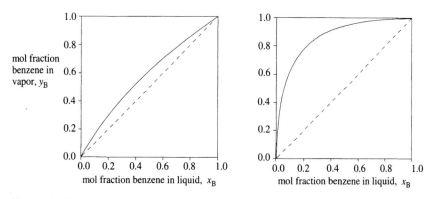

Figure 4.68. Vapor–liquid equilibrium diagrams for benzene mixed with other compounds.

If a mixture is not ideal, the phase-equilibrium maps are not as symmetric as the maps in Figures 4.67 and 4.68. In fact, the equilibrium line may cross the diagonal. Maps of vapor–liquid equilibrium data for nonideal mixtures are shown in Exercises 4.21 and 4.26.

Let's use the map in Figure 4.67 to analyze benzene/toluene systems. As we did with the absorber, let's begin with a closed system. Consider a two-phase system of benzene and toluene at equilibrium in a closed vessel as shown in Figure 4.69. A known quantity of liquid benzene is added to the vessel. Indicate on the benzene/toluene phase equilibrium map this system before and after the liquid benzene is added.

The system begins at a point on the equilibrium line and then moves horizontally to the right such as shown in Figure 4.69. Why horizontally? Because only the liquid composition changes initially; the vapor composition is unchanged. Why to the *right*? Because the mol fraction of benzene in the liquid *increases*.

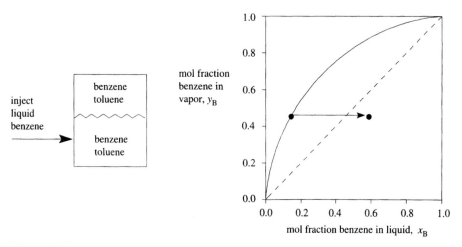

Figure 4.69. Analysis of a closed benzene/toluene system.

How will the system evolve from this point? It will return to equilibrium by decreasing the mol fraction of benzene in the liquid and thus increasing the mol fraction of benzene in the vapor. How does this translate into a direction on our map? The system proceeds to the left and upward. At what angle? Recall how we analyzed the two-phase system of air/benzene over oil/benzene. Consider the extreme case of an infinite amount of liquid and a tiny amount of vapor. x_B would decrease only infinitesimally and y_B would increase a lot; the path would be almost straight up. What about an infinite amount of vapor and a tiny amount of liquid? Yes – the path would be almost horizontal to the left. And what about a system with V moles of vapor and L moles of liquid? Take a guess. That's correct – the system would move to the equilibrium line at a slope of $-L/V$ (see Figure 4.70). You can verify this by applying a mole balance to the closed system, similar to the mass balance we applied to the absorber.

Our motivation for creating the new map of the vapor–liquid equilibrium data was to analyze the multistage countercurrent distillation process diagrammed in Figure 4.64. Let's redraw this process (Figure 4.71) so it is similar to the multistage countercurrent absorber in Figure 4.41.

Multistage distillation is similar to multistage absorption, except there is only one input, which enters somewhere between the top and bottom stage (not necessarily the middle stage). An energy balance on the distillation process in Figure 4.71 would reveal problems with the top and bottom stages. Consider the bottom stage – a liquid

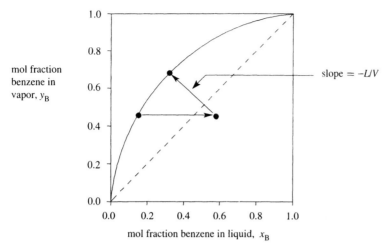

Figure 4.70. Reequilibration of the closed system after liquid benzene is injected.

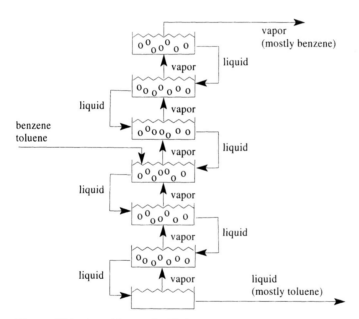

Figure 4.71. A multistage distillation process.

stream enters and divides into a liquid stream and a vapor stream. We need to add energy to the bottom stage to vaporize some of the liquid. Consequently, below the bottom stage is an evaporator, or boiler. An overall energy balance of the system would reveal that we now need to remove the energy added by the evaporator. We withdraw energy at the top to condense some of the vapor that enters. Above the top stage lies a condenser. The multistage distillation process is shown in Figure 4.72.

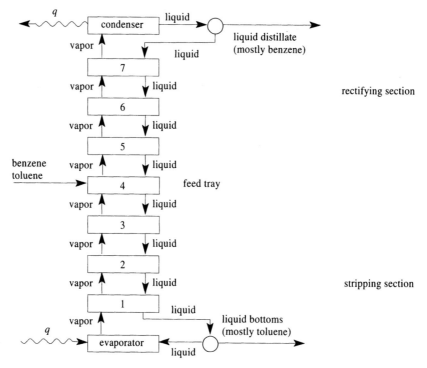

Figure 4.72. A multistage distillation process.

4.8   Distillation Columns

An actual multistage distillation process consists of a series of *trays* with holes in the bottom of each to allow the vapor to bubble up through the liquid. The liquid flows over the edge of a tray onto the tray below. The trays are round and are stacked in a tube, forming a column. For this reason distillation units are commonly called *distillation columns*. Some additional nomenclature is introduced in Figure 4.72. The tray to which the input is injected is called the *feed tray*. The trays above the feed tray (including the condenser) comprise the *rectifying section*. The trays below the feed tray (including the evaporator) comprise the *stripping section*. Finally, the output from the top of the column is the *distillate* and the output from the bottom of the column are the *bottoms*.

Distillation is an integral component in the chemical process industry. About 90–95% of all separations are done by distillation. It is estimated that there are 40,000 distillation columns in operation in the United States alone, representing a capital investment of $8 billion (Humphrey 1995).

The graphical method for analyzing distillation columns is similar to the graphical method for analyzing multistage countercurrent absorbers: Find a reference point, draw an operating line, and step off the stages. The difference is that we must perform this method twice – once for each section of the column.

Let's start with the stripping section. The reference point is the composition of the bottoms stream. Note that the bottoms composition is the same as the composition of the liquid entering the evaporator. And because the vapor stream is the only stream that leaves the evaporator, its composition must be the same as the entering liquid. Our reference point is thus *on the diagonal* of the vapor–liquid equilibrium map.

By analogy with the absorber, we expect the operating line to have slope L/V. Examine the distillation column and convince yourself that the ratio L/V is the same for every stage in the stripping section, including the evaporator. Thus, we expect that the coordinates of x_B entering a stage and y_B leaving that stage lie on the operating line. Is the ratio L/V less than one or greater than one? That is, is the liquid flow rate into a stage smaller or larger than the vapor flow rate? Sketch a border around the entire stripping section. Liquid enters the section from the feed tray and two streams leave: the vapor to the feed tray and the liquid bottoms. The vapor flow to the feed tray must be less than the liquid flow from the feed tray. Thus $L > V$ and $L/V > 1$. The operating line has a slope greater than the diagonal. We thus draw the operating line for the stripping section (see Figure 4.73). Note that it starts near the lower left of the map, near the coordinates of pure toluene.

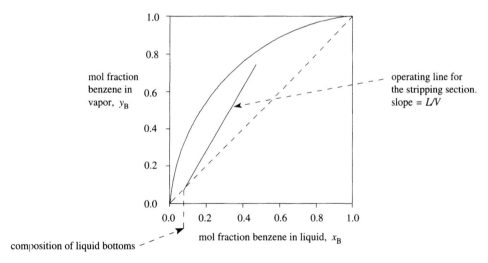

Figure 4.73. The operating line for the stripping section.

We use the stripping-section operating line to step off the stages as shown in Figure 4.74. The stages are numbered on the equilibrium line. Where do we stop stepping? The answer comes from the analysis of the rectifying section.

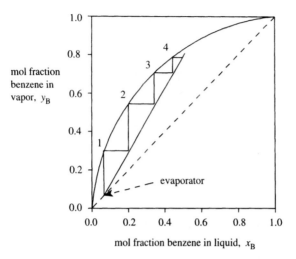

Figure 4.74. Analysis of the stripping section.

The rectifying section is analyzed in a manner similar to the stripping section; the graphical analysis of the rectifying section is shown in Figure 4.75. Note that the reference point is on the diagonal near the coordinates of pure benzene. Also note that the slope of the operating line is less than the slope of the diagonal.

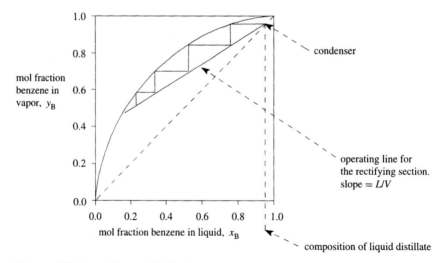

Figure 4.75. A multistage distillation process.

Figures 4.74 and 4.75 are combined in Figure 4.76. How does one know when to switch from the stripping-section operating line to the rectifying-section operating line? That's the location of the feed tray. In the analysis shown in Figure 4.76, the feed tray is the third tray in the column. (We will assume in this text that the feed is a vapor–liquid equilibrated stream. In this case we have a feed with $x_B = 0.35$ and $y_B = 0.70$.)

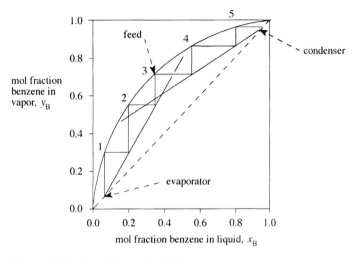

Figure 4.76. A multistage distillation process.

What if the feed has a higher concentration of benzene, say $x_B = 0.5$ and $y_B = 0.8$? The analysis is shown in Figure 4.77. The feed tray is now tray 5. Note that we now have more trays in the stripping section (four) and fewer trays in the rectifying section (one). That makes sense: Because the feed has a higher concentration of benzene, it is easier to make the benzene-rich product and more difficult to make the toluene-rich product.

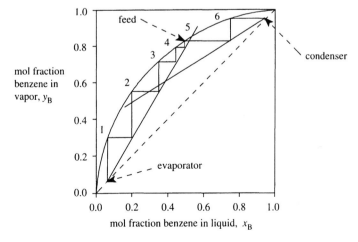

Figure 4.77. A multistage distillation process.

As stated above, we consider in this text only feeds in which the vapor and liquid are in equilibrium. Thus the feed point will lie on the equilibrium line. The more general cases are treated in the chemical engineering course on separation processes. In the examples shown here, we have started counting trays at the bottom (or top) and the feed composition conveniently matches one of the intersections with the equilibrium line. In other cases (such as the exercises at the end of this chapter) the feed composition is less conveniently matched to the diagram. In these cases, you might consider starting at the feed composition and stepping off trays upward toward the condenser and downward toward the evaporator.

Note that we would need *many* more trays to distill feed mixtures with benzene concentrations greater than shown in Figure 4.77. The stripping-section operating line and the equilibrium line intersect at a *pinch point*. It would take an infinite number of trays to reach the pinch point. Moreover, it is impossible to distill a feed with a composition above the pinch point. Does this mean one cannot produce 99% toluene from a 90/10 benzene/toluene feed? How might one redesign a column to accommodate a 90/10 benzene/toluene mixture? We need to move the pinch point. We cannot change the equilibrium line.[3] So we change the operating line by changing the ratio L/V. In fact, we can decrease this ratio until $L/V \approx 1$ for both the stripping and rectifying section. The analysis in Figure 4.76 is repeated for $L/V \approx 1$ in both sections in Figure 4.78.

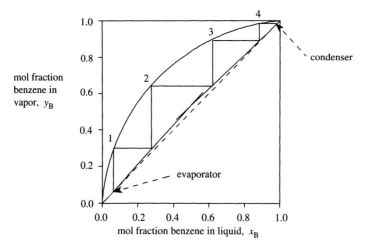

Figure 4.78. A multistage distillation process.

---

[3] Actually, we *can* change the equilibrium line, by changing the pressure. This is the basis of separations called *pressure-swing processes*. But *how* will the equilibrium line change? That's a subject of the course in chemical engineering thermodynamics. For now, you would have to return to the lab and measure more data.

Figure 4.78 indicates that we could accommodate feed of any composition with a four-tray distillation column. That seems better than the columns analyzed in Figures 4.76 and 4.77. Why bother with L/V ratios other than 1? Well, consider this: If $L/V = 1.001$ in the stripping section, for example, and we are drawing 1 mol/min of bottoms, what is the flow rate of liquid down the column in the stripping section? Roughly, L is about 1,000 mol/min. Thus although you only need four trays, the column must have an enormous diameter to handle 1,000 mol/min of liquid–vapor countercurrent flow. Instead, for $L/V = 2$, and we again draw 1 mol/min of bottoms, the liquid flow rate in the column is only 1 mol/min. Changing L/V from 1.001 to 2 doubles the number of trays, but it reduces the diameter by a factor of a thousand! Since the cost of a distillation column is roughly proportional to the amount of steel it contains, changing L/V from 1.001 to 2 will reduce the capital cost by a factor of 500. The operating cost is also reduced – it takes a lot more steam to evaporate 1,000 mol/min than it does to evaporate 1 mol/min.

This graphical method makes it easy to design a column (to specify the number of trays) or analyze the effect of changing the conditions in an existing column – feed composition, bottoms composition, distillate composition, or L/V ratios in the stripping and rectifying sections. Can you imagine analyzing distillation columns mathematically? Neither could Warren McCabe and Ernest Thiele when they were graduate students at MIT in 1925. They developed the graphical procedure shown in Figures 4.73 through 4.78, known as the McCabe–Thiele method. Nowadays distillation columns are analyzed with commercial software, which is efficient but also quite opaque. (Chemical engineering students in the senior design course use such software and routinely propose distillation columns with two trays, 3 m in height, and 100 m in diameter.) The performance of a column, such as the change in composition from tray to tray, or the existence of pinch points, is easily visualized with a McCabe–Thiele diagram. If you use commercial software, prepare a McCabe–Thiele plot of your results. (Your manager will ask to see it anyway. And when you become a manager, always ask to see the McCabe–Thiele diagram.) In Chapter 5 we will examine the internal workings of distillation columns to design the tray diameters and tray spacings.

4.9 Summary

Processes can be analyzed using experimental data. Because experimental data are often presented graphically, analysis based on empirical data often uses graphical methods.

Maps are useful devices for representing data. You need not be familiar with every map of technical data, just like you need not know every street map in the world.

When you see a new street map, you can immediately find your way about because you have learned how to read street maps. If you develop some basic skills in reading engineering maps, you will seldom be lost. The procedures for reading street maps apply to engineering maps. When you encounter a new map, label the landmarks – such as "vapor, liquid, and solid" regions or an equilibrium line. Get your bearings, label extremes on the map – left upper corner, lower right corner, etc. Then find your present location and your destination on the map. Finally, explore alternate routes to get from here to there.

While applying graphical analysis to analyze the separation processes, we developed several graphical devices, such as operating lines, tie lines, and the McCabe–Thiele method. These devices are *design tools* for analyzing separation processes, analogous to the operating equations obtained by mathematical analysis.

REFERENCES

CRC Handbook of Chemistry and Physics, CRC Press, West Palm Beach, FL. A new edition is issued every year.

Blanch, H. W., and Clark, D. S. 1995. *Biochemical Engineering*, Marcel-Dekker, New York.

Humphrey, J. L. 1995. "Separation Processes: Playing a Critical Role," *Chemical Engineering Progress*, October, pp. 31–35.

King, C. J. 1971. *Separation Processes*, McGraw-Hill, New York.

Lightfoot, E. N. 1988. "Recovery of Potentially Valuable Biologicals from Dilute Solution." In *Chemical Engineering Education in a Changing Environment*, ed. S. I. Sandler and B. A. Finlayson, AIChE Publications, New York.

McCabe, W. L., and Thiele, E. W. 1925. *Industrial & Engineering Chemistry* **17**: 605.

Nystrom, J. M. 1984. *Product Purification and Downstream Processing*, 5th Biennial Executive Forum, A. D. Little, Boston.

Sandler, S. I. 1989. *Chemical and Engineering Thermodynamics*, Wiley, New York.

Sherwood, T. K., Pigford, R. L., and Wilke, C. R. 1975. *Mass Transfer*, McGraw-Hill, New York.

EXERCISES

4.1 Plot the data given in (A), (B), and (C) to determine the functional relation between the two variables. That is, obtain an equation to express the second variable as a function of the first variable. Each set of data will yield a straight line when plotted on one of three types of graph paper – normal, semilog, or log–log. (See Appendix D.)

(A) Fluid mechanics. The frictional force on a flat plate (F) as a function of the velocity (v) of the fluid flowing past the plate.

v (cm/sec)	F (dynes)
4.0	1.35
5.0	1.8
10.	5.3
20.	15.
45.	50.
70.	98.

(B) The Leibnitz experiment to measure the conversion of potential energy to heat. The temperature change of a water bath (ΔT) is measured as a function of the mass (M) that falls a given distance.

M (kg)	ΔT (mK)
10.	12.
25.	30.
40.	49.
55.	64.5
62.5	75.
75.	90.

(C) Chemical kinetics. The concentration of reactant A in a batch reactor as a function of time, t.

t (min)	[A] (mol/m^3)
0.	90.
10.	50.
30.	15.
40.	8.5
55.	3.5
70.	1.5

Phase Diagrams

4.2 An air stream is contaminated by 100 ppm (on a molar basis) of compound X. We wish to remove compound X from the air by condensation. 100 ppm (parts per million) $= 0.01$ mol% $= 10^{-4}$ mol fraction.

(A) If the pressure of the air is 1 atm, what is the partial pressure of compound X?

(B) If the temperature of the air is 300 K, what is the temperature of compound X?

(C) Consider the phase diagram of compound X shown below. Locate the point on the graph that corresponds to 100 molar ppm of compound X in air at 300 K and 1 atm.

(D) If the contaminated air is cooled at constant pressure (1 atm), at what temperature will compound X start to condense?

(E) To what temperature must one cool the air (at total pressure of 1 atm) to reduce the contamination level to 10 ppm?

(F) If the contaminated air is compressed at constant temperature (300 K), at what pressure will compound X start to condense?

(G) To what pressure must one compress the air (at 300 K) to reduce the compound X level to 10 ppm?

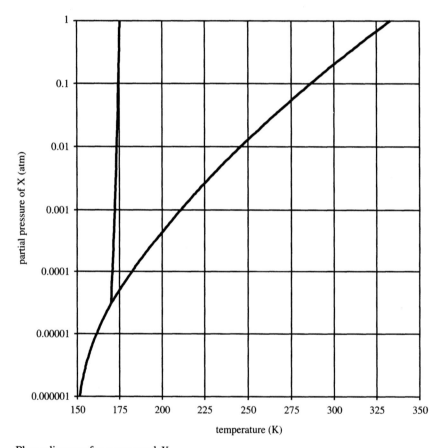

Phase diagram for compound X.

4.3 The phase diagram for the chemical A is shown below. Design a process to transform the gas at 300°C and 2 atm to a solid at 0°C and 1 atm without forming any liquid at any point in the process. Label the function of each unit and specify the temperature and pressure of each stream. Begin your process with a stream of chemical A at 300°C and 2 atm. Your process need not include peripheral equipment for energy efficiency.

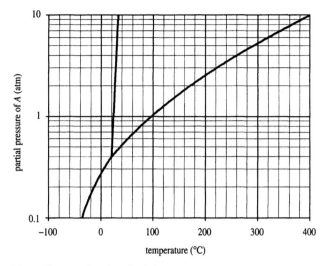

Phase diagram for chemical *A*.

(This exercise appeared on an exam. It was estimated that it could be completed in 15 minutes.)

4.4 We wish to reduce the partial pressure of benzene in air from 10 torr to 1 torr. The phase map below shows the path for cooling the contaminated air at constant total pressure, as discussed in Section 4.4.

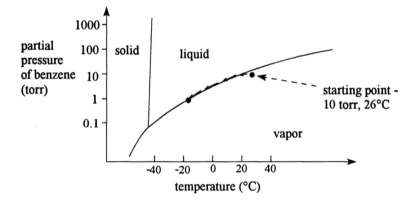

Use the phase map below to plot the same process as plotted above – cooling the contaminated air at constant total pressure to reduce the benzene in the vapor phase by 90 mol%.

The *x*-axis, molar volume, is a logarithmic scale. For reference, at 1 atm the vapor/liquid region spans from 0.09 liter/mol (liquid molar volume) to 22.4 liter/mol (vapor molar volume). A point within the two-phase region corresponds to the average molar volume, calculated as

average molar volume = (liquid mol fract)(liquid molar volume)

+ (vapor mol fract)(vapor molar volume).

A point midway between the two borders is a two-phase mixture with about 5% vapor, 95% liquid.

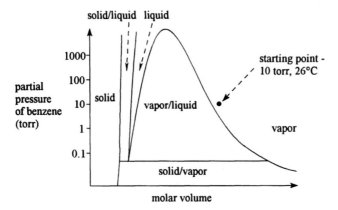

4.5   In Section 4.4 we analyzed two processes to reduce the partial pressure of benzene in air from 10 torr to 1 torr. One process reduced the temperature at constant pressure and another increased the pressure at constant temperature. Consider here a process to reduce the temperature and total pressure at constant total volume. Plot a process to reduce the partial pressure of benzene in air from 10 torr to 1 torr on the two phase maps below. See the previous exercise for a note on average molar volumes.

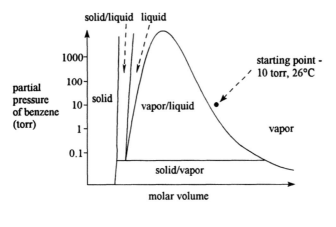

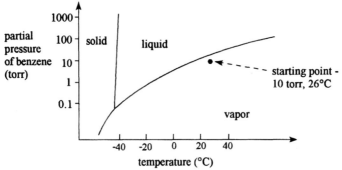

4.6 Air with 100% humidity has the maximum partial pressure of water at a given temperature. Air with 100% humidity is saturated with water vapor.

(A) Indicate air at 29°C (84°F) with 25% humidity with an "*A*" on the pressure–temperature phase diagram for H_2O.

(B) Clouds are two-phase mixtures of water vapor and water liquid suspended in air. Indicate a cloud at 20°C (68°F) with a "*B*" on the pressure–temperature phase diagram.

(C) A cloud at ground level is fog. Early-morning fog usually disappears as the sun rises and warms the air. Assume the fog is 0.02 mol fraction H_2O (liquid plus vapor) in air at 10°C (50°F) and 1 atm. Show the transition from fog to no fog on the diagram. At what temperature does the fog disappear?

(D) On a certain day in November the air over Ithaca, New York is clear, 4°C (39°F), with 50% humidity. Later that night, this same air cools to −4°C (25°F). Do you predict snow overnight? Unlike TV weather reporters, *you* will get no credit for guessing "*yes*" or "*no.*" You must justify your prediction in words or by drawing on the pressure–temperature phase diagram.

(E) A cold, dry mass of polar air moving south mixes with a warm, wet mass of gulf air moving north. The polar air has a temperature of −10°C and 0.% humidity. The gulf air has a temperature of 22°C and 95.% humidity. Assume equal portions of polar air and gulf air are mixed. Calculate the temperature of the mixture and the mol fraction of H_2O in the mixture.

(F) Calculate the amount of rain that will fall, as the depth of water that will accumulate on the ground (in inches). Assume there are 14.7 lbs of air above each square inch of ground.

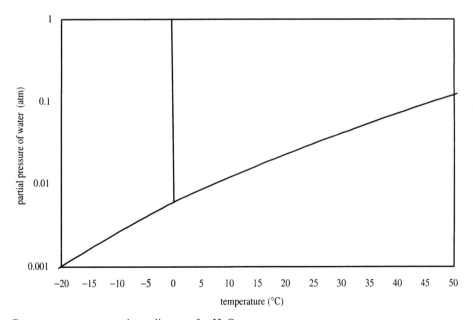

Pressure–temperature phase diagram for H_2O.

(Parts (A)–(D) of this exercise appeared on an exam. It was estimated they could be completed in 15 minutes.)

Absorbers and Strippers

4.7 Consider three different configurations of absorbers for removing an organic pollutant, P, from air. For this particular system, the equilibrium line is straight,

$$\left(\frac{\text{kg of } P}{\text{kg of air}}\right)_{eq} = \frac{1}{10}\left(\frac{\text{kg of } P}{\text{kg of oil}}\right)_{eq},$$

$$w_{P/\text{air}} = \frac{w_{P/\text{oil}}}{10}.$$

In each of the three configurations, the amount of P in the air is reduced to 1.0 wt% P from 4.0 wt% P.

(A) Consider a single-stage absorber. The oil and air are well mixed and the streams that exit the unit are in equilibrium with each other. Use a graphical method to compute the flow rate of oil.

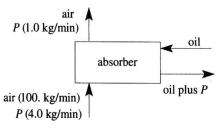

(B) Consider the three-stage absorber shown on the left below. The streams that exit each absorber are in equilibrium with each other. Note that the flow rate of oil into each stage is not necessarily the same. Use a graphical method to compute the total flow rate of oil.

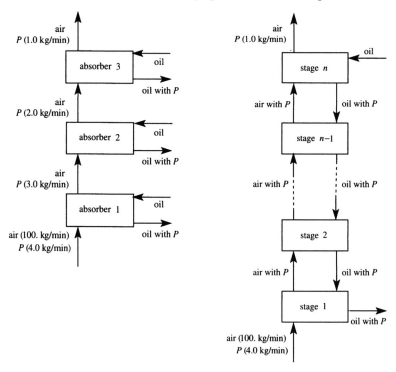

(C) Consider a countercurrent multistage absorber as shown on the right of the preceding figure. The flow rate of oil is 10 kg/min. Again, the streams that exit each absorber are in equilibrium with each other. Use a graphical method to calculate the number of stages required.

(D) Which design, (A), (B), or (C), uses less oil? Hypothesize as to why.

4.8 Your lungs can be modeled as an absorber that transfers O_2 from the air to your blood.

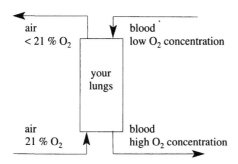

In this simple model we will assume:

- The lungs are a single-stage absorber.
- The flows through the lungs are continuous and at steady state.
- The O_2/air and O_2/blood streams leaving the lungs are in equilibrium.

The map below is a graphical analysis of lungs for a normal breathing rate and a normal pulse rate. Because you will use this map to graph your answers to parts (A)–(D) of this exercise, you are advised to photocopy and enlarge the map.

(A) Indicate the following on the map:

    1. mol fraction of O_2 in the inlet air,
    2. mol fraction of O_2 in the outlet air,
    3. mol fraction of O_2 in the inlet blood,
    4. mol fraction of O_2 in the outlet blood.

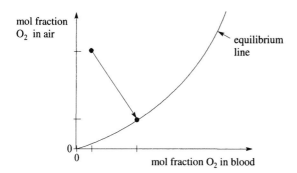

(B) Sketch on the map a graphical analysis of hyperventilation (in which the breathing rate doubles but the pulse rate is unchanged). Assume the mol fraction of O_2 in the inlet blood is unchanged.

(C) You are transported to a planet whose atmosphere has one half the O_2 concentration as Earth. Draw a graphical analysis on the map to determine what you must do to achieve the same mol fraction of O_2 in blood you had on Earth.

(D) In an emergency, a saline solution (water with salt) is administered intravenously to someone who has suffered a severe loss of blood. Consider a person who has lost half his or her blood. This person's blood has half the capacity to absorb oxygen. Draw a graphical analysis on the map to determine what this person must do to achieve the same mol fraction of O_2 in diluted blood.

(E) Your lungs also serve as a stripper that transfers CO_2 from your blood to the air. Sketch a graphical analysis of this stripper on the map below. The concentration of CO_2 in the air is essentially zero (0.03 mol%). Assume that the blood leaving the lungs contains CO_2. Indicate on your sketch:

1. mol fraction of CO_2 in the outlet air,
2. mol fraction of CO_2 in the inlet blood,
3. mol fraction of CO_2 in the outlet blood.

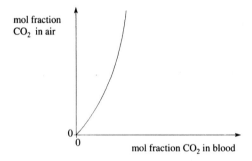

We are indebted to Professor Richard Seagrave for suggesting the concept for this exercise. Similar interesting applications to biological systems can be found in Seagrave's text, *Biological Applications of Heat and Mass Transfer*, Iowa State University Press, 1971.

(This exercise appeared on an exam. It was estimated it could be completed in 20 minutes.)

4.9 Consider an absorber with the flow rates and compositions given in the diagram below.

(A) What flow rate of oil is needed if the absorber has only one equilibrium stage?

(B) What is the minimum flow rate of oil?

(C) Calculate the number of stages needed if the flow rate of oil is 25% higher than the minimum determined in part (B).

(D) What flow rate of oil is needed if the absorber has four equilibrium stages?

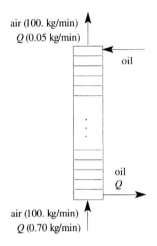

air (100. kg/min)
Q (0.05 kg/min)

oil

oil
Q

air (100. kg/min)
Q (0.70 kg/min)

The graph below is used for parts (A)–(D) of this exercise. Because it will be difficult to draw the solutions to all four parts on a single graph you are advised to photocopy and enlarge the graph below. Or, you can download an enlarged graph from www.cheme.cornell.edu/~tmd/Graphs.

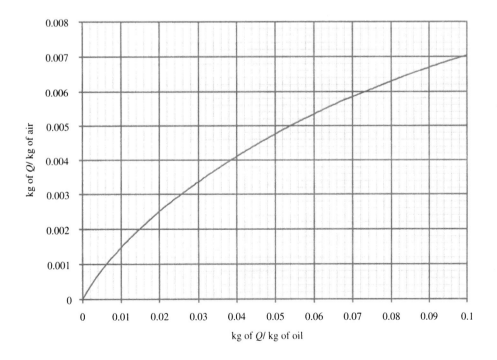

4.10 **(A)** A countercurrent multistage absorber shown below removes K from air. How many stages are required *and* what is the mass fraction of K in the oil effluent? (The equilibrium plot appears at the end of the exercise.)

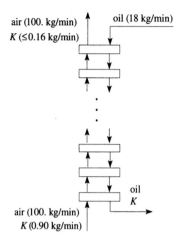

(B) A different countercurrent multistage absorber shown below removes K from air. In this case, the oil entering the absorber is not pure; the oil contains some K. How many stages are required?

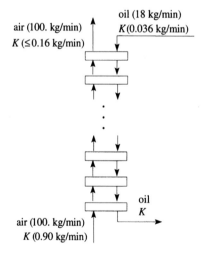

(Parts (A) and (B) of this exercise appeared on an exam. It was estimated that they could be completed in 20 minutes.)

(C) A yet-different countercurrent multistage absorber removes K from air. Pure oil enters the column at the top. Oil leaves the column at the bottom and at an intermediate stage. Given the flow rates and compositions shown below, how many stages are required?

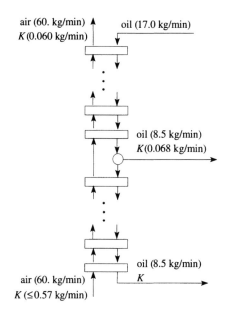

(Part (C) of this exercise appeared on another exam. It was estimated that it could be completed in 20 minutes.)

The graph below is used for parts (A), (B), and (C). Because it will be difficult to draw the solutions to all three parts on a single graph you are advised to photocopy and enlarge the graph below. Or, you can download the graph from www.cheme.cornell.edu/ ~tmd/Graphs.

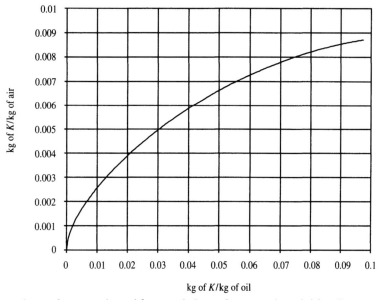

4.11 Groundwater is contaminated frequently by anthropogenic activities. For example, storage tanks may leak chemicals into the soil, which are then incorporated into an aquifer. An example is the leakage of gasoline into groundwater from storage tanks underneath filling stations. Public health, safety, and environmental concerns have spawned a number of designs for removing these trace chemicals.

Two designs are considered here. Both involve the adsorption of benzene onto activated carbon, a substance similar to charcoal and relatively inexpensive to produce and regenerate. In the first scheme, the benzene adsorbs directly from the contaminated water onto the charcoal.

First Scheme:

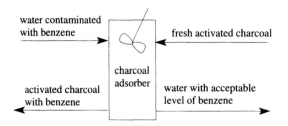

The adsorber above is a well-mixed slurry of water and charcoal. The concentration of benzene in the water and the concentration of benzene adsorbed on the charcoal are uniform throughout the adsorber.

A second scheme uses air to first strip the benzene from the groundwater. The contaminated air then passes through a bed of activated carbon, which adsorbs the benzene from the air.

Second Scheme:

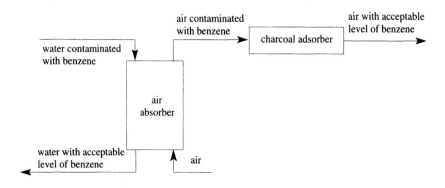

The adsorber in the second scheme is a packed bed of charcoal. The air is not mixed in the adsorber and therefore the concentration of benzene in the air varies with position in the adsorber, as well as with time, because the adsorber is charged initially with fresh charcoal. Until the charcoal saturates with benzene, the charcoal absorber reduces the benzene/air concentration by a factor of 1,000. When the outlet concentration of benzene reaches a critical level, breakthrough has occurred and the charcoal is replaced.

Use the graphical data below to calculate the capacity and cost of each scheme. Which scheme has the lower capital cost? Which scheme has the lower operating cost? Calculate the profit (or loss) associated with each scheme. Which scheme is best overall?
Here are some specifications:

· The process must treat 100 gallons of water per minute.
· The water is contaminated with 20 ppm (by weight) benzene.

- The water leaving each process has a residual concentration of 20 ppb (by weight) benzene.
- The air leaving the second scheme contains essentially zero benzene.
- The equipment is depreciated by a ten-year, straight-line method.

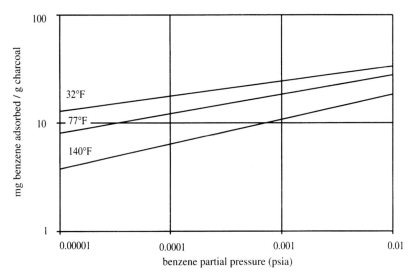

Gas-phase adsorption isotherms for benzene over activated charcoal. (Adapted from Calgon Corporation, Pittsburgh, PA, a subsidiary of Merck & Co., Inc.)

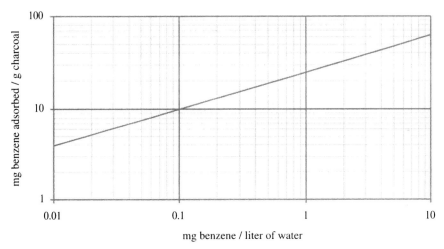

Liquid-phase adsorption isotherm for benzene over activated charcoal. (Adapted from Westates Carbon, Inc., Los Angeles, CA.)

Here are some facts obtained from colleagues at Levine-Fricke – Engineers, Hydrogeologists, and Applied Scientists, an environmental engineering firm in Emeryville, California:

- A water/charcoal adsorber with a capacity of 50 gallons of water per minute holds 1,800 pounds of activated carbon and costs $9,800, which includes the cost of pipes, pumps, valves, controls, etc.

- Activated carbon costs $2.50 per pound. (Most of the cost is for transportation.)
- An air absorber with a capacity of 25 gallons of water per minute costs $20,000, which includes the cost of pipes, pumps, valves, controls, etc.
- The air absorbers use 350 cubic feet of air to strip the benzene from 25 gallons of water.
- An air/charcoal adsorber with a capacity of 350 cubic feet of air per minute holds 2,200 pounds of activated carbon and costs $6,800, which includes the cost of pipes, pumps, valves, controls, etc.

Finally, here are some conversion factors:

1 gal = 3.8 liter,   1 kg = 2.2 lb,   1 ft^3 = 28.3 liter,   1 mol of benzene = 78 g.

Note that standard pressure is 1 atm.

4.12 Examine the process in the second scheme of Exercise 4.11. *Sketch* on a plot similar to the one below a graphical analysis of the multistage air/water column.

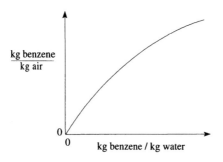

Your sketch need only be *qualitative*. Ignore all *quantitative* compositions and flow rates. Just sketch a generic analysis of a unit in which clean air strips benzene from contaminated water.

Label on your sketch the following:

- the equilibrium line,
- the operating line,
- the inlet *and* outlet compositions of the air,
- and the inlet *and* outlet compositions of the water.

Finally, indicate how many stages your analysis predicts.

(This exercise appeared on an exam. It was estimated that it could be completed in 15 minutes.)

4.13 Use the thermodynamic data in the six plots below to design a process to separate E and X from a dilute water solution. Your process must also separate E from X, although the E may contain up to 10 wt% X and the X may contain up to 10 wt% E.

Specifications:

One input stream:	3 wt% E, 2 wt% X, 95 wt% H_2O.
Two product streams:	(1) >90 wt% E, <10 wt% X, no H_2O,
	(2) >90 wt% X, <10 wt% E, no H_2O.

Boiling points (in °C, at 1 atm)

E	X	H_2O	oil	N_2
100	100	100	320	−196

Thermodynamic data (all at 20°C, 1 atm):

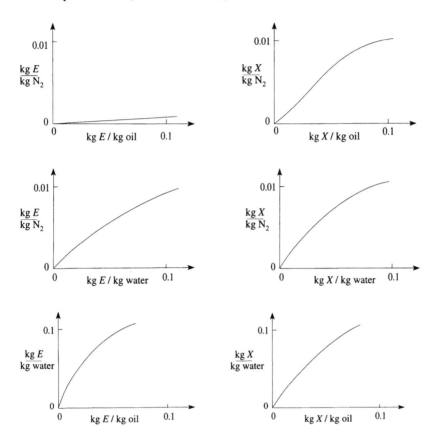

You may assume:

    Water does not dissolve in the oil; the oil does not dissolve in water.

    The oil does not evaporate into N_2 at 20°C; water does (1.4 wt% H_2O vapor in N_2 at 20°C).

Guidelines:

    Minimize the number of units. Minimize waste.

    Label each stream with qualitative compositions.

    Label each separator with the temperature and the physical basis for the separation (e.g., liquid/solid).

    You may omit heat exchangers.

    You need *not* specify the operating parameters (number of stages, L/V ratios) for the separators.

(This exercise appeared on an exam. It was estimated that it could be completed in 25 minutes.)

Flash Drums, Tie Lines, and the Lever Rule

4.14 The unit below treats a mixture of tetrachloromethane (CCl_4) and chlorobenzene (C_6H_5Cl) by flash distillation.

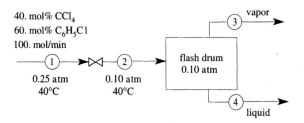

40. mol% CCl_4
60. mol% C_6H_5Cl
100. mol/min

Use the CCl_4–C_6H_5Cl phase diagram below to determine the compositions (mol fractions CCl_4 and C_6H_5Cl) and flow rates (mol/min) of streams 3 and 4. Also, label on the CCl_4–C_6H_5Cl phase diagram the phases corresponding to each region and indicate the coordinates of (1) the stream before the expansion valve, (2) the vapor stream, and (3) the liquid stream.

4.15 Use the phase diagram for CCl_4–C_6H_5Cl mixtures below to determine the flow rate and composition of stream 4 leaving the flash drum shown on the following page. Also determine the pressure in the flash drum.

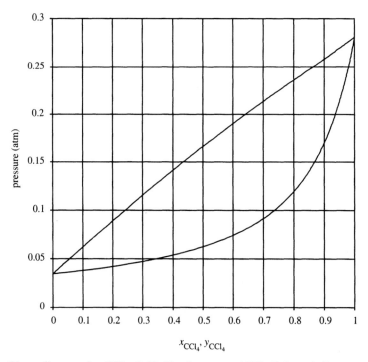

Phase diagram for CCl_4–C_6H_5Cl mixtures at 40°C. Enlarged diagram available at www.cheme.cornell.edu/~tmd/Graphs.

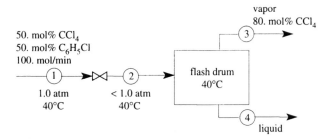

4.16 The flash drum below separates a 50/50 mixture of acetonitrile (ACN) and 2-methylpyridine (MP) at 98°C and 1 atm.

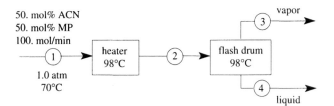

(A) Determine the composition of stream 3 *and* the composition of stream 4.

(B) *Use the lever rule* to determine the flow rate of stream 3 *and* the flow rate of stream 4.

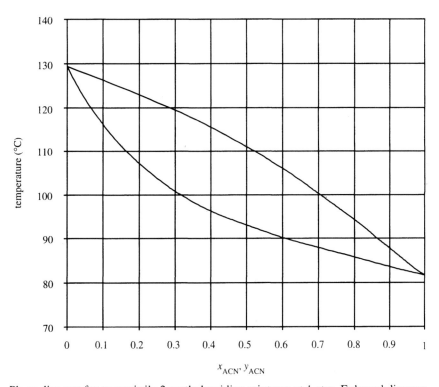

Phase diagram for acetonitrile 2-methylpyridine mixtures at 1 atm. Enlarged diagram available at www.cheme.cornell.edu/-tmd/Graphs.

A second flash drum is added.

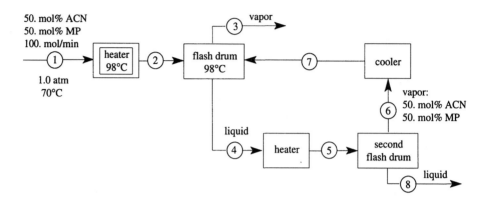

(C) Determine the temperature in the second flash drum.

(D) Determine the composition of stream 8.

(E) Calculate the flow rate of stream 8 *and* the flow rate of stream 7. You may use the lever rule and/or mass balances. Note that the flow rate of stream 4 may be changed by the addition of the second flash drum.

(Parts (A)–(D) of this exercise appeared on an exam. It was estimated they could be completed in 12 minutes.)

4.17 A mixture of M and K is treated with the flash drum shown below.

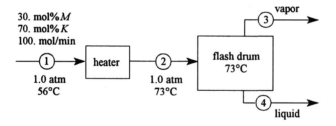

Use the phase diagram for $M-K$ mixtures to determine the compositions (mol fractions M and K) and flow rates (mol/min) of streams 3 and 4. (Justify *briefly* the number of significant figures in your answers.) Also, label on the $M-K$ phase diagram the phases corresponding to each region and indicate the coordinates of (1) stream 1, (2) the vapor-phase stream, and (3) the liquid-phase stream.

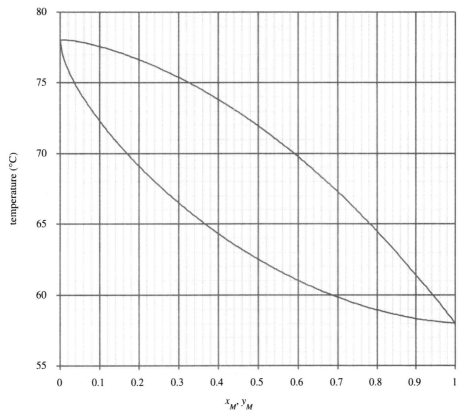

Phase diagram for $M-K$ mixtures at 1 atm. Enlarged diagram available at www.cheme.cornell. edu/~tmd/Graphs.

4.18 The flash drum diagrammed below produces an equimolar vapor of M and K.

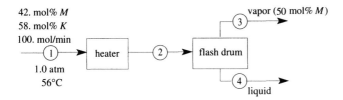

(A) What is the composition of stream 4 (in mol% M)?

(B) What is the flow rate of stream 4 (in mol/min)?

(C) What is the temperature in the flash drum?

(D) The temperature is changed to maximize the concentration of M in stream 3. If the input stream contains 42 mol% M, what is the maximum possible concentration of M in stream 3?

(This exercise appeared on an exam. It was estimated that it could be completed in 15 minutes.)

4.19 The process diagrammed below produces an equimolar mixture of *M* and *K* at 100. mol/min from either stream 3 (vapor) or stream (4), depending on the composition of the input (stream 1), which varies over time.

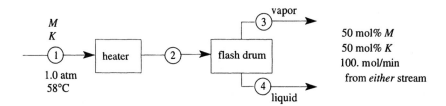

Assuming the equimolar product is the vapor stream (stream 3),

(A) What is the range of compositions permitted in stream 1?

(B) What is the composition of stream 4 (in mol% *M*)?

(C) What is the temperature in the flash drum?

Assuming the equimolar product is the liquid stream (stream 4),

(D) What is the range of compositions permitted in stream 1?

(E) What is the composition of stream 3 (in mol% *M*)?

(F) What is the temperature in the flash drum?

Assume now that the equimolar product is leaving via stream 4 (liquid). Assume further that the vapor flow rate (stream 3) is 190. mol/min.

(G) What is the flow rate and composition of stream 1?

4.20 Use the experimental data given on the following page to determine the flow rate (mol/min) and composition of stream 7 in the process diagrammed below. Both flash drums operate at 1.0 atm.

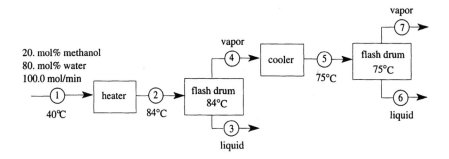

The data below were measured with methanol/water mixtures at 1.0 atm.

T (°C)	$x_{methanol}$	$y_{methanol}$	T (°C)	$x_{methanol}$	$y_{methanol}$
100.0	0.0	0.0	79.1	0.268	0.648
98.4	0.012	0.068	78.1	0.294	0.666
96.9	0.020	0.121	76.5	0.352	0.704
95.8	0.026	0.159	75.3	0.402	0.734
95.1	0.033	0.188	74.2	0.454	0.760
94.1	0.036	0.215	73.2	0.502	0.785
92.2	0.053	0.275	72.0	0.563	0.812
90.0	0.074	0.356	70.9	0.624	0.835
88.6	0.087	0.395	69.2	0.717	0.877
86.9	0.108	0.440	68.1	0.790	0.910
85.4	0.129	0.488	67.2	0.843	0.930
83.4	0.164	0.537	66.9	0.857	0.939
82.0	0.191	0.572	65.7	0.938	0.971
			65.0	1.0	1.0

(Experimental data from the compilation by Gmehling, J., and Onken, U. 1977. *Vapor–Liquid Equilibrium Data Collection*, Dechema, Frankfurt, Germany, vol. 1, p. 60.)

Distillation Columns

4.21 A distillation column that separates acetonitrile ($CH_3C{\equiv}N$) and water operates at a pressure of 1 atm with parameters shown by the McCabe–Thiele diagram shown below.

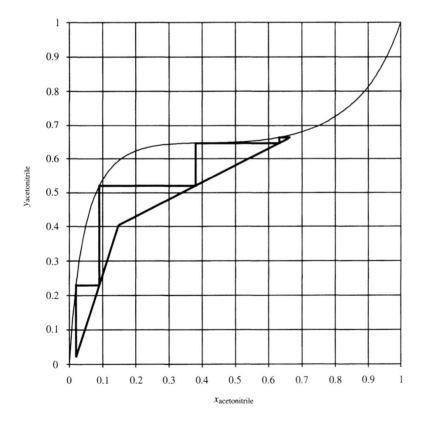

(A) How many stages does this distillation column have?

(B) What is the mol% acetonitrile in the acetonitrile-rich product?

(C) What is the mol% acetonitrile in the water-rich product?

(D) What is the mol% acetonitrile in the liquid and vapor streams that leave the feed stage (that is, the stage into which the input is fed)?

(E) What is the ratio of liquid to vapor flow rates in the rectifying section (the upper section) of the column?

(F) Shown below is a phase diagram for acetonitrile–water mixtures at 1 atm as a function of temperature and composition. Label the four regions of the diagram.

(G) The distillation column described on the previous page is preceded by a heater to bring the acetonitrile–water mixture (20 mol% acetonitrile) to match the conditions of the feed stage. Use the phase diagram below to determine the temperature in the preheater.

(H) What is the ratio of liquid to vapor in the stream leaving the preheater?

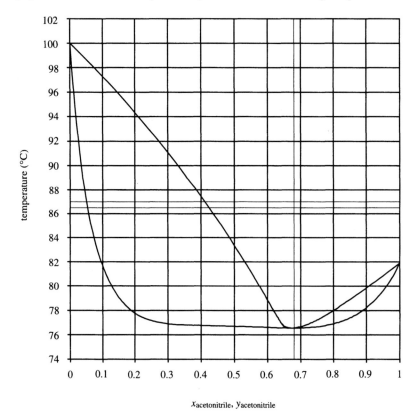

$x_{\text{acetonitrile}}, y_{\text{acetonitrile}}$

Phase diagram for ACN-water mixtures. Enlarged diagram available at www. cheme.cornell.edu/~tmd/Graphs.

(An exercise similar to this appeared on an exam. It was estimated it could be completed in 20 minutes.)

(Diagrams generated from data compiled by Gmehling, J., and Onken, U. 1977. *Vapor–Liquid Equilibrium Data Collection*, Dechema, Frankfurt, Germany, vol. 1, p. 78.)

4.22 The temperature–composition phase diagram for acetonitrile (ACN)/2-methylpyridine (MP) mixtures in Exercise 4.16 can be plotted as the x–y diagram shown below.

(A) Mark the point that corresponds to pure ACN.

(B) Mark the point that corresponds to a vapor–liquid mixture of ACN and MP at equilibrium at 98°C.

(C) Mark the point that corresponds to a vapor–liquid mixture *not* at equilibrium: vapor with 80% ACN above liquid with 70% ACN.

(D) After the system in (C) reaches equilibrium, will the vapor contain less than 80% ACN or more than 80% ACN? Justify your reasoning on the diagram below.

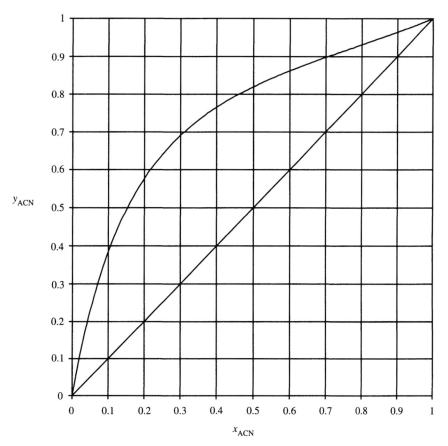

x–y diagram for two-phase systems of acetonitrile/2-methylpyridine mixtures at 1 atm. Enlarged diagram available at www.cheme.cornell.edu/~tmd/Graphs.

(This exercise appeared on an exam. It was estimated it could be completed in 12 minutes.)

4.23 A typical flash drum has one liquid–vapor equilibrium stage. The effectiveness of a flash drum is improved by adding stages, as shown on the following page. The liquid feed is split into two streams: A liquid stream is delivered to the top stage and the other stream is converted entirely to vapor and bubbled through the bottom stage.

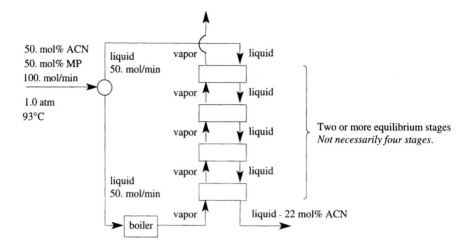

A multistage flash drum may be analyzed with an operating line on an x–y diagram, similar to the McCabe–Thiele analysis of a distillation column. But note the following differences:

(A) The liquid leaving the bottom stage has a different composition from the vapor entering the bottom stage. Locate the lower point of the operating line on the x–y diagram for ACN/MP mixtures given in Exercise 4.22.

(B) All stages have the same L/V ratio, the ratio of liquid flow rate to vapor flow rate. Draw the operating line on the x–y diagram.

Use the operating line to answer parts (C) and (D).

(C) Determine the number of equilibrium stages required.

(D) Determine the composition of the vapor output.

Sketch a different operating line to answer part (E).

(E) Given a feed with 50.% ACN, what is the maximum concentration of ACN in the vapor output? You may use any number of stages; you may use any L/V ratio.

(This exercise appeared on an exam. It was estimated it could be completed in 20 minutes.)

4.24 The column shown below distills a water–ethanol mixture.

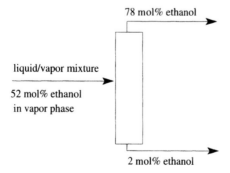

Refer to the ethanol–water liquid–vapor equilibrium below.

(A) What is the minimum number of stages this column could have?

(B) What is the maximum ratio of liquid/vapor flow for the stripping portion (the stages below the feed)?

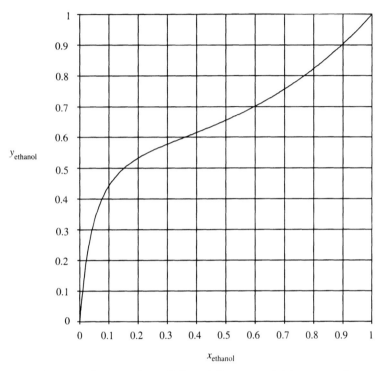

Ethanol–water mixtures at 1 atm. Enlarged diagram available at www.cheme. cornell.edu/~tmd/Graphs.

(This exercise appeared on an exam. It was estimated that it could be completed in 15 minutes.)

(Diagram generated from data compiled by Gmehling, J., Onken, U., and Arlt, W. 1982. *Vapor–Liquid Equilibrium Data Collection – Organic Hydroxy Compounds: Alcohols*, Dechema, Frankfurt, Germany, vol. 1, p. 165.)

4.25 Design a distillation column to separate a mixture of ethanol and water into a water-rich stream (98 mol% water) and an ethanol-rich stream (80 mol% ethanol). The feed to the distillation column is a liquid–vapor mixture equilibrated such that $y_{ethanol} = 0.443$ and $x_{ethanol} = 0.10$. Use the ethanol–water vapor–liquid diagram from Exercise 4.24 to determine

- the number of stages,
- the reflux ratio (L/V) for the stripping section,
- the reflux ratio (L/V) for the rectifying section,
- the stage into which the feed enters, and
- if the feed needs to be heated or cooled before it enters the feed stage.

For uniformity, please number your stages from the bottom of the distillation column.

Although there are a multitude of viable designs for the column, avoid the absurd. For example, a column with reflux ratio of 1 (no output) or a column with maximum throughput (requiring an infinite number of stages) would be considered absurd.

4.26 Design a distillation column to separate a mixture of 1,2-dichloroethane (CH_2ClCH_2Cl) and 1-propanol ($CH_3CH_2CH_2OH$) into a DCE-rich stream (74 mol% DCE) and a propanol-rich stream (99 mol% propanol). The input to the distillation column is a liquid–vapor mixture equilibrated such that $x_{DCE} = 0.30$. Use the DCE–propanol vapor–liquid diagram to determine operating conditions and the number of stages. You need not find the absolute optimal conditions; just find a workable solution. Specify

- the number of stages,
- the reflux ratio (L/V) for the stripping section,
- the reflux ratio (L/V) for the rectifying section,
- the stage into which the feed enters, and
- if the feed needs to be heated or cooled before it enters the feed stage.

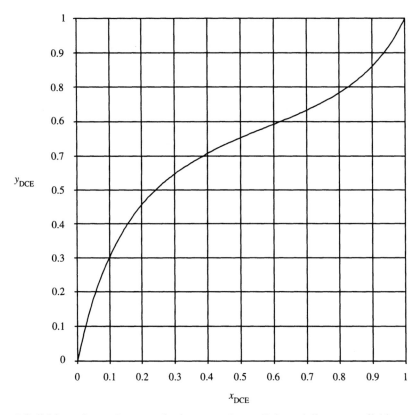

1,2-dichloroethane –1-propanol mixtures at 1 atm. Enlarged diagram available at www.cheme.cornell.edu/~tmd/Graphs.

(Diagram generated from data compiled by Gmehling, J., Onken, U., and Arlt, W. 1982. *Vapor–Liquid Equilibrium Data Collection – Organic Hydroxy Compounds: Alcohols*, Dechema, Frankfurt, Germany, vol. 1, part 2c, p. 477.)

4.27 From which mixture would it be easiest to produce 99.9% *A* by distillation? Assume the feed is a liquid/vapor mixture at equilibrium, with 50 mol% *A* in the vapor phase. Justify your choice, either with words or a graphical analysis.

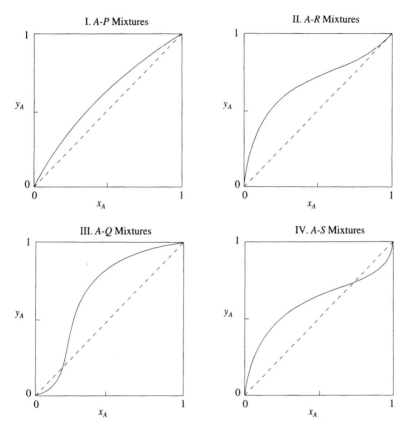

4.28 Use the experimental data given in Exercise 4.20 to specify the operating conditions in the distillation column below. The column has nine trays. Determine operating conditions that will maximize L/V in the lower trays (stripping section) and minimize L/V in the upper trays (rectifying section). Also, specify the temperature of stream 1 so the vapor–liquid composition of the input matches that of the feed tray.

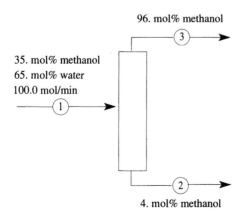

4.29 Tabulated below are vapor–liquid equilibrium data for nitric acid/water mixtures at 1.0 atm. Use these data to plot a phase diagram (suitable for analyzing a flash drum) and an x–y liquid–vapor diagram (suitable for McCabe–Thiele analysis). Label the axes and, where appropriate, label the regions of the plot (liquid, vapor, etc.).

T (°C)	$x_{nitric\ acid}$	$y_{nitric\ acid}$	T (°C)	$x_{nitric\ acid}$	$y_{nitric\ acid}$
100.	0.0	0.0	117.0	0.474	0.651
104.	0.067	0.003	115.0	0.515	0.762
104.5	0.072	0.003	113.0	0.530	0.764
106.5	0.102	0.010	112.6	0.540	0.768
107.0	0.110	0.012	111.5	0.557	0.857
108.5	0.135	0.020	108.8	0.574	0.864
109.5	0.141	0.023	106.0	0.606	0.936
110.5	0.162	0.035	102.9	0.651	0.942
111.5	0.181	0.042	97.5	0.700	0.960
112.0	0.181	0.042	96.1	0.719	0.972
114.5	0.217	0.082	95.8	0.723	0.984
115.5	0.233	0.096	95.5	0.738	0.986
117.5	0.282	0.165	92.0	0.755	0.983
119.2	0.348	0.297	91.0	0.802	0.983
119.4	0.341	0.259	87.2	0.853	0.984
120.0	0.383	0.375	86.9	0.878	0.988
119.9	0.374	0.375	82.8	0.991	0.997
118.5	0.450	0.564	83.0	1.0	1.0

(Experimental data from Gmehling, J., Onken, U., and Rarey-Nies, J. R. 1988. *Vapor–Liquid Equilibrium Data Collection – Aqueous Systems – Supplement 2*, Dechema, Frankfurt, Germany, vol. 1, part 1B, p. 384.)

4.30 **(A)** The multistage distillation column shown below separates a mixture of F and Z.

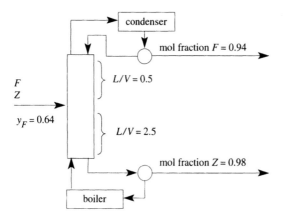

How many stages are needed? At what stage does the feed enter?

The vapor–liquid equilibrium data for F–Z mixtures are plotted at the end of this exercise.

(B) The distillation column shown on the following page provides an intermediate output stream.

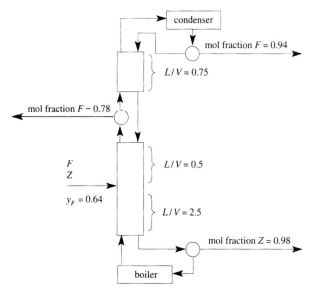

How many stages are needed? At what stage does the feed enter? Above what stage does the intermediate stream leave? (**Hint:** Contrary to usual practice, begin stepping off stages at the top.)

The vapor–liquid equilibrium data for $F-Z$ mixtures are plotted below. Because you need to analyze two columns, you are advised to work your solutions from enlarged photocopies of the diagram, or from diagrams available at www.cheme.cornell.edu/~tmd/Graphs.

(This exercise appeared on an exam. It was estimated that it could be completed in 25 minutes.)

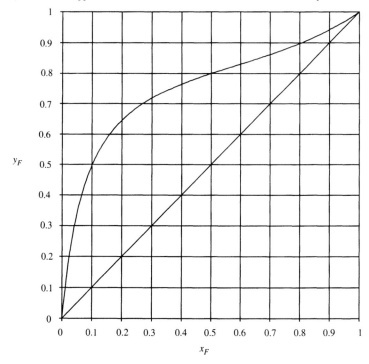

$F-Z$ mixtures at 1 atm.

Process Design

4.31 Design a process that starts with 20 mol% acetonitrile and yields two streams with compositions ≥98 mol% acetonitrile and ≤2 mol% acetonitrile. Indicate the compositions of all streams in your process and the number of stages in the distillation column(s). You may wish to use the liquid–vapor equilibrium data for acetonitrile/water mixtures at 1.0 atm (Exercise 4.21) as well as the data at 0.2 atm (shown below).

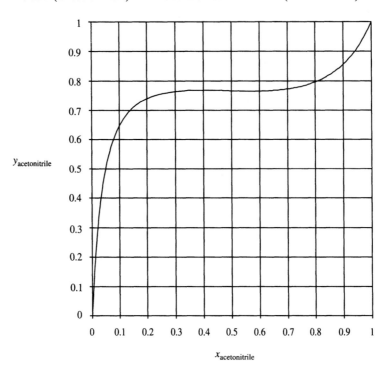

$y_{acetonitrile}$

$x_{acetonitrile}$

Acetonitrile/water mixtures at 0.2 atm. Enlarged diagram available at www.cheme.cornell.edu/~tmd/Graphs.

(Diagram generated from data given in Gmehling, J., and Onken, U. 1977. *Vapor–Liquid Equilibrium Data Collection – Organic Hydroxy Compounds: Alcohols*, Dechema, Frankfurt, Germany, vol. 1, p. 79.)

4.32 Shown below is a phase diagram for a two-component system as a function of composition and temperature. Recall that a coordinate in the region labeled "liquid" corresponds to a liquid with a homogeneous composition. Likewise, a coordinate in the region labeled "vapor" corresponds to a vapor with a homogeneous composition. A coordinate in the "two-phase" region corresponds to a liquid–vapor mixture. The compositions of the two phases are given by a tie line. The quantities of the two phases are given by the lever rule.

A phase diagram for immiscible liquids is more complicated. Shown on the following page is the phase diagram for water and 1-butanol. A liquid mixture with very little butanol is homogeneous. Likewise, a liquid mixture with very little water is homogeneous. For an intermediate composition below 70°C, the mixture separates into two liquid phases – one phase is mostly water and one phase is mostly butanol. The other two-phase regions are liquid–vapor mixtures. The compositions of the two phases are given by a tie line and the quantities of the two phases are given by the lever rule.

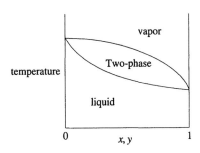

(A) Label on the diagram the boiling point of pure water.

(B) Label on the diagram the boiling point of pure butanol.

(C) Complete the following table for a mixture of 0.90 mol butanol and 0.10 mol water. If only one phase is present, leave the last three columns blank.

T (°C)	1 or 2 phases?	First phase			Second phase		
		Liquid or vapor?	Mol fraction butanol	Total moles	Liquid or vapor?	Mol fraction butanol	Total moles
60							
80							
105							
120							

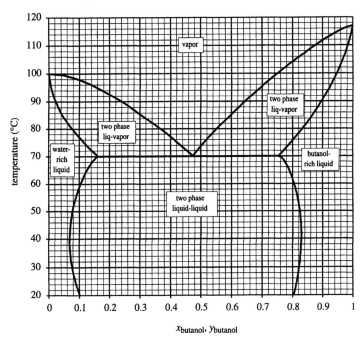

Enlarged diagram available at www.cheme.cornell.edu/~tmd/Graphs.

(D) Complete the following table for a mixture of 0.65 mol butanol and 0.35 mol water. If only one phase is present, leave the last three columns blank.

T (°C)	I or 2 phases?	First phase			Second phase		
		Liquid or vapor?	Mol fraction butanol	Total moles	Liquid or vapor?	Mol fraction butanol	Total moles
60							
80							
105							
120							

(E) Design a process that separates a mixture of 50 mol% butanol and 50 mol% water at 60°C to yield two products: (1) a water-rich stream with at most 5 mol% butanol and (2) a butanol-rich stream with at least 95 mol% butanol. *Note*: It is feasible to separate the phases of any two-phase system, such as liquid–vapor, or liquid–liquid. You may ignore energy efficiency; you need not include heat exchangers. Many designs are workable. Explore different options and choose your best.

- Label the function of each unit.
- Indicate the temperature in each unit.
- Indicate *quantitatively* the mol% of butanol in each stream.

Some guidelines:

- Minimize the number of units.
- Minimize waste.

(This exercise appeared on an exam. It was estimated that it could be completed in 35 minutes.)

4.33 W and R are separated by the process below. Use the W–R phase diagram and the x_W-y_W liquid–vapor equilibrium diagram to complete parts (A)–(D).

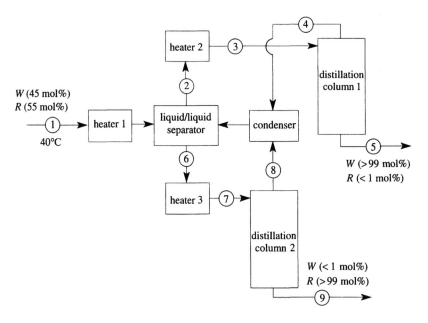

(A) What temperature in heater 1 will optimize the liquid/liquid separation?

(B) Label streams 2 and 6 on the $W-R$ phase diagram below.

In both distillation columns the feed enters at the top stage. The distillate stream from each column is thus vapor in equilibrium at the temperature of the preheater. It is not necessary to optimize the distillation columns. Just specify parameters so each column will produce the output streams 5 and 9.

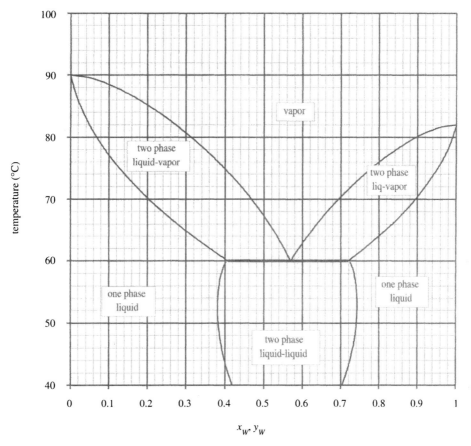

$W-R$ phase diagram at 1 atm. Enlarged diagram available at www.cheme.cornell.edu/~tmd/ Graphs.

(C) Design distillation column 1 by completing the following steps:
Specify the temperature for heater 2.
Indicate stream 4 on the $W-R$ phase diagram.
Indicate stream 3 (feed stream) on the x_W-y_W liquid–vapor equilibrium diagram.
Specify the number of equilibrium stages.
Specify the *approximate* L/V ratio.

(D) Design distillation column 2 by completing the following steps:
Specify the temperature for heater 3.
Indicate the composition of stream 8 on the $W-R$ phase diagram.
Indicate stream 7 (feed stream) on the x_W-y_W liquid–vapor equilibrium diagram.

Specify the number of equilibrium stages.

Specify the *approximate L/V* ratio.

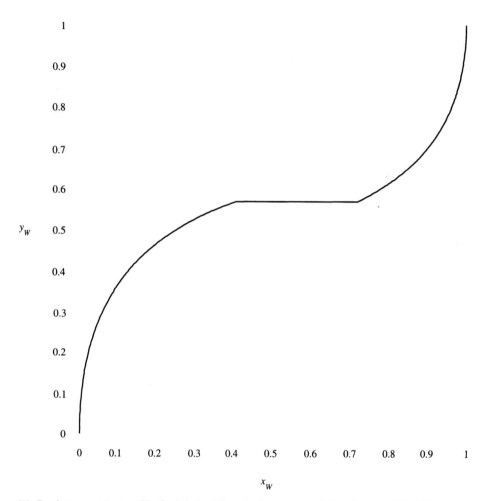

W–R mixtures at 1 atm. (Vertical dashed lines indicate $x_W = 1\%$ and $x_W = 99\%$.) Enlarged diagram available at www.cheme.cornell.edu/~tmd/Graphs.

(This exercise appeared on an exam. It was estimated that it could be completed in 30 minutes.)

4.34 Design a process to separate a solid composed of 50 mol% *P* and 50 mol% *Q* into two products: a stream with ≥ 98 mol% *Q* and a stream with ≤ 2 mol% *Q*. Your process should minimize waste and the number of units. You may assume that liquid/solid mixtures can be separated perfectly. However, solid/solid mixtures cannot be separated.

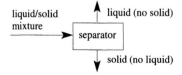

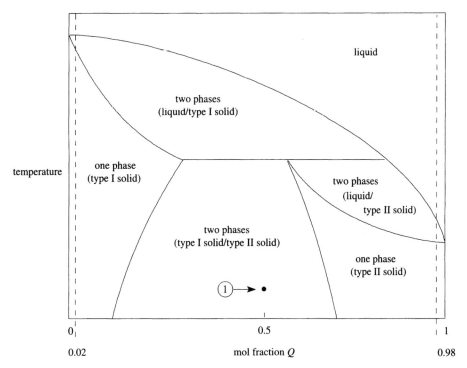

Phase diagram for $P-Q$ mixtures. Enlarged diagram available at www.cheme.cornell.edu/
~tmd/Graphs.

Indicate the compositions and temperatures of every stream by writing the stream number on a copy of the diagram above. For example, the process begins at point ①. You need not write any compositions or temperatures on your flowsheet.

(This exercise appeared on a exam. It was estimated that it could be completed in 25 minutes.)

4.35 Design a process to produce pure Na from a liquid mixture of 65 weight% Na, and 35 weight% K. Your process should maximize the amount of Na produced and minimize the number of units needed. You may assume that liquid/solid mixtures can be separated perfectly. However, solid/solid mixtures cannot be separated.

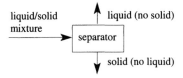

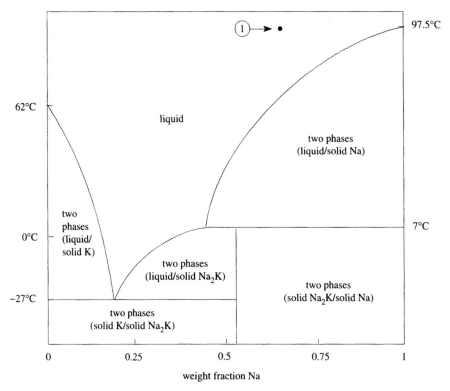

Phase diagram for Na–K mixtures. Available at www.cheme.cornell.edu/~tmd/Graphs.

Indicate the compositions and temperatures of every stream in your process by writing the stream number on a copy of the diagram above. For example, the process begins at point ①. You need not write any compositions or temperatures on your flowsheet.

(This exercise appeared on a exam. It was estimated that it could be completed in 25 minutes.)

4.36 Magnesium sulfate crystals exist in five compositions:

$MgSO_4 \cdot 12H_2O$ (0.36 wt fraction $MgSO_4$),

$MgSO_4 \cdot 7H_2O$ (0.48 wt fraction $MgSO_4$),

$MgSO_4 \cdot 6H_2O$ (0.53 wt fraction $MgSO_4$),

$MgSO_4 \cdot H_2O$ (0.87 wt fraction $MgSO_4$),

$MgSO_4$ (1.0 wt fraction $MgSO_4$).

The phases present in a mixture of $MgSO_4$ and H_2O are shown on the map below. The region labeled "liquid solution" is a single phase. All other regions are two phase, either liquid/solid or solid/solid.

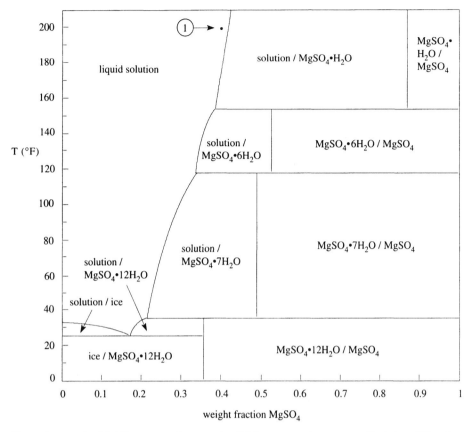

Phase diagram for MgSO₄-water mixtures. Available at www.cheme.cornell.edu/~tmd/Graphs.

Design a process to produce $MgSO_4 \cdot 6H_2O$ (0.53 wt fraction $MgSO_4$) from a liquid solution with 0.40 wt fraction $MgSO_4$ at 200°F. Your process should maximize the amount of $MgSO_4 \cdot 6H_2O$ produced and minimize the number of units needed.

Indicate the compositions and temperatures of every stream in your process by writing the stream number on the phase diagram. For example, the process begins at point ①. You need not write any compositions or temperatures on your flowsheet. You may omit heat exchangers.

You may assume that liquid/solid mixtures can be separated perfectly. However, solid/solid mixtures cannot be separated.

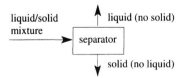

(Exercise adapted from Felder, R. M., and Rousseau, R. W. 1986. *Elementary Principles of Chemical Processes*, 2nd ed., Wiley, New York, p. 259. Figure adapted from Perry, R. H., and Green, D. W. 1984. *Perry's Chemical*

Engineers' Handbook, 6th ed., McGraw-Hill, New York, pp. 19–26. Copyright McGraw-Hill, Inc. Reproduced by permission.)

(This exercise appeared on an exam. It was estimated that it could be completed in 15 minutes.)

4.37 We used tie lines with liquid–vapor phase diagrams to calculate the compositions and flow rates of the two phases that separated. Tie lines can also be used to analyze the mixing of two miscible liquids. The composition of two miscible liquids, such as water and methanol, can be represented by a linear scale.

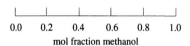

Example: Use a tie line and the lever rule to determine the composition of the mixer effluent.

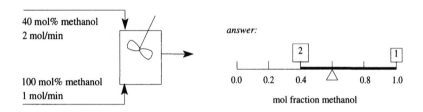

The tie line is drawn between 0.4 and 1.0 mol fraction methanol. Amounts representing the respective flow rates are placed at these points. This "lever" is balanced by placing the fulcrum at 0.6 mol fraction methanol. The composition of the effluent is thus 60 mol% methanol.

(A) Use a tie line and the lever rule to determine the flow rate of stream 1.

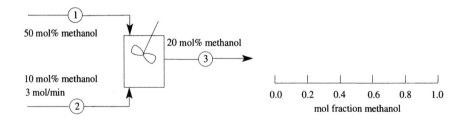

Mixtures of three miscible liquids can be represented by a ternary diagram such as the one shown below for water–methanol–ethanol mixtures. Any coordinate on this ternary diagram corresponds to a single liquid phase.

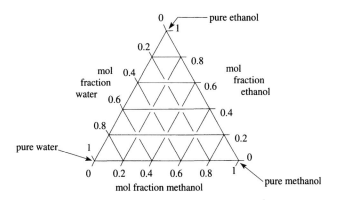

The mol fraction of ethanol is given by horizontal lines.

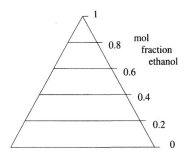

The mol fraction of methanol is given by these lines.

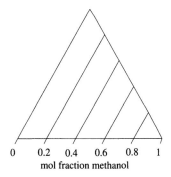

The mol fraction of water is given by these lines:

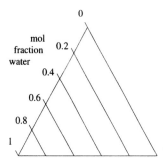

The point below corresponds to a mixture of 10% water, 30% ethanol, and 60% methanol.

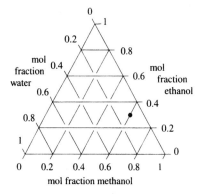

mol fraction methanol

Use the ternary diagram at the end of the exercise to complete parts (B) and (C).

(B) Use a tie line to calculate the composition of stream 3 in the process below. Note that on a ternary diagram of three miscible fluids, tie lines can be drawn at any angle.

20 mol% water
10 mol% methanol
70 mol% ethanol
1 mol/min

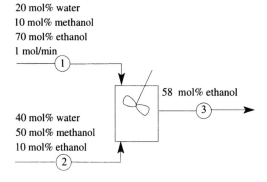

58 mol% ethanol

40 mol% water
50 mol% methanol
10 mol% ethanol

(C) Apply the lever rule to the tie line drawn in part (B) to determine the flow rate of stream 2 (in mol/min).

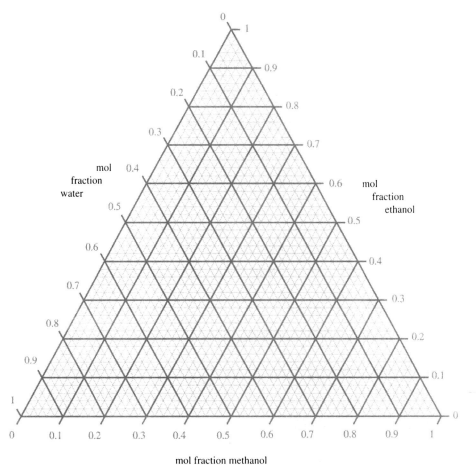

mol fraction methanol

Ternary diagram for water–methanol–ethanol mixtures. Enlarged diagram available at www.cheme.cornell.edu/~tmd/Graphs.

(This exercise appeared on an exam. It was estimated that it could be completed in 20 minutes.)

4.38 Mixtures of partially immiscible liquids separate into two liquid phases for some compositions. For example, hexane and methylcyclopentane (mcp) are miscible for all compositions. Also, aniline and mcp are miscible for all compositions. However, aniline and hexane are immiscible. The ternary phase diagram for mcp–hexane–aniline at 45°C and 1 atm is sketched below.

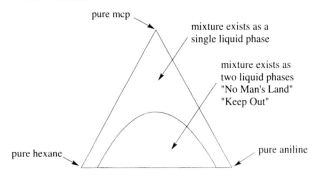

The compositions of the two liquid phases in equilibrium are obtained by a horizontal tie line, as shown below. In this example, a mixture of 15% mcp, 60% hexane, and 25% aniline separates into two liquid phases; one phase is 15% mcp, 75% hexane, and 10% aniline and the other phase is 15% mcp, 10% hexane, and 75% aniline.

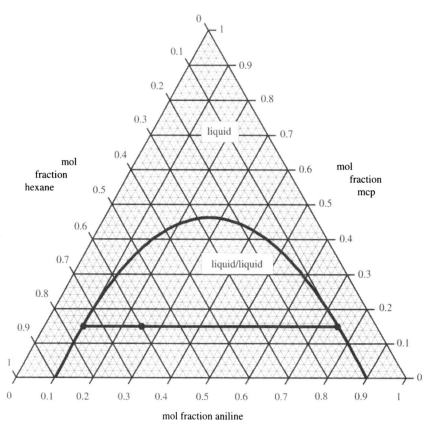

Ternary diagram for mcp–hexane–aniline mixtures.

Design a process that uses these starting materials:

- 100 mol/min of 50 mol% hexane, 50 mol% mcp,
- 100 mol% aniline (unlimited flow rate)

and yields a product of *at least* 75 mol% hexane; the remainder is mcp and aniline in any ratio. The product should have a flow rate of *at least* 35 mol/min. You may ignore the fate of all other output streams in your process.

Your process should use liquid–liquid separation, not distillation. Specify the *approximate* flow rates and compositions of each stream. You may specify a composition by indicating a point on the ternary diagram. For example, the aniline stream is indicated by a $\boxed{1}$ on the diagram above.

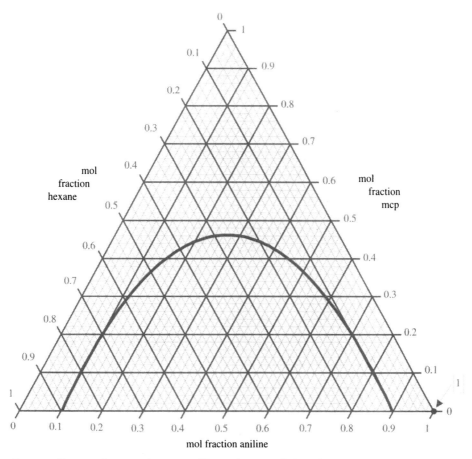

Ternary diagram for mcp–hexane–aniline mixtures. Enlarged diagram available at www.cheme.cornell.edu/~tmd/Graphs.

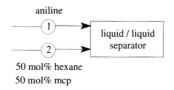

(This exercise appeared on an exam. It was estimated that it could be completed in 25 minutes.)

4.39 The two preceding exercises use ternary diagrams plotted on equilateral triangles to analyze three-component mixtures that are completely miscible (Exercise 4.37) and three-component mixtures that separate into two liquid phases (Exercise 4.38).

Below is a different representation of three-component mixtures: *n*-heptane (H), methylcyclo hexane (MCH), and aniline (A). The weight fraction of H is the *x* axis; the weight fraction of MCH is the *y* axis. The weight fraction of A is implied; a mixture with 40% H and 50% MCH is assumed to have 10% A.

There are three regions on the map: two regions of single-phase liquids separated by a two-phase liquid–liquid region. The dashed lines in the two-phase region are tie lines. Additional tie lines may be drawn by interpolating between the dashed lines.

(A) Label on the map the points (or regions) that correspond to:

> pure *n*-heptane (H),
> pure methylcyclohexane (MCH),
> pure aniline (A),
> aniline-rich liquid,
> aniline-poor liquid.

(B) Sketch the tie line for two-phase mixtures that contain no *n*-heptane.

(C) A mixture with 10% H, 40% MCH, and 50% A will separate into two liquid phases. Determine the compositions of the *two* liquid phases.

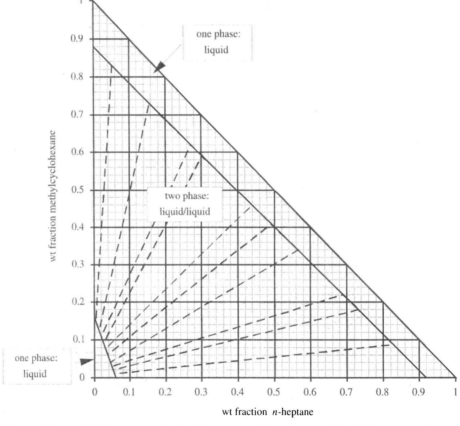

Ternary diagram for heptane–methylcyclohexane–aniline mixtures. Available at www.cheme. cornell.edu/~tmd/Graphs.

(This exercise appeared on an exam. It was estimated that it could be completed in 12 minutes.)

4.40 The process below treats a mixture of methylcyclohexane (MCH) and aniline (A) by liquid–liquid extraction with *n*-heptane (H) and then distillation to remove the H, to produce A with less than 2% MCH. The liquid–liquid separator has one equilibrium stage.

   (A) Use mass balances and the ternary diagram for Exercise 4.39 (available at www.cheme. cornell.edu/~tmd/Graphs) to determine the composition *and* flow rate of stream 4.

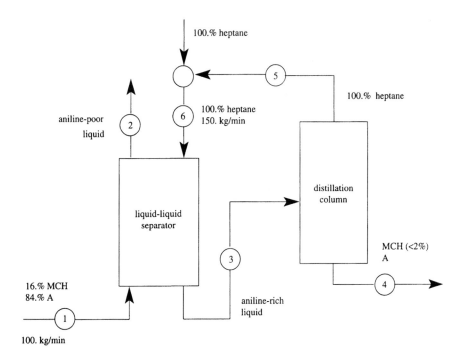

   (B) Add units to recover useful components from stream 2. That is, add units to minimize waste.

   - You are *not permitted* to separate MCH and A by distillation.
   - Stream 4 must have less than 2% MCH.
   - Specify the compositions of *all* streams that you add.

   (Hint: Draw a border around your entire process to analyze your design.)

   (This exercise appeared on an exam. It was estimated that it could be completed in 30 minutes.)

4.41 Shown below is a phase diagram for mixtures of *G* and *H*. Graphical methods can be applied to diagrams of this type to calculate a mass balance. For example, one may apply the lever rule to a tie line spanning the two-phase region to calculate the relative sizes of the liquid and vapor phases.

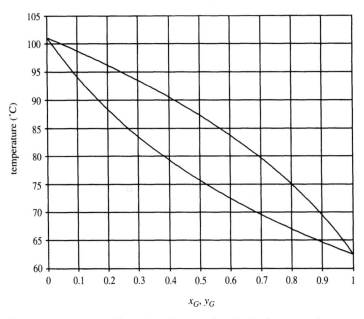

Temperature–composition phase diagram for *G–H* mixtures at 1 atm.

Another useful map is the enthalpy–composition map shown below. Applying the lever rule on an enthalpy–composition map provides a graphical *energy* balance.

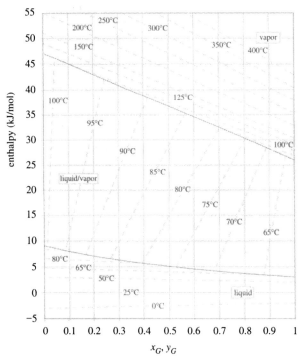

Enthalpy–composition phase diagram for *G–H* mixtures at 1 atm. Enlarged diagram available at www.cheme.cornell.edu/ ~tmd/Graphs.

Examine the enthalpy–composition map for G–H mixtures at 1 atm. The dashed lines in the liquid region and the vapor region show the temperature at a given composition and enthalpy. The dashed lines in the liquid–vapor region are tie lines. Unlike the previous phase diagrams, the tie lines in the liquid–vapor region are not horizontal. Liquid and vapor at equilibrium have the same temperature but not the same enthalpy.

(A) Follow steps (i)–(v) to determine the enthalpy (in kJ/mol) of liquid and vapor phases in equilibrium at 87.5°C.

   (i) Draw a tie line on the temperature–composition phase diagram at 87.5°C. What are the liquid and vapor compositions?

   (ii) On the enthalpy–concentration diagram, mark on the liquid border the composition of the liquid phase. Similarly, mark on the vapor border the composition of the vapor phase.

   (iii) Connect these two points with a tie line. Your tie line should lie midway between the tie lines at 85°C and 90°C.

   (iv) What is the enthalpy of the liquid phase (in kJ/mol)?

   (v) What is the enthalpy of the vapor phase (in kJ/mol)?

(B) A liquid with 20 mol% G at 0°C is heated to a vapor at 125°C. Follow steps (i)–(iii) to calculate the change in enthalpy.

   (i) What is the enthalpy of 20 mol% G at 0°C (in kJ/mol)?

   (ii) What is the enthalpy of 20 mol% G at 125°C (in kJ/mol)?

   (iii) What is the change in enthalpy (in kJ/mol)?

(C) One mol of 50 mol% G at 125°C is mixed with 1 mol of 90 mol% G at 400°C. Follow steps (i)–(iv) to calculate the temperature and composition of the resulting mixture.

   (i) Locate the coordinates of the two components to be mixed and connect the points with a tie line.

   (ii) Use the lever rule to locate the fulcrum at the mixture composition.

   (iii) What is the composition of the mixture?

   (iv) What is the temperature of the mixture?

(D) One mol of H liquid at 0°C is mixed with 1 mol of G vapor at 200°C. Follow steps (i)–(v) to calculate the temperature and composition of the resulting mixture.

   (i) Locate the coordinates of the two components to be mixed and connect the points with a tie line.

   (ii) Use the lever rule to locate the fulcrum at the mixture composition.

   (iii) The mixture should have two phases. Locate the liquid–vapor tie line nearest to the mixture. If necessary, draw a new tie line parallel to a predrawn tie line. What is the mixture temperature?

   (iv) Follow the liquid–vapor tie line to the liquid border. What is the composition of the liquid phase?

   (v) How many moles (total) are in the liquid phase?

(This exercise appeared on an exam. It was estimated that it could be completed in 25 minutes.)

4.42 The previous exercise illustrates the utility of enthalpy concentration diagrams for performing energy balances graphically. Graphical energy balances are particularly useful for substances that release heat upon mixing, such as water and sulfuric acid.

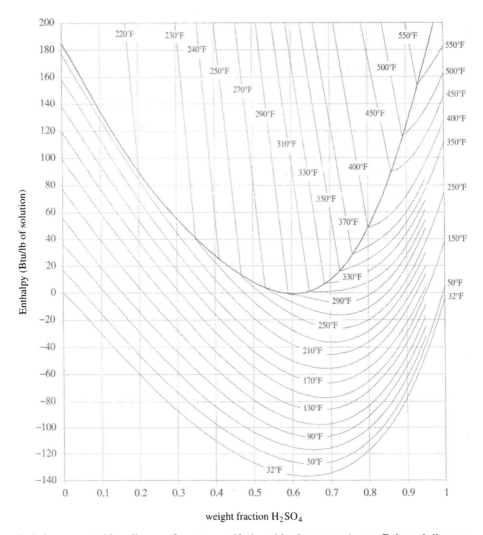

Enthalpy–composition diagram for water–sulfuric acid mixtures at 1 atm. Enlarged diagram available at www.cheme.cornell.edu/~tmd/Graphs.

(Diagram adapted from Hougen, O. A., and Watson, K. M. 1946. *Chemical Process Principles Charts*, Wiley, New York.)

The sulfuric acid–water phase diagram shows two regions: liquid and liquid/vapor. The regions are separated by the dark arc. Below this line is liquid – one phase. Above this line is liquid/vapor mixtures – two phases, "keep out."

The arcs in the liquid region show the temperature at a given composition and enthalpy. The straight lines in the liquid–vapor region are tie lines. Unlike other phase diagrams, the tie lines are not horizontal. Given a composition and enthalpy in the two-phase region, a tie line connects the liquid and vapor borders. The contact at the vapor border is off the map. Each tie line corresponds to a specific temperature; liquid and vapor at equilibrium are at the same temperature.

(A) A liquid with 20 wt% H_2SO_4 at 32°F is to be warmed to 130°F. Calculate the change in enthalpy.

  (i) What is the enthalpy of 20 wt% H_2SO_4 at 32°F (in Btu/lb)?
  (ii) What is the enthalpy of 20 wt% H_2SO_4 at 130°F (in Btu/lb)?
  (iii) What is the change in enthalpy (in Btu/lb)?

(B) 100 lbs of 20 wt% H_2SO_4 at 32°F are mixed with 100 lbs of 90 wt% H_2SO_4 at 250°F. Calculate the temperature and composition of the resulting mixture.

  (i) Locate the coordinates of the two components to be mixed and connect the points with a tie line.
  (ii) Use the lever rule to locate the fulcrum at the mixture composition.
  (iii) What is the composition of the mixture?
  (iv) What is the temperature of the mixture? (Mixing sulfuric acid and water releases heat. The mixture's temperature may be higher than the average of the initial temperatures.)

(C) 30 lbs of water at 150°F are mixed with 70 lbs of 100% H_2SO_4 at 150°F. Calculate the temperature and composition of the resulting mixture.

  (i) Locate the coordinates of the two components to be mixed and connect the points with a tie line.
  (ii) Use the lever rule to locate the fulcrum at the mixture composition.
  (iii) The mixture should be a two-phase one. Locate the liquid–vapor tie line nearest to the mixture. If necessary, draw a new tie line parallel to a predrawn tie line. What is the mixture temperature?
  (iv) Trace the liquid–vapor tie line to the liquid border. What is the composition of the liquid phase?

5

Dimensional Analysis and Dynamic Scaling

WE HAVE NOW CONSIDERED two methods for modeling and analyzing processes: mathematical modeling based on laws and graphical analysis based on empirical measurements. In mathematical modeling we started with a general principle and substituted the specifics of the process. The general principle might be the conservation of mass, (*rate of mass in*) = (*rate of mass out*) at steady state, or a constitutive equation, such as $PV = nRT$. For empirical analysis we first measured data, such as the solid–liquid–vapor phase diagram of a pure substance or a vapor-liquid–composition diagram for a mixture of two components. We did not concern ourselves with why these data were the way they were. We did not dwell on why some systems are nonideal or why some systems have an azeotrope. We just used the data. We never extrapolated from the measured data. That is, if we measured vapor–liquid–composition data for mixtures of benzene–toluene at 80°C as a function of pressure, we used these data only to analyze units that separated benzene–toluene mixtures at 80°C.

What if we wanted to operate the unit at 60°C? We would have to return to the lab and remeasure the data at 60°C. We had no means of extrapolating the data to different conditions. As you will learn later, thermodynamics provides the tools needed to extrapolate equilibrium data. But lacking a knowledge of thermodynamics, is there a way to extrapolate? More generally, how does one extrapolate in systems too complex to yield an analytical solution? How does one design a large unit too complex to be analyzed mathematically? Should one build the unit and hope for the best? That approach would cost too much in time and money. Our government and the military use this approach because they have a lot of time and money. You will not have this luxury.

Instead, we use a third means of modeling: We will study a small working example and extrapolate the process to industrial size. Or, as was stated eloquently by Leo Baekeland (1863–1944), inventor of Bakelite, the first synthetic plastic,

Commit your blunders on a small scale and make your profits on a large scale.

But is it possible to extrapolate the behavior of a model to predict the behavior of a large unit? One of the earliest to recognize this connection was Leonardo da Vinci (~1500) (Zlokarnik 1991: 5):

> Vitruvius says that small models are of no avail for ascertaining the effects of large ones; and I here propose to prove that this conclusion is a false one.

A change in dimensions is known as scaling. The scaled dimension is usually length (*i.e.*, size) and/or time. Extrapolation of a small unit to a large unit or extrapolation of a slow process to a fast process is called *scale-up*. Similarly, extrapolation of a small process to a microscopic process or extrapolation of a fast process to a slow process is called *scale-down*.

So how does one extrapolate from the small to the large? Are there formal rules? Let's look at some examples. We divide the examples into two types: static systems and dynamic systems.

Static systems – systems with no motion – include maps and molecular models. We scale-up a map to calculate distances between cities, for example. Similarly, we scale-down the dimensions of an atomic model to calculate interatomic distances. The rule of scaling in static systems is obvious: The size scales linearly. This is geometric scaling. Examples are shown in Figures 5.1 and 5.2.

Figure 5.1. A two-dimensional static model, a map. Copyright Cornell University. Reproduced by permission.

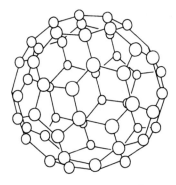

Figure 5.2. A drawing of a three-dimensional static model, an atomic model of the spherical carbon compound, Buckminster fullerene (C_{60}).

Compare the scaling of maps and molecular models to the scaling of dynamic systems. Consider lizards for example. Figure 5.3 is a picture of a small lizard, shown

Figure 5.3. A small lizard, about 10 cm long. Copyright Deborah Duncan. Reproduced by permission.

close to its actual size. This design works quite well for a small lizard. What if you wanted to scale-up this design to make a very large lizard – for example a lizard the size of a house? Should you scale the dimensions linearly? Linear scaling was used in most of the 1950s science fiction B movies (see Figure 5.4). Why does the large lizard in Figure 5.4 look unrealistic? Perhaps because we know that large lizards are proportioned differently than small lizards. Large lizards tend to be shaped like dinosaurs (Figure 5.5), a design that worked well for about 100 million years.

Why do small lizards and large lizards have different proportions? Consider how various features of a lizard scale with size. The mass of a lizard is proportional to the volume of a lizard, which is proportional to the height cubed:

$$body\ mass \propto (height)^3. \tag{5.1}$$

However, the compressive strength of a structural element, such as a bone, is proportional to its cross-sectional area, which is proportional to the height squared:

$$bone\ strength \propto (height)^2. \tag{5.2}$$

If all the dimensions of a lizard were scaled linearly, the mass of a large lizard would exceed the capacity of its skeletal system. As dynamic systems get larger, the length

Figure 5.4. A common lizard scaled linearly to gigantic size. Scene from *The Giant Gila Monster* (1959). Movie Still Archives.

Figure 5.5. A large reptile, triceratops, about 10 m long. Copyright Deborah Duncan. Reproduced by permission.

dimensions cannot scale linearly. Scaling in dynamic systems is based on dimensional analysis, the subject of this chapter. But before we take up dimensional analysis we need to know what a dimension is.

5.1   Units and Dimensions

Any physical quantity has dimensions – not just the usual dimensions of length and time, but others too. Base dimensions, listed in Table 5.1, are denoted by uppercase letters. For temperature we use Θ because T was already used for time (Θ is the Greek letter for Q, but at least "theta" sounds like "t"). The last two items in Table 5.1 will not be encountered again in this text.

Table 5.1. Base dimensions

Base dimension	Symbol
length	L
mass	M
time	T
temperature	Θ
amount	N
electric charge	Q
luminous intensity	I

Each base dimension has a *base unit.* The base unit for each dimension depends on the system of units. Two systems dominate: Système International (SI) and English. The base units are listed in Table 5.2. Any physical quantity can be expressed as a product of a number and a combination of base units.

Table 5.2. The SI and English systems of units

Base dimension	SI (mks)	English
length, L	meter, m	foot, ft
mass, M	kilogram, kg	pound-mass, lb_m
time, T	second, s	second, s
temperature, Θ	Celsius, °C, or Kelvin, K	Fahrenheit, °F, or Rankine, °R
amount, N	mole	mole

The existence of two systems of units causes frustration, tedium, and error. It is important to develop a rigorous method of converting between systems. The guidelines are:

- Multiply by factors of one: For example, 2.54 cm/1 inch = 1.
- Keep meticulous accounting.

From base units we create *multiple units*. Multiple units are, for example, milligram and kilometer. Why multiple units? Because people like to use numbers between 0.1 and 100 and SI units are sometimes awkward. For example, consider length. One describes a marathon as a 10 kilometer run, as opposed to a 10,000 meter run. Similarly, when describing the distance between two atoms, it is convenient to say a C—H bond length is 0.11 nanometers, as opposed to 1.1×10^{-10} m. One describes a feature on a silicon wafer as approximately 1 μm, as opposed to 1×10^{-6} m. The common prefixes are listed in Table 5.3. For the dimension of time, multiple units include hour, day, year, and century. English units of length include inch, foot, yard, and mile. The English system is fraught with special units for each dimension.

Table 5.3. Prefixes for multiple units

Prefix	Symbol	Multiplier
mega	M	10^6
kilo	k	10^3
centi	c	10^{-2}
milli	m	10^{-3}
micro	μ	10^{-6}
nano	n	10^{-9}
pico	p	10^{-12}

The third and final class of units are *derived units*. Derived units are the products of base units and have special names. Some important derived units are listed in Table 5.4.

Table 5.4. Derived units and dimensions

Quantity	Dimensions	Units in SI
volume	L^3	m^3
velocity	L/T	m/s
acceleration	L/T^2	m/s^2
momentum ($p = mv$)	ML/T	$kg \cdot m/s$
force ($f = ma$)	ML/T^2	$kg \cdot m/s^2 \equiv$ newton (N)
pressure (force/area)	M/LT^2	$kg/m \cdot s^2 = N/m^2 \equiv$ pascal (Pa)
energy ($1/2mv^2$)	ML^2/T^2	$kg \cdot m^2/s^2 = N \cdot m \equiv$ joule (J)
power	ML^2/T^3	$kg \cdot m^2/s^3 = J/s \equiv$ watt (W)

We conclude this topic with two rules. The first rule is

Dimensional consistency: All quantities in a sum must have the same dimensions.

Example: The total energy in a system is the sum of the kinetic and potential energy:

$$total\ energy = kinetic\ energy + potential\ energy, \tag{5.3}$$

$$E = \frac{1}{2}mv^2 + mgh, \tag{5.4}$$

$$\frac{ML^2}{T^2} [=] M\left(\frac{L}{T}\right)^2 + M\left(\frac{L}{T^2}\right)L, \tag{5.5}$$

$$\frac{ML^2}{T^2} [=] \frac{ML^2}{T^2} + \frac{ML^2}{T^2}. \tag{5.6}$$

The symbol "[=]" means "has dimensions of." The second rule is

All quantities in a sum must have the same units.

Thus

$$distance = 1\,km + 1\,foot + 1\,mm \neq 3 \tag{5.7}$$

$$= 1\,km + 1\,foot\left(\frac{0.3048\,m}{1\,foot}\right)\left(\frac{1\,km}{1000\,m}\right) + 1\,mm\left(\frac{1\,km}{10^6\,mm}\right)$$

$$= 1.000306\,km. \tag{5.8}$$

5.2   Dimensional Analysis of a Pendulum

To illustrate the principles of dynamic scaling, we consider the motion of a pendulum. A simple pendulum is depicted in Figure 5.6. Can we extrapolate the behavior of a

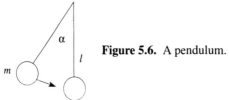

Figure 5.6. A pendulum.

small pendulum to a large pendulum, such as a person swinging on the end of a trapeze? What determines the period of a pendulum? How does it depend on mass? If a child and an adult are side by side on identical swings, will they have the same period? Must they swing at the same amplitude to have the same period? Our everyday experience with swings and trapezes suggests how pendulums scale with size and mass. Can we describe the scaling mathematically?

We begin by listing in Table 5.5 the physical quantities that describe a pendulum. We also assign a mnemonic symbol to each quantity and list each quantity's dimensions.

Table 5.5. The parameters of a pendulum

Physical quantity	Symbol	Dimensions
period of oscillation	t_p	T
length of pendulum	l	L
mass of pendulum	m	M
gravitational acceleration	g	L/T^2
amplitude	α	(none)

Why does amplitude have no dimensions? Recall that amplitude is an angle measured in radians. The amplitude of an angle is the ratio of the length of a radius divided by the arc swept out by the radius. The dimensions of amplitude are thus L/L, which is dimensionless.

Have we neglected anything? How about the viscosity of the medium through which the pendulum moves? This would be important, for example, in designing an underwater pendulum. We neglect it for now.

We wish to use a small model of a pendulum to predict the period of oscillation, t_p, from the physical parameters. We seek an equation of the form

$$t_p = \text{some function of } l, m, g, \text{ and } \alpha \tag{5.9}$$

or

$$t_p = f(l, m, g, \alpha). \tag{5.10}$$

This function must be dimensionally consistent. Whatever combination of parameters we derive for the right side of Eq. (5.10), each term must have dimensions of time, T. That is, the function we seek will have some general form

$$t_p = (\text{term 1}) + (\text{term 2}) + (\text{term 3}) + \cdots \tag{5.11}$$

and each term must have the dimensions of time. This means that no term may contain the parameter m. Why? Because there is no complementary term to cancel its dimension of mass. Thus, the period of oscillation must be independent of mass. The function we seek simplifies to

$$t_p = f(l, g, \alpha). \tag{5.12}$$

Of course, it is possible that we have neglected a parameter whose base dimensions include M. We discount that possibility and continue.

What about the length, l? Does the pendulum's behavior depend on length? Yes, we may include l because the constant g also contains the dimension of length:

$$l [=] L, \qquad g [=] L/T^2. \tag{5.13}$$

So, we must divide l by g:

$$\frac{l}{g} [=] \frac{L}{L/T^2} [=] T^2. \tag{5.14}$$

Thus, *if* length is to appear in our function, it must appear in ratio with g. The function we seek thus has two variables:

$$t_p = f\left(\frac{l}{g}, \alpha\right). \tag{5.15}$$

Furthermore, the function is determined in part by dimensional consistency. Each term on the right side must have dimensions of time:

$$t_p [=] T, \qquad \frac{l}{g} [=] T^2, \qquad \left(\frac{l}{g}\right)^{1/2} [=] T. \tag{5.16}$$

Therefore,

$$t_p = \left(\frac{l}{g}\right)^{1/2} f(\alpha). \tag{5.17}$$

243

Because α has no dimensions, dimensional consistency cannot provide any information on $f(\alpha)$. Possibilities include

$$f(\alpha) = \alpha^3, \text{ or}$$
$$= \sin \alpha, \text{ or}$$
$$= e^{\alpha}, \text{ or}$$
$$= \text{a constant.} \tag{5.18}$$

To proceed further, one must measure the behavior of a pendulum for different values of α and l. Note that our experimental agenda has been shortened by dimensional analysis; we need not waste time measuring the effect of the pendulum mass. Furthermore, our analysis of the data is simplified. Dimensional analysis has assured us that we may plot t_p versus $(l/g)^{1/2}$, for a constant amplitude, and the data *for all pendulums* will lie on a single curve.

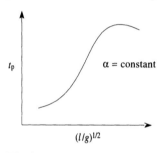

Figure 5.7a. A possible plot of a pendulum period as a function of $(l/g)^{1/2}$.

Plotting data measured with various amplitudes could generate a family of curves.

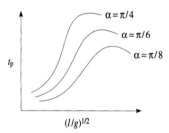

Figure 5.7b. A possible plot of a pendulum period as a function of $(l/g)^{1/2}$ for various amplitudes.

Upon measuring the pendulum behavior as a function of α and l one finds experimentally that the period of oscillation is independent of the amplitude, for small amplitudes ($\alpha < \pi/2 \, (90°)$). That is, the curves for different amplitudes are all the same. Furthermore, the relationship between t_p and $(l/g)^{1/2}$ is linear, with a slope of 2π, as shown in Figure 5.7c.

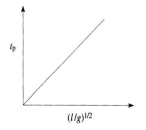

Figure 5.7c. The plot of an actual pendulum period as a function of $(l/g)^{1/2}$.

Any pendulum is thus described by a simple function:

$$t_{\mathrm{p}} = 2\pi \left(\frac{l}{g} \right)^{1/2} . \tag{5.19}$$

Let's summarize what we have found for the pendulum. Dimensional analysis can simplify the analysis of a model. But dimensional analysis is limited; eventually we must go to the lab. However, our experimental agenda is shortened by dimensional analysis. In addition, our analysis of the data will be simplified because we know how quantities are grouped.

It is worth noting that our analysis of the pendulum was aided by some key experiments that preceded ours. Specifically, we benefited from experiments that determined physical constants. In 1604, Galileo Galilei sought a formula for the time a mass m falls a distance h:

$$t_{\mathrm{fall}} = f(m, h, \ldots). \tag{5.20}$$

Galileo discovered that t_{fall} was independent of mass and proposed a constant to relate the dependence of t_{fall} on h. The constant is g. Other key experiments have provided the universal gravitational constant, Planck's constant, Boltzmann's constant, and the speed of light.

5.3   Dimensional Analysis of Walking

Let's apply dimensional analysis to another phenomenon – walking. Since all humans are constructed similarly, we should be able to predict a person's walking velocity based on physical parameters. Clearly this is a system that would be difficult to analyze at a fundamental level.

We begin by forming a list of the possible parameters of walking. The first parameter is what we seek – velocity. We start a table similar to that of the pendulum analysis.

Table 5.6a. The parameters of walking

Parameter	Symbol	Dimensions
velocity	v	L/T

What else might play a role? Tall people seem to walk very fast, at least compared to children. So we might expect height to play a role. However, not everyone is proportioned the same. More precisely, the length of a person's leg is the pertinent parameter. We add this to our table.

Table 5.6b. The parameters of walking

Parameter	Symbol	Dimensions
velocity	v	L/T
leg length	l	L

We might also expect a person's mass to affect walking, so we include mass.

Table 5.6c. The parameters of walking

Parameter	Symbol	Dimensions
velocity	v	L/T
leg length	l	L
mass	m	M

Walking involves lifting one's mass during the step. Thus one would expect gravity is involved. That is, a person walking on the Moon has a distinctly different gait than on Earth, even in the same space suit. Our table now looks like Table 5.6d.

Table 5.6d. The parameters of walking

Parameter	Symbol	Dimensions
velocity	v	L/T
leg length	l	L
mass	m	M
gravity	g	L/T^2

We note also that people have different styles of walking. Some take small steps and some take long steps. And although this tends to scale with leg length, it is truly an independent parameter. People with identical leg lengths have different stride lengths. If the length of stride is redundant with the leg length, this will be revealed in the experimental data. For example, a dimensionless group such as (leg length)/(stride length) will be found to be superfluous. For now we add stride length to the table.

Table 5.6e. The parameters of walking

Parameter	Symbol	Dimensions
velocity	v	L/T
leg length	l	L
mass	m	M
gravity	g	L/T^2
stride length	s	L

We now apply dimensional analysis to guide us in measuring and analyzing data on walking. We could proceed in the haphazard fashion that yielded a functional relationship for the pendulum. Like the pendulum, dimensional analysis of walking is simple enough to guess the answer. But let's use this opportunity to introduce a more rigorous method, a method we will find useful for more complex systems.

Recall that we started the pendulum analysis by looking for a formula that predicted the period, t_p, in terms of four physical parameters:

$$t_p = f(l, m, g, \alpha). \tag{5.10}$$

We applied dimensional analysis to simplify Eq. (5.10) to the form in Eq. (5.17):

$$t_p = \left(\frac{l}{g}\right)^{1/2} f(\alpha). \tag{5.17}$$

From experimental data we determined the form of the function to be

$$t_p = 2\pi \left(\frac{l}{g}\right)^{1/2}. \tag{5.19}$$

We can rearrange Eq. (5.19) into a dimensionless form to obtain

$$2\pi = \frac{g^{1/2}t_p}{l^{1/2}}. \tag{5.21}$$

We obtained Eq. (5.17) by finding a combination of l, m, g, and α that had dimensions of time. Alternatively, we could have obtained a dimensionless equation by finding a combination of t_p, l, m, g, and α that was dimensionless. Although adding the parameter t_p may seem to complicate the analysis, this approach lends itself well to a general algorithm. We will use this dimensionless approach to analyze walking.

A dimensionless term is called a Π group. We seek the Π group(s) that describe walking. A Π group for walking will have the general form

$$\Pi = v^a l^b m^c g^d s^e, \tag{5.22}$$

where exponents a, b, c, d, and e are to be determined. We express Π in terms of its dimensions, as listed in Table 5.6e:

$$\Pi [=] \left(\frac{L}{T}\right)^a L^b M^c \left(\frac{L}{T^2}\right)^d L^e \tag{5.23}$$

$$[=] L^{a+b+d+e} T^{-a-2d} M^c. \tag{5.24}$$

Because Π must be dimensionless, the exponent on each term in Eq. (5.24) must be zero. This requirement yields three equations:

Length, L: $a + b + d + e = 0$, $\tag{5.25}$

Time, T: $-a - 2d = 0$, $\tag{5.26}$

Mass, M: $c = 0$. $\tag{5.27}$

Because Eq. (5.27) demands that c be zero, and c is the exponent of m in Eq. (5.22), there is no contribution from mass. The velocity of walking is independent of mass, similar to the pendulum. (This is not too surprising when one considers the similarity of motion in the two systems.) What remains are two equations and four unknowns. As you will learn later, there are infinite combinations of a, b, d, and e that will satisfy these two equations. We need to specify something. We introduce the Buckingham Pi Theorem, which states

Number of Dimensionless Groups = Number of Parameters

$$- \textit{Number of Dimensions}, \tag{5.28}$$

or in other words,

Number of Dimensionless Groups = Number of Unknowns

$$- \textit{Number of Equations}. \tag{5.29}$$

Applying Eq. (5.29) to our analysis of walking yields

$$N_{\Pi} = 4 - 2 = 2. \tag{5.30}$$

Fine. So how does one determine these two dimensionless groups? We need another rule:

For each Π group one must choose a "core variable."

A core variable is a parameter that appears in only one Π group. It forms the core of a dimensionless group. Because our goal is to find a function in terms of dimensionless groups, such as

$$\Pi_1 = f(\Pi_2, \Pi_3, \ldots, \Pi_n), \tag{5.31}$$

it is convenient to designate the variable one wishes to predict as a core variable. For walking, we wish to predict the velocity. The other core variable should be an independent variable – a parameter one would vary in the experiments. For walking, stride length is a suitable choice for the second core variable.

We now derive the two Π groups, which we will designate Π_1 and Π_2. Π_1 will have velocity as its core variable and Π_2 will have stride length as its core variable.

Π_1: This dimensionless group contains v and not s. Thus the exponent on s must be zero; $e = 0$. We are free to choose any number for the exponent on v. It is usually convenient to choose 1; we set $a = 1$. Substituting $a = 1$ and $e = 0$ into Eqs. (5.25) and (5.26) we obtain

$$1 + b + d + 0 = 0, \tag{5.32}$$

$$-1 - 2d = 0. \tag{5.33}$$

As stated in Chapter 3, methods for solving systems of equations are beyond the intended scope of this text. But this system is trivial. From Eq. (5.33) we see $2d = -1$, or $d = -1/2$. Substituting into Eq. (5.32) we have $1 + b - 1/2 = 0$, or $b = -1/2$. Thus

$$\Pi_1 = v^1 l^{-1/2} m^0 g^{-1/2} s^0 = \frac{v}{(gl)^{1/2}}. \tag{5.34}$$

This first dimensionless group appears in the analysis of many physical systems. In general, this group appears when an object's motion causes something to be lifted in a gravitational field. Walkers lift their bodies with every step. Ships create bow waves. This dimensionless group first appeared in William Froude's analysis of a bow wave's contribution to a ship's resistance. This group (actually its square) is named for Froude. We will adopt the conventional form for this first Π group, and thus,

$$\Pi_1 = \frac{v^2}{gl} = \text{Froude number} \equiv \text{Fr}. \tag{5.35}$$

Squaring the dimensionless group will not affect the results of our analysis. However, adopting a conventional form will allow us to compare our analysis of walking to the analyses of similar systems.

Π_2: This dimensionless group contains s and not v. Thus the exponent on v must be zero; $a = 0$. Again we are free to choose any number for the exponent on the core variable, s. Again we choose 1; we set $e = 1$. When we substitute $a = 0$ and $e = 1$ into Eqs. (5.25) and (5.26) we obtain

$$0 + b + d + 1 = 0, \tag{5.36}$$

$$-0 - 2d = 0. \tag{5.37}$$

Again the solution is trivial. From Eq. (5.37) we have $d = 0$. Substituting $d = 0$ into Eq. (5.36) we have $b = -1$. Thus

$$\Pi_2 = v^0 l^{-1} m^0 g^0 s^1 = \frac{s}{l}. \tag{5.38}$$

We have determined that a dimensionless group containing stride length is a function of a dimensionless group containing velocity:

$$\Pi_2 = f(\Pi_1), \tag{5.39}$$

$$\frac{s}{l} = f\left(\frac{v^2}{gl}\right). \tag{5.40}$$

It is conventional to say that the *reduced* stride length is a function of the *reduced* velocity. That is, the square of the velocity has been scaled by gl. Similarly, stride length has been scaled by the leg length. With these scalings, data for all walkers should lie on a universal correlation, shown graphically in Figure 5.8. That is, a person walking with a stride length of 8′ may be dynamically similar to a person walking with a stride length of 6′ if their strides are scaled similarly. Similarly scaled strides would have the same reduced stride length, s/l, and thus would have the same reduced velocity, v^2/gl. But the person with the longer stride would have a higher velocity.

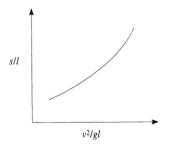

Figure 5.8. A possible universal correlation between reduced velocity and reduced stride length.

What is the correlation between reduced stride length and reduced velocity? We cannot determine that from dimensional analysis. To determine the correlation we must perform experiments.

Data obtained by first-year engineering students at Cornell, shown in Figure 5.9, have an upward trend. Let's examine the physical situations that correspond to the

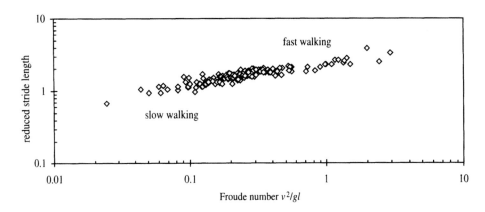

Figure 5.9. Experimental data for reduced stride length versus Froude number for first-year engineers at Cornell (1993–1996).

various data points. Data measured on fast-walking students lie in the upper right portion of the plot, $s/l > 2$ and $v^2/gl > 0.8$. Slow-paced walkers correspond to the data in the lower left portion of the plot, $s/l < 1.5$ and $v^2/gl < 0.3$. The data in the middle were measured on students walking at their "natural" pace, including (surprisingly) people walking backwards at their natural pace. There is probably another domain at high s/l corresponding to running. The curve might shift because the phenomenon of running is different from walking; at times both feet are off the ground.

The dimensional analysis of walking demonstrates an important concept in dynamic similarity. The magnitude of a dimensionless group indicates the "character" of the phenomenon. That is, if you are told that a walker had a Froude number of 2.0, you could be reasonably confident that the walker was "speed walking." Morticia Addams's style of walking (as portrayed by Carolyn Jones in the 1960s TV series "The Addams Family," and Angelica Huston in the 1990s movies of the same title) is "slow walking." At the other extreme, the style of walking known as "trucking" (introduced in the late 1960s and exhibited by many of the cartoon characters of Robert Crumb, creator of Mr. Natural) would be typified by high values of s/l, perhaps greater than 3.

Figure 5.10. Trucking. From *The Apex Treasury of Underground Comics*, Don Donahue and Susan Goodrick, eds., 1974. Copyright 1974, Quick Fox Publishers.

The analysis of walking has been extended to include another biped – the ostrich – as well as to include quadrupeds such as dogs and elephants. As shown in Figure 5.11, the data for a diverse range of walkers lie on one general curve.

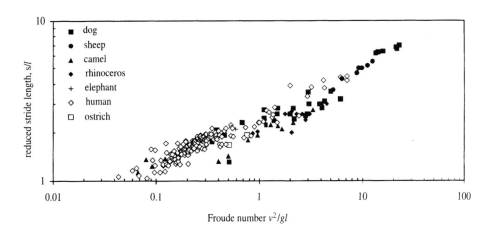

Figure 5.11. Experimental data for reduced stride as a function of Froude number for quadrupeds and bipeds. (Adapted from Figure 3.10 of *Dynamics of Dinosaurs and Other Extinct Giants*, R. McN. Alexander, Columbia University Press, New York, 1989.)

It is noted (Alexander 1989, 1992) that quadrupeds – from ferrets to rhinoceroses – change gait from trotting to galloping at a Froude number of about 2.55, although the absolute velocities of each are different. Galloping is more strenuous than trotting. Elephants don't gallop, but buffalo do. The dimensional analysis of bone strength suggests the types of gaits an animal can attain. The largest of the dinosaurs, apatosaurus (34 metric tons) had a reduced bone strength similar to an elephant, suggesting that its fastest gait was a trot. Triceratops (6–9 metric tons), however, had a reduced bone strength more similar to an African buffalo, suggesting it was capable of galloping. If triceratops did gallop, the correlation in Figure 5.11 suggests it could gallop at 9 m/sec. (The best human sprinters can achieve 11 m/sec.) The analysis of dinosaur running is an excellent example of the use of dynamic scaling to predict results for experimentally inaccessible systems. With dynamic analysis, for example, one may predict if triceratops could outrun a predator such as an allosaurus.

5.4   Dimensional Analysis of a Solid Sphere Moving through a Fluid

Let's apply dimensional analysis to another phenomenon – the terminal velocity of a sphere (Figure 5.12).

 Figure 5.12. A solid sphere falling through a fluid.

This phenomenon is common in chemical engineering processes, for instance

· Silt settling in water. How long does it take to clarify muddy water?
· Dust in air. Will ash from your company's smokestack fall on your company's property, in the nearby town, or in the ocean? How long would a nuclear winter last?
· What force is needed to tow a sphere through a fluid?

In these examples it is impractical to measure the actual process. The particles are either too small (approximately 1 μm for some dust) and/or the terminal velocity is too slow (nuclear winter is predicted to last 1–10 years). We would much rather take measurements on more convenient systems, such as a plastic sphere of diameter 1 cm falling through water. We will apply dimensional analysis to reveal how to scale the results of our experiments to interesting but experimentally inaccessible real processes.

To aid in identifying the parameters of this system, let's examine the physics of the phenomenon. Why does a particle fall? Well, gravity has something to do with this. But why does a particle fall and not rise? A sphere of polyethylene will fall in air but rise in water. Perhaps the density of the particle compared to the density of the fluid is important. Why does a falling particle reach a terminal velocity? Why doesn't the acceleration due to gravity increase the velocity without limit (with due respect to relativity)? The terminal velocity suggests there is an equal and opposing force: friction owing to the surrounding fluid.

Again we begin by constructing a table. Again the first choice is obvious – the terminal velocity of the sphere.

Table 5.7a. The parameters of a sphere moving through a fluid

Parameter	Symbol	Dimensions
terminal velocity	v	L/T

We need to determine the parameters that affect the terminal velocity. The parameters can be divided into two groups: parameters that describe the sphere and parameters that describe the fluid. First we consider the sphere. A key parameter is the sphere's diameter. Another is the buoyancy of the sphere, given by the difference between the sphere density and fluid density. Our table now looks like Table 5.7b.

Table 5.7b. The parameters of a sphere moving through a fluid

Parameter	Symbol	Dimensions
terminal velocity	v	L/T
sphere diameter	D	L
buoyancy	$\rho_{sph} - \rho_{fluid}$	M/L^3

Is there anything else about the sphere? How about its mass? Yes, the mass has an effect, but the mass is redundant with the buoyancy. How about the "smoothness" of the sphere's surface? We will ignore that effect for now and assume that the sphere is perfectly smooth.

We now turn to the properties of the fluid. The frictional force exerted by the fluid is related to the viscosity of the fluid. What are the dimensions of viscosity? When we consult a reference book for a typical viscosity, we see that viscosity has units of "poise." That doesn't help much. Sometimes it is useful to use a fundamental law to determine the units of a quantity. For example, to remember the units of force one can use "$F = ma$." For energy one can use "$E = (1/2) mv^2$." For viscosity, a law dear to the hearts of chemical engineers (but probably unknown to first-year engineers) is

Newton's law of viscosity:

$$-(viscosity)\frac{d(velocity)}{d(distance)} = \frac{shear\,force}{area},$$ (5.41)

$$(viscosity)\frac{L/T}{L} [=] \frac{ML/T^2}{L^2},$$ (5.42)

$$(viscosity) [=] \frac{M}{LT}.$$ (5.43)

We add viscosity to our table.

Table 5.7c. The parameters of a sphere moving through a fluid

Parameter	Symbol	Dimensions
terminal velocity	v	L/T
sphere diameter	D	L
buoyancy	$\rho_{sph} - \rho_{fluid}$	M/L^3
fluid viscosity	μ	M/LT

The next parameter is subtle. When the fluid passes the sphere, the inertia of the fluid changes. Imagine a stationary sphere in a moving fluid. The fluid must pass around the particle, which changes the fluid's inertia, as sketched in Figure 5.13. The fluid's inertia is proportional to the mass of the fluid and the change in velocity of the fluid. The change in velocity is already accounted for by the velocity of the sphere. However, we need to account for the mass of the fluid. Again we use the fluid density. Finally, we add physical constants. In this case, we add the constant for gravitational acceleration.

Table 5.7d. The parameters of a sphere moving through a fluid

Parameter	Symbol	Dimensions
terminal velocity	v	L/T
sphere diameter	D	L
buoyancy	$\rho_{sph} - \rho_{fluid}$	M/L^3
fluid viscosity	μ	M/LT
fluid density	ρ_{fluid}	M/L^3
gravitational acceleration	g	L/T^2

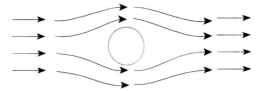

Figure 5.13. Fluid moving around a solid sphere.

We realize it may seem that Table 5.7 was constructed in an arbitrary manner. How did we know to include the density of the fluid? What would have happened if we used the density of the sphere instead of the buoyancy? The answer is that we would still be able to calculate dimensionless groups for this system. However, when we went to the lab we would find that the dimensionless groups were not related by a single function – a plot of the results would yield considerable scatter instead of a single line. Why did we use specific parameters, such as diameter instead of radius? Either parameter would have been acceptable. We chose diameter so we could compare our results with those of previous workers.

We now determine the dimensionless groups. Each will have the general form:

$$\Pi = v^a D^b (\rho_{sph} - \rho_{fluid})^c \mu^d \rho_{fluid}^e g^f \tag{5.44}$$

$$[=] \left(\frac{L}{T}\right)^a L^b \left(\frac{M}{L^3}\right)^c \left(\frac{M}{LT}\right)^d \left(\frac{M}{L^3}\right)^e \left(\frac{L}{T^2}\right)^f \tag{5.45}$$

$$[=] M^{c+d+e} T^{-a-d-2f} L^{a+b-3c-d-3e+f}. \tag{5.46}$$

Again for Π to be dimensionless, each exponent in Eq. (5.46) must be zero. This requirement yields three equations:

Mass, M: $c + d + e = 0,$ (5.47)

Time, T: $-a - d - 2f = 0,$ (5.48)

Length, L: $a + b - 3c - d - 3e + f = 0.$ (5.49)

We have six unknowns (a to f) and three equations. We thus have $6 - 3 = 3\Pi$ groups. To solve Eqs. (5.47)–(5.49) we must fix three of the unknowns, which we do by selecting three core variables. What do the guidelines suggest for core variables?

· What do we want to know? The dependent variable is *velocity*.
· Which parameters do we plan to vary in our experiments?

We will eventually go to the lab and drop spheres of different diameters and densities through fluids with different densities and viscosities. Candidates for the independent parameters are the sphere diameter, the buoyancy, the fluid viscosity, and the fluid density. However, to compare our analysis with previous work, we should consider some alternate choices. Consider the following rationalization. We wish to describe the terminal velocity of a sphere falling (or rising) through a fluid. But we can also analyze a more general system. That is, we are concerned only with the resistive force on a sphere moving through a fluid. It does not matter how the sphere is propelled through the fluid. For terminal velocity, the sphere is propelled by its buoyancy. However, the system applies also to a mutually buoyant sphere ($\rho_{sph} = \rho_{fluid}$) being towed through a fluid or a sphere fixed in place with a fluid flowing past it. We can describe systems with broader range if we do not regard velocity as a *dependent* variable. We may wish to vary the velocity, for example, by towing a sphere through the fluid. The choice of core variables is chiefly for convenience. The Π groups derived with one valid set of core variables can be related to the Π groups derived with other valid sets of core variables.

The traditional core variables for the terminal velocity of a sphere are $(\rho_{sph} - \rho_{fluid})$, g, and μ. We now use these to derive the Π groups.

Π_1 contains $(\rho_{sph} - \rho_{fluid})$, but not g, and not μ. Thus, the exponents on the parameters g and μ in Eq. (5.44) must be zero; set $f = 0$ and set $d = 0$. The exponent of the core variable $(\rho_{sph} - \rho_{fluid})$ may be any nonzero number; $c = 1$ is convenient. Substituting these exponents into Eqs. (5.47)–(5.49) we get

Mass, M: $1 + 0 + e = 0 \Rightarrow e = -1$, $\qquad\qquad$ (5.50)

Time, T: $-a - 0 - 2(0) = 0 \Rightarrow a = 0$, $\qquad\qquad$ (5.51)

Length, L: $0 + b - 3(1) - 0 - 3(-1) + 0 = 0 \Rightarrow b = 0$. $\qquad$ (5.52)

In summary, for Π_1 we have

$$\left.\begin{array}{l} a = 0 \\ b = 0 \\ c = 1 \\ d = 0 \\ e = -1 \\ f = 0 \end{array}\right\} \quad \Pi_1 = (\rho_{sph} - \rho_{fluid})^1 (\rho_{fluid})^{-1} = \frac{\rho_{sph} - \rho_{fluid}}{\rho_{fluid}}. \qquad (5.53)$$

Thus the buoyancy is scaled by the density of the fluid. One may think of this dimensionless group as a reduced buoyancy, but that parlance is not used by fluid mechanicians and consequently this dimensionless group has no name. Nevertheless we will call it reduced buoyancy and accept the stigma that comes with this convenience.

Π_2 contains g, but not $(\rho_{sph} - \rho_{fluid})$ and not μ. Thus, we set the exponents of $(\rho_{sph} - \rho_{fluid})$ and μ to zero; from Eq. (5.44) $c = 0$ and $d = 0$. The exponent on the core variable g can be any nonzero number. Again, $f = 1$ is convenient. We substitute these exponents into Eqs. (5.47)–(5.49):

Mass, M: $0 + 0 + e = 0 \Rightarrow e = 0$, $\qquad\qquad$ (5.54)

Time, T: $-a - 0 - 2(1) = 0 \Rightarrow a = -2$, $\qquad\qquad$ (5.55)

Length, L: $-2 + b - 3(0) - 0 - 3(0) + 1 = 0 \Rightarrow b = 1$. $\qquad$ (5.56)

In summary, for Π_2 we have

$$\left.\begin{array}{l} a = -2 \\ b = 1 \\ c = 0 \\ d = 0 \\ e = 0 \\ f = 1 \end{array}\right\} \Pi_2 = v^{-2}D^1g^1 = \frac{Dg}{v^2}. \tag{5.57}$$

The reciprocal of this dimensionless group is v^2/Dg, the Froude number. Similar to the analysis of walking where we arbitrarily squared the Π group, we will arbitrarily invert Π_2 to obtain the Froude number as our second dimensionless group. We would have obtained the Froude number if we had set the exponent of g to -1. Inverting Π_2 will not affect our analysis, but it will allow comparison with analyses of similar systems.

Because this is our second encounter with the Froude number it warrants further inspection. Like the other two dozen or so common dimensionless groups in chemical engineering the Froude number can be expressed as a ratio of forces. Specifically, the Froude number can be expressed as the ratio of inertial force to gravitational force. As we will see later, this provides insight to the phenomenon.

$$\Pi_2 = \text{Fr} = \frac{v^2}{Dg} = \frac{(mass)(v^2/D)}{(mass)g} = \frac{inertial\ force}{gravitational\ force}. \tag{5.58}$$

Π_3 contains μ, but not $(\rho_{\text{sph}} - \rho_{\text{fluid}})$ and not g. Thus, from Eq. (5.44) set $c = 0$, set $f = 0$, and set d to any nonzero number. Again, $d = 1$ is convenient. Substituting into Eqs. (5.47)–(5.49) yields

Mass, M: $0 + 1 + e = 0 \Rightarrow e = -1,$ \hfill (5.59)

Time, T: $-a - 1 - 2(0) = 0 \Rightarrow a = -1,$ \hfill (5.60)

Length, L: $-1 + b - 3(0) - 1 - 3(-1) + 0 = 0 \Rightarrow b = -1.$ \hfill (5.61)

In summary, for Π_3 we have

$$\left.\begin{array}{l} a = -1 \\ b = -1 \\ c = 0 \\ d = 1 \\ e = -1 \\ f = 0 \end{array}\right\} \Pi_3 = v^{-1}D^{-1}\mu^1\rho_{\text{fluid}}^{-1} = \frac{\mu}{Dv\rho_{\text{fluid}}}. \tag{5.62}$$

The reciprocal of Π_3 is also a famous dimensionless group – perhaps *the* most famous dimensionless group – the Reynolds number, Re. Similar to the Froude number, the

Reynolds number can also be expressed as a ratio of forces:

$$\mathrm{Re} = \frac{Dv\rho_{\mathrm{fluid}}}{\mu} = \frac{inertial\ force}{viscous\ force}. \tag{5.63}$$

One encounters the Reynolds number throughout chemical engineering. We will return later to the physical significance of the Froude number and the Reynolds number.

Let's summarize to this point our analysis of a sphere moving through a fluid. We combined six parameters to form three dimensionless groups. What is the functional relationship between these groups? That is, what is the function f such that

$$\frac{v^2}{Dg} = f\left(\frac{\rho_{\mathrm{sph}} - \rho_{\mathrm{fluid}}}{\rho_{\mathrm{fluid}}}, \frac{Dv\rho_{\mathrm{fluid}}}{\mu} \right)? \tag{5.64}$$

Dimensional analysis can take us no further. To determine the function f we must appeal to theory and/or experiment. The theory lies outside the realm of this text. Moreover, there is no theory that encompasses the functional form of f for all values of Fr, reduced buoyancy, and Re. To proceed further we must measure terminal velocities of spheres falling through fluids.

Before proceeding, let's consider the effect of choosing different core variables. What would be the result of choosing the obvious candidates for core variables: v, $(\rho_{\mathrm{sph}} - \rho_{\mathrm{fluid}})$, and μ? If we use an exponent of 1 for each core variable one obtains, for Π_1,

$$\Pi_1 = \frac{v}{\sqrt{Dg}}. \tag{5.65}$$

This is the square root of the Froude number. For Π_2, one obtains

$$\Pi_2 = \frac{\rho_{\mathrm{sph}} - \rho_{\mathrm{fluid}}}{\rho_{\mathrm{fluid}}}. \tag{5.66}$$

This is the reduced buoyancy, the same result as before. Finally, for Π_3, one obtains

$$\Pi_3 = \frac{\mu}{D^{3/2}\rho_{\mathrm{fluid}}g^{1/2}}. \tag{5.67}$$

The square of this group is the Galileo number:

$$\Pi_3^2 = \frac{\mu^2}{D^3 \rho_{\mathrm{fluid}}^2 g} = N_{\mathrm{Ga}}. \tag{5.68}$$

The Galileo number is the ratio of viscous effects to gravitational effects. It is also the Froude number divided by the square of the Reynolds number,

$$N_{\mathrm{Ga}} = \frac{\mathrm{Fr}}{\mathrm{Re}^2}. \tag{5.69}$$

To determine experimentally the function in Eq. (5.64), first-year engineers at Cornell observed spheres of various materials (steel, brass, aluminum, glass, lucite, nylon, teflon, and polypropylene) falling through various fluids (water, vegetable oil, and glycerol). The students measured each sphere's mass and diameter, the distance each sphere fell, and the time elapsed. The students then used spreadsheets to calculate the parameters of Table 5.7d in SI units, and finally they combined these parameters to calculate the three dimensionless groups: Fr, Re, and the reduced buoyancy.

To determine the relation in Eq. (5.64) we need to plot the data. Plotting the walking data was straightforward because there were only two dimensionless groups. Because three dimensionless groups describe spheres moving through fluids, plotting is more difficult. We have at least two options. We could prepare a three-dimensional plot with axes Re, Fr, and reduced buoyancy. The correlation would then be a surface in this three-dimensional space. We tried this and it gets ugly – trust us. Another option is to plot, for example, reduced buoyancy versus Re for values of constant Fr. We would then have a family of curves for different values of Fr. This is shown in Figure 5.14.

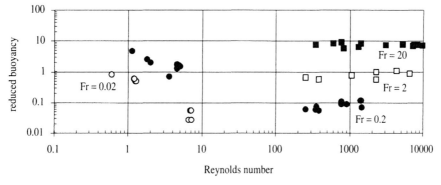

Figure 5.14. Data for a sphere moving through a fluid, plotted for Froude numbers of 0.02, 0.2, 2, and 20, ±10%.

The trends in the data sets in Figure 5.14 suggest that the correlation can be simplified. Note that if the Froude number increases by a factor of 10, the reduced buoyancy increases by a factor of 10. This suggests that one can divide the reduced buoyancy by the Froude number and all the data will collapse onto one universal curve. The data in Figure 5.14 are replotted as reduced buoyancy/Fr versus Re, shown in Figure 5.15. The data seem to lie on a single correlation.

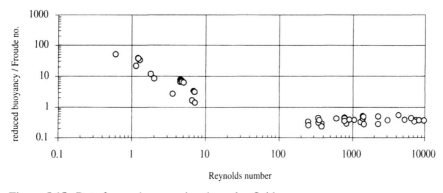

Figure 5.15. Data for a sphere moving through a fluid.

Figures 5.14 and 5.15 contain only a portion of the experimental data. Let's see if the correlation in Figure 5.15 holds for data with arbitrary values of Fr. Figure 5.16 shows all the data plotted in the manner of Figure 5.15, but with the reduced buoy-

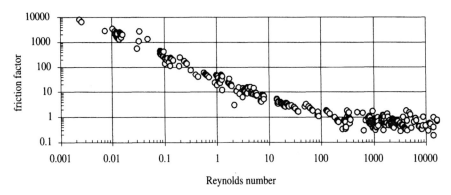

Figure 5.16. Data for a sphere moving through a fluid.

ancy/Fr multiplied by 4/3. The factor of 4/3 appears when one derives the relation with fluid mechanics; 4/3 comes from the volume of a sphere, $(4/3)\pi r^3$. The reduced buoyancy/Fr multiplied by 4/3 is known as the friction factor, as given by

$$friction\ factor = \frac{4}{3}\frac{\rho_{sph} - \rho_{fluid}}{\rho_{fluid}}\frac{1}{Fr}. \tag{5.70}$$

It should come as no surprise that the correlation in Figure 5.16 has been studied before. The data measured by the first-year engineers agree well with the correlation determined previously, shown in Figure 5.17.

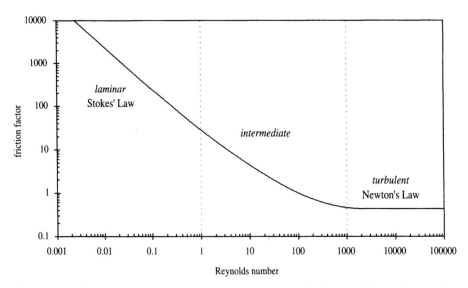

Figure 5.17. Correlation for a sphere moving through a fluid. (Adapted from Bird, R. B., Stewart, W. E., and Lightfoot, E. N. 1960. *Transport Phenomena*, Fig. 6.3-1.)

The correlation between friction factor and Reynolds number in Figure 5.17 has three distinct regions. There is a linear region for Re < 1, a curved region for 1 < Re < 1,000, and a flat region for Re > 1,000.

For Re < 1, the plot is linear with a negative slope. Recall the definition of the Reynolds number:

$$\text{Re} = \frac{Dv\rho_{\text{fluid}}}{\mu}. \tag{5.63}$$

A low Reynolds number occurs at low velocity, low fluid density, a small sphere, or high fluid viscosity. Let's derive the equation that corresponds to this regime. Because the line is straight we know that the functional form is

$$y = mx = b. \tag{5.71}$$

We choose two convenient points on the line to calculate the slope: $(\text{Re}, \text{F}) = (0.5, 50)$ and $(0.005, 5000)$. Because this is a log–log plot, one must be careful calculating the slope. Specifically, we calculate the distance between the logarithms of these coordinates, not the distances between the coordinates:

$$m = slope = \frac{rise}{run} = \frac{\log_{10} 5000 - \log_{10} 50}{\log_{10} 0.005 - \log_{10} 0.5}$$

$$= \frac{(\log_{10} 5 + \log_{10} 10^3) - (\log_{10} 5 + \log_{10} 10^1)}{(\log_{10} 5 + \log_{10} 10^{-3}) - (\log_{10} 5 + \log_{10} 10^{-1})}$$

$$= \frac{3 - 1}{-3 - (-1)} = -1. \tag{5.72}$$

Substitute the slope into Eq. (5.71) to get

$$y = (-1)x + b, \tag{5.73}$$

$$\log_{10} \text{F} = (-1)\log_{10} \text{Re} + b \tag{5.74}$$

$$= \log_{10} \text{Re}^{-1} + b. \tag{5.75}$$

Raise both sides of the equation to the power of 10:

$$10^{\log_{10} \text{F}} = 10^{[\log_{10} \text{Re}^{-1} + b]} \tag{5.76}$$

$$= 10^{\log_{10} \text{Re}^{-1}} 10^b, \tag{5.77}$$

$$\text{F} = \text{Re}^{-1} 10^b. \tag{5.78}$$

The constant b may be calculated by inserting a point on the correlation into Eq. (5.78), such as $(\text{Re}, \text{F}) = (0.5, 50)$. Doing so one calculates $10^b = 25$. Thus

$$\frac{4}{3} \frac{\rho_{\text{sph}} - \rho_{\text{fluid}}}{\rho_{\text{fluid}}} \frac{1}{\text{Fr}} = \frac{25}{\text{Re}} \quad \text{for} \quad \text{Re} < 1. \tag{5.79}$$

Equation (5.79) is a form of Stokes's law. It is not a universal law, such as is the Conservation of Mass, because it is restricted to flows with Re < 1. Stokes's law is a constrained law.

For Re > 1,000, the plot is essentially flat. Thus the friction factor is independent of the Reynolds number:

$$\frac{4}{3} \frac{\rho_{sph} - \rho_{fluid}}{\rho_{fluid}} \frac{1}{Fr} \approx 0.44 \quad \text{for} \quad Re > 1,000. \tag{5.80}$$

Equation (5.80) is known as Newton's Law of Resistance, not to be confused with Newton's Laws of Motion, or Newton's Law of Viscosity, or Newton's Law of Cooling.

In the intermediate region, $1 < Re < 1,000$, the system is in transition. We won't attempt to extract an equation from our data. It is generally accepted that this region can be approximated by

$$\frac{4}{3} \frac{\rho_{sph} - \rho_{fluid}}{\rho_{fluid}} \frac{1}{Fr} = \frac{18}{Re^{3/5}} \quad \text{for} \quad 1 < Re < 1,000. \tag{5.81}$$

The intermediate region is thus approximated by a straight line of slope $-3/5$.

The fluid streamlines around the sphere are distinctly different for each region. Figure 5.18 shows fluid streamlines for flow around a cylinder, which is similar to flow around a sphere. (It's not easy to photograph streamlines around a sphere.) At low Reynolds number, the fluid passes smoothly past the object; the streamlines gently warp around the cylinder. This type of flow is called *laminar* and Stokes's law applies. At higher Reynolds number eddies develop behind the object. At even higher Reynolds number, the eddies break up and no longer form steady streamlines; the pattern in Figure 5.18f changes with time. For Re > 1,000 the flow is turbulent and the friction factor becomes independent of Re.

Let's use our universal correlation to extrapolate outside of the range of objects we measured, for example to microscopic dust settling in air. This is an important consideration in a "clean room" for the manufacture of integrated circuits. If a person who has recently inhaled tobacco smoke exhales into the room, how long does it take for the smoke particles to settle out of the air? Or, imagine a large dust cloud injected into the upper atmosphere, for example, after a volcanic eruption, a large asteroid impact, or global thermonuclear war. If the dust cloud is above the cleansing action of rain, how long before the dust particles settle to Earth?

Assume that microscopic dust settling in air is in the low Reynolds number regime. We start with Stokes's law, Eq. (5.79), to derive an equation for the terminal velocity and substitute the definitions for the Froude number and Reynolds number:

$$\frac{4}{3} \left(\frac{\rho_{sph} - \rho_{fluid}}{\rho_{fluid}} \right) \left(\frac{Dg}{v^2} \right) = \frac{25}{\left(\frac{Dv\rho_{fluid}}{\mu} \right)}. \tag{5.82}$$

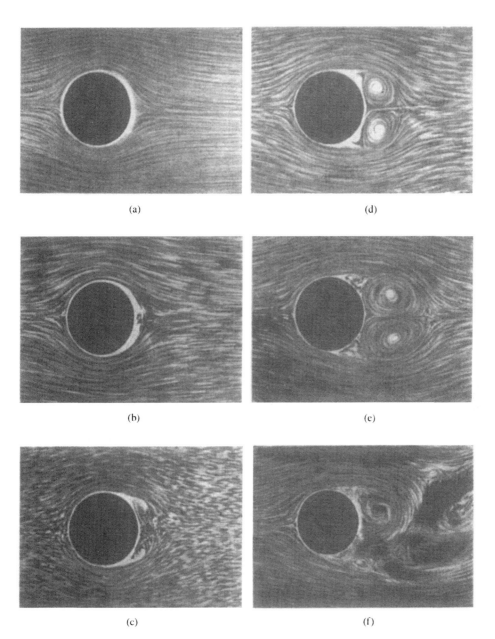

(a) (d)

(b) (e)

(c) (f)

Figure 5.18. Fluid streamlines for a fluid moving around a cylinder. (From Batchelor, G. K. 1967. *Introduction to Fluid Mechanics*, Fig. 5.11.3, on Plate 10. Copyright Cambridge University Press. Reproduced by permission.)

Solving for velocity we get

$$v = \frac{4}{75} \frac{gD^2}{\mu}(\rho_{sph} - \rho_{fluid}). \tag{5.83}$$

We will approximate the prefactor of $4/75$ by $1/18$, which is the result obtained by mathematical modeling of a sphere at low Reynolds number. For solids falling through air Eq. (5.83) can be simplified further. Because $\rho_{sph} \approx 2{,}000$ kg/m^3 and $\rho_{fluid} \approx 1$ kg/m^3, $\rho_{sph} - \rho_{fluid} \approx \rho_{sph}$, which yields

$$v = \frac{1}{18} \frac{g D^2 \rho_{sph}}{\mu}. \tag{5.84}$$

Equation (5.84) is a more common form of Stokes's law. Substitute into Stokes's law the constants for a 1 μm particle in air.

$$v = \frac{1}{18} \frac{(9.8 \text{ m/s}^2)(10^{-6} \text{ m})^2(2000 \text{ kg/m}^3)}{1.8 \times 10^{-6} \text{ Pa·s}} = 6 \times 10^{-4} \text{ m/s} = 4 \text{ cm/min}. \tag{5.85}$$

The dust settles quite slowly. How long does it take a 1 μm particle to settle from a height of 50 km? Because distance equals velocity $\times$ time,

$$time = distance \times \frac{1}{velocity} \tag{5.86}$$

$$= 50 \text{ km} \left(\frac{1}{4 \text{ cm/min}} \right) \left(\frac{1{,}000 \text{ m}}{1 \text{ km}} \right) \left(\frac{100 \text{ cm}}{1 \text{ m}} \right) \left(\frac{1 \text{ hr}}{60 \text{ min}} \right) \left(\frac{1 \text{ day}}{24 \text{ hr}} \right) \left(\frac{1 \text{ year}}{365 \text{ days}} \right)$$

$$= 2.5 \text{ years}. \tag{5.87}$$

Now, without performing a calculation similar to Eqs. (5.85)–(5.87), estimate how long it takes a 10 μm particle to settle from a height of 50 km. Because Eq. (5.84) tells us the velocity scales as D^2, the velocity of a 10 μm particle is 100 times greater than that of a 1 μm particle. The 10 μm particle will settle in about 9 days.

Before we conclude we must check our initial assumption that the terminal velocity of a 1 μm particle is in the laminar regime. Calculate the Reynolds number:.

$$Re = \frac{D v \rho}{\mu} = \frac{(1 \times 10^{-6} \text{ m})(6 \times 10^{-4} \text{ m/s})(2{,}000 \text{ kg/m}^3)}{1.8 \times 10^{-6} \text{ Pa·s}} = 0.7. \tag{5.88}$$

The Reynolds number is less than one, so it was legitimate to apply Stokes's law. However, for a particle of diameter 10 μm, Re $= 700$. This lies outside the valid range of Stokes's law. So, will a 10 μm particle fall faster or slower than predicted by Stokes's law?

5.5   Design Tools from Dimensional Analysis

Correlations between dimensionless groups are *design tools*. Just as we derived operating equations to analyze the desalinator and studied the McCabe–Thiele method to analyze distillation columns, the correlation between the Reynolds number and the friction factor in Figure 5.17 can be used to analyze spheres moving through fluids. Similarly, the correlation between reduced stride and Fr in Figure 5.11 is a design tool for devices that walk. The process of developing design tools from dimensional analysis is diagrammed in Figure 5.19. You may wish to compare Figure 5.19 to the equivalent schematics for mathematical analysis (Figure 3.15) and graphical methods (Figure 4.48).

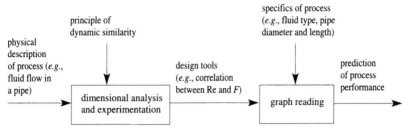

Figure 5.19. Schematic of the process to create and apply dimensional-analysis design tools.

There are myriad correlations between dimensionless groups. We will examine a few more examples in this chapter. Again, it is not necessary to memorize each correlation. Rather, it is important to know how to apply the principles of dynamic similarity to create new design tools.

5.6 Dynamic Scaling – The Alaskan Pipeline

In the three previous studies – the pendulum, walking, and spheres through fluids – we followed the same steps: dimensional analysis, then experiment, and finally analysis of the data. In the analysis of walking and the terminal velocity of spheres we measured many data to obtain functional relationships. We can use these relations to extrapolate to dimensionally similar phenomena such as dinosaurs walking or dust settling in air. What if we don't care to derive the functional relation? That is, assume we know the magnitude of the dimensionless groups of the system of interest. If we are to perform a single experiment on a model system to predict the behavior of the real system at these specific conditions, what should be the specific conditions in the model? Dynamic similarity tells us that the magnitudes of the dimensionless groups must be the same for the model system and the real system.

We wish to specify pumps to send oil through the Alaskan pipeline (Figure 5.20).

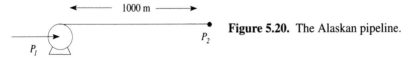

Figure 5.20. The Alaskan pipeline.

Assume that pumps are stationed every 1,000 m and the desired flow is 4×10^5 barrels/day. What pressure must each pump deliver?

$$P_1 - P_2 = \Delta P = ? \tag{5.89}$$

It would be impractical to build a full-scale system to measure the pressure drop; the pipe is ~1 m in diameter. Let's instead build a model. But this time our approach will be different. Rather than measure a lot of data with the model to find a functional relation we will take only one measurement with the model. The values of the dimensionless groups for the model will be set equal to the values of the dimensionless groups for the Alaskan pipeline. The model and the Alaskan pipeline are thus dynamically similar. To do this we need to (1) derive the dimensionless groups for flow through a pipe and (2) match the dimensionless groups for the actual system and the model.

Step 1. Prepare a table of the parameters.

The first parameter is what we seek, the pressure drop of the fluid. We construct Table 5.8a.

Table 5.8a. The parameters of fluid flow through a pipe

Parameter	Symbol	Dimensions
pressure drop	ΔP	M/LT^2

Next, we turn to the physical parameters of the pipe. One might expect the pipe diameter to be important; it is harder to force fluid through a small pipe than a large pipe. Similarly, for a given diameter of pipe, a fluid flows faster through a short pipe than a long pipe. We add pipe diameter and pipe length to the table.

Table 5.8b. The parameters of fluid flow through a pipe

Parameter	Symbol	Dimensions
pressure drop	ΔP	M/LT^2
pipe length	l	L
pipe diameter	d	L

Next, we consider the properties of the fluid, starting with the dynamic property, the fluid velocity, v. From our study of spheres moving through fluids we expect viscosity to be important. Also, the inertia of the fluid, ρv, will play a role. We already have the fluid velocity in our table. We need the fluid density.

Table 5.8c. The parameters of fluid flow through a pipe

Parameter	Symbol	Dimensions
pressure drop	ΔP	M/LT^2
pipe length	l	L
pipe diameter	d	L
velocity	v	L/T
viscosity	μ	M/LT
density	ρ_{fluid}	M/L^3

Step 2. Write a general expression for the dimensionless group, Π:

$$\Pi = (\Delta P)^a l^b d^c v^d \mu^e \rho_{\text{fluid}}^f \tag{5.90}$$

$$[=] \left(\frac{M}{LT^2} \right)^a L^b L^c \left(\frac{L}{T} \right)^d \left(\frac{M}{LT} \right)^e \left(\frac{M}{L^3} \right)^f \tag{5.91}$$

$$[=] M^{a+e+f} T^{-2a-d-e} L^{-a+b+c+d-e-3f}. \tag{5.92}$$

Step 3. Set all exponents equal to zero to satisfy the requirement that Π is dimensionless:

Mass, M: $a + e + f = 0,$ (5.93)

Time, T: $-2a - d - e = 0,$ (5.94)

Lenght, L: $-a + b + c + d - e - 3f = 0.$ (5.95)

Step 4. How many Π groups are there? We apply the Buckingham Π Theorem:

6 variables $-$ 3 equations $= 3$ Π groups.

Step 5. Choose three core variables, one for each Π group:

- ΔP – This is the parameter whose value we seek; it would be convenient if this variable appears only once, so we can solve for it, $\Delta P = \ldots$.
- l – This is what we will vary if ΔP is too high or too low, so again it would be convenient if it appears only once.
- μ – This is a fluid property.

As foretold earlier in this chapter, there are rules for a valid set of core variables.

Rule 1. *All dimensions of the system must be represented in the core variables.*

The dimensions of flow through a pipe are M, L, and T. M is represented by ΔP and μ. L is represented by all three, and T is represented by ΔP and μ.

Rule 2. *The core variables must not form a dimensionless group.*

This requires more effort to check. The process of calculating the Π groups will reveal when this rule is violated; one will be unable to find a set of exponents for every Π group.

Step 6. Derive the Π groups.

Π_1 contains ΔP and not l or μ. Set $a = 1$, and $b = 0$, and $e = 0$. Then

Mass, M: $1 + 0 + f = 0 \Rightarrow f = -1,$ (5.96)

Time, T: $-2(1) - d - 0 = 0 \Rightarrow d = -2,$ (5.97)

Length, L: $-1 + 0 + c + (-2) - 0 - 3(-1) = 0 \Rightarrow c = 0.$ (5.98)

In summary, for Π_1

$$\left.\begin{array}{l} a=1 \\ b=0 \\ c=0 \\ d=-2 \\ e=0 \\ f=-1 \end{array}\right\} \Pi_1 = (\Delta P)^1 v^{-2} (\rho_{\text{fluid}})^{-1} = \frac{\Delta P}{v^2 \rho_{\text{fluid}}}. \tag{5.99}$$

This dimensionless group is the Euler number (pronounced "Oiler" number), Eu, another useful group in chemical engineering:

$$\text{Eu} = \frac{\Delta P}{v^2 \rho_{\text{fluid}}} = \frac{\left(\frac{\Delta P}{v}\right)d}{v \rho_{\text{fluid}} d} = \frac{friction\, force}{inertial\, force}. \tag{5.100}$$

Π_2 contains l and not ΔP or μ. Set $b=1$, and $a=0$, and $e=0$. Then

Mass, M: $0+0+f=0 \Rightarrow f=0,$ (5.101)

Time, T: $-2(0)-d-0=0 \Rightarrow d=0,$ (5.102)

Length, L: $-0+1+c+0-0-3(0)=0 \Rightarrow c=-1.$ (5.103)

In summary, for Π_2

$$\left.\begin{array}{l} a=0 \\ b=1 \\ c=-1 \\ d=0 \\ e=0 \\ f=0 \end{array}\right\} \Pi_2 = l^1 d^{-1} = \frac{l}{d}. \tag{5.104}$$

Π_3 contains μ and not ΔP or l. Set $e=1$, and $a=0$, and $b=0$. Then

Mass, M: $0+1+f=0 \Rightarrow f=-1,$ (5.105)

Time, T: $-2(0)-d-1=0 \Rightarrow d=-1,$ (5.106)

Length, L: $-0+0+c+(-1)-1-3(-1)=0 \Rightarrow c=-1.$ (5.107)

In summary, for Π_3

$$\left.\begin{array}{l} a=0 \\ b=0 \\ c=-1 \\ d=-1 \\ e=1 \\ f=-1 \end{array}\right\} \Pi_3 = d^{-1} v^{-1} \mu^1 (\rho_{\text{fluid}})^{-1} = \frac{\mu}{d v \rho_{\text{fluid}}}. \tag{5.108}$$

The inverse of Π_3,

$$\Pi_3^{-1} = \frac{d v \rho_{\text{fluid}}}{\mu}, \tag{5.109}$$

should look familiar. It is the Reynolds number, with a slight twist. The pertinent length is the diameter of the pipe. In our first encounter with the Reynolds number, the length scale was represented by the diameter of the sphere. Recall that the Reynolds number is the ratio

$$\text{Re} = \frac{d v \rho}{\mu} = \frac{\textit{inertial effects}}{\textit{viscous effects}}. \tag{5.63}$$

As we will see later the magnitude of Re again indicates the characteristics of the fluid flow.

Note that the Reynolds number is just as appropriate for the third dimensionless group as is the particular form we derived. We arrived at the particular group in Eq. (5.108) because we set $e = 1$ for convenience. It is equally appropriate to set $e = -1$ and thus to arrive at the Reynolds number.

We now design a model system to predict the characteristics of the real system. To do so, the magnitude of the dimensionless numbers must be the same for each system:

$$\text{Eu}_{\text{model}} = \text{Eu}_{\text{real}}, \tag{5.110}$$

$$\Pi_{2,\text{model}} = \Pi_{2,\text{real}}, \tag{5.111}$$

$$\text{Re}_{\text{model}} = \text{Re}_{\text{real}}. \tag{5.112}$$

This is *dynamic similarity*. Let's design a model system manageable enough to build in one's kitchen.

Let's model the Alaskan pipeline with water flowing through a conveniently sized glass tube. First we convert the Alaskan pipeline flow rate from English units to SI units:

$$\left(\frac{4 \times 10^5 \,\text{bbl}}{\text{day}}\right)\left(\frac{0.159 \,\text{m}^3}{1 \,\text{bbl}}\right)\left(\frac{1 \,\text{day}}{24 \,\text{hr}}\right)\left(\frac{1 \,\text{hr}}{60 \,\text{min}}\right)\left(\frac{1 \,\text{min}}{60 \,\text{s}}\right) = 0.7 \,\text{m}^3/\text{s}. \tag{5.113}$$

How does one calculate the average fluid velocity from the flow rate? Let's examine the dimensions of each. Flow rate has dimensions of L^3/T. Fluid velocity has dimensions of L/T. We must account for a difference in dimensions of L^2, the dimensions of area. It is thus dimensionally consistent to write

$$\textit{flow rate} = (\textit{fluid velocity}) \times (\textit{cross sectional area of pipe}). \tag{5.114}$$

Note that dimensional consistency cannot guarantee you are correct, but it can reveal when you are wrong. We calculate

$$\textit{flow rate} = v \pi \left(\frac{d}{2}\right)^2, \tag{5.115}$$

$$v = \frac{4 \,(\textit{flow rate})}{\pi d^2} = \frac{4(0.8 \,\text{m}^3/\text{s})}{\pi \,(1 \,\text{m})^2} = 1 \,\text{m/s}. \tag{5.116}$$

We prepare a table of the parameters of each system.

Table 5.9. The parameters of fluid flow through a pipe

Parameter	Symbol	Alaska	Kitchen
pressure drop	ΔP	?	(to be measured)
pipe length	l	1,000 m	(to be calculated)
pipe diameter	d	1 m	0.003 m (3 mm I.D. tube)
velocity	v	1 m/s	(to be calculated)
fluid viscosity	μ	10 Pa·s (oil)	10^{-3} Pa·s (water)
fluid density	ρ_{fluid}	800 kg/m^3 (oil)	1,000 kg/m^3 (water)

How does one calculate the length of the glass tube and the velocity of the water? Equate the dimensionless groups. To find l_{model}, use Π_2:

$$\Pi_{2,\text{model}} = \Pi_{2,\text{Alaska}}, \tag{5.117}$$

$$\left(\frac{l}{d}\right)_{\text{model}} = \left(\frac{l}{d}\right)_{\text{Alaska}}, \tag{5.118}$$

$$\frac{l_{\text{model}}}{0.003\,\text{m}} = \frac{1{,}000\,\text{m}}{1\,\text{m}}, \tag{5.119}$$

$$l_{\text{model}} = 3\,\text{m}. \tag{5.120}$$

To find v_{model}, use Re:

$$\text{Re}_{\text{model}} = \text{Re}_{\text{Alaska}}, \tag{5.121}$$

$$\left(\frac{d v \rho_{\text{fluid}}}{\mu}\right)_{\text{model}} = \left(\frac{d v \rho_{\text{fluid}}}{\mu}\right)_{\text{Alaska}}, \tag{5.122}$$

$$\frac{(0.003\,\text{m})v_{\text{model}}(1{,}000\,\text{kg/m}^3)}{10^{-3}\,\text{Pa·s}} = \frac{(1\,\text{m})(1\,\text{m/s})(800\,\text{kg/m}^3)}{10\,\text{Pa·s}}, \tag{5.123}$$

$$v_{\text{model}} = 2.7 \times 10^{-2}\,\text{m/s}. \tag{5.124}$$

Both v_{model} and l_{model} are reasonable. If either parameter were not reasonable (a length too long or a velocity too fast) we could choose a different fluid and/or a different pipe diameter.

We now measure ΔP with our model system. What volumetric flow rate should we have in our model?

$$\text{flow rate} = v \times \text{area} = (2.7 \times 10^{-2}\,\text{m/s}) \times \pi \left(\frac{0.003\,\text{m}}{2}\right)^2 \left(\frac{100\,\text{cm}}{1\,\text{m}}\right)^3 \left(\frac{60\,\text{s}}{1\,\text{min}}\right)$$

$$= 11\,\text{cm}^3/\text{min}. \tag{5.125}$$

We adjust the kitchen faucet to obtain the desired flow rate. How does one measure the pressure drop? Drill two holes in the glass tube separated by a distance of 3.0 m and attach vertical tubes. Measure the difference in the water height in the vertical tubes.

We observe in our model system a difference of 2.9 cm water, which we convert to SI units:

$$2.9 \, \text{cm water} \left(\frac{98 \, \text{Pa}}{1 \, \text{cm water}} \right) = 280 \, \text{Pa}. \tag{5.126}$$

How do we use this experimental result from our model to predict the pressure drop in the real system? Use the Euler number:

$$\text{Eu}_{\text{model}} = \text{Eu}_{\text{Alaska}}, \tag{5.127}$$

$$\left(\frac{\Delta P}{v^2 \rho} \right)_{\text{model}} = \left(\frac{\Delta P}{v^2 \rho} \right)_{\text{Alaska}}, \tag{5.128}$$

$$\frac{290 \, \text{Pa}}{(2.7 \times 10^{-2} \, \text{m/s})^2 (1{,}000 \, \text{kg/m}^3)} = \frac{\Delta P_{\text{Alaska}}}{(1 \, \text{m/s})^2 (800 \, \text{kg/m}^3)}, \tag{5.129}$$

$$\Delta P_{\text{Alaska}} = 3.2 \times 10^5 \, \text{Pa}. \tag{5.130}$$

Although SI units are convenient for calculation, I have no sense for a pressure of 3.2×10^5 Pa. So I convert to more convenient metric units and English units:

$$3.2 \times 10^5 \, \text{Pa} \left(\frac{1 \, \text{atm}}{1.01 \times 10^5 \, \text{Pa}} \right) = 3.2 \, \text{atm} \left(\frac{14.7 \, \text{psi}}{1 \, \text{atm}} \right) = 46 \, \text{psi}. \tag{5.131}$$

Thus, as promised, one experiment performed on a model has yielded the pressure drop for the actual system. However, we did not obtain a functional relationship. If any of the parameters of the real system change, we must perform another experiment with the model to predict the new pressure drop.

5.7 Flow through a Pipe

In the previous section we derived three Π groups to describe fluid flow in pipes:

$$\frac{\Delta P}{v^2 \rho} \equiv \text{Eu} = \frac{friction \; force}{inertial \; force}, \tag{5.100}$$

$$\frac{l}{d} \equiv reduced \; length \tag{5.104}$$

$$\frac{dv\rho}{\mu} \equiv \text{Re} = \frac{inertial \; force}{viscous \; force}. \tag{5.109}$$

For real pipes, we need an additional parameter to describe the flow – the tube roughness. Roughness is an inherent property of the pipe material; glass tubes are smoother than concrete drains. Roughness is also caused by deposits that accumulate in a pipe, known as scaling. Because roughness varies considerably, there is no precise way to characterize it. A sensible descriptor of roughness is the average height of features along the tube wall, k. Some typical values of k are given in Table 5.10.

Table 5.10. Roughness factors
for flow through a pipe

Material	k (mm)
glass	0.002
drawn copper	0.002
cast iron	0.3
concrete	2.0

Source: Perry and Chilton, 1973; 5.21.

The addition of another parameter requires a fourth dimensionless group. To find this fourth group we could return to the formal method used in the previous section. But because the dimensions of k are simple, we can guess the new Π group. How does one compensate for k's dimension of L? Divide by a parameter with dimension of L. What is the logical parameter to combine with k? The two candidates are l and d. Because l is a core variable, it can appear in only one Π group. So we divide k by the tube diameter to form k/d. And this makes sense. The roughness should be normalized by the diameter of the tube. Whereas 1 mm of roughness may be insignificant in a pipe of diameter 1 m ($k/d = 0.001$), it would be significant in a pipe of diameter 1 cm ($k/d = 0.1$).

In our analysis of the terminal velocity of a sphere we found we could combine two of the Π groups to form a composite dimensionless group, the friction factor:

$$friction\ factor = \frac{4}{3}\frac{\rho_{sph} - \rho_{fluid}}{\rho_{fluid}}\frac{1}{Fr}. \tag{5.70}$$

A similar analysis of flow through pipes reveals that the Euler number and the reduced length can be combined into one dimensionless group, also a called a *friction factor*. Again, a seemingly arbitrary factor appears, again as a result of theoretical modeling. Beware: Because the factor of 1/8 is arbitrary it is omitted from some plots. The Fanning friction factor is given by

$$friction\ factor = \frac{1}{8}\frac{d}{l}Eu. \tag{5.132}$$

As shown in Figure 5.21, roughness increases the friction factor, but only in the region of high Reynolds number. What type of fluid flow is typical of this region? Turbulence. Why should roughness have the largest effect here?

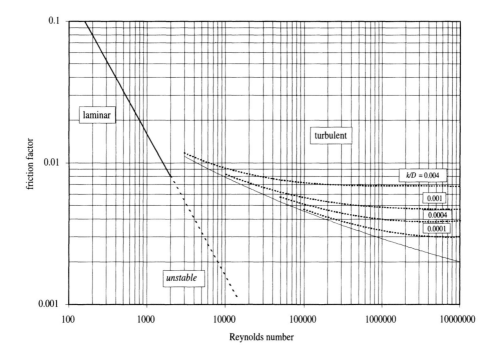

Figure 5.21. The friction factor (Eq. 5.132) as a function of Re, with effect of pipe roughness, k. (Adapted from Bird, R. B., Stewart, W. E., and Lightfoot, E. N. 1960. *Transport Phenomena*, Fig. 6.2-2.)

The region intermediate to laminar and turbulent is labeled "unstable." The flow pattern in this region depends on the history of the flow. If a laminar flow increases velocity gently, it can remain laminar for Re well above 1,000. However, once perturbed to turbulence, the flow remains turbulent.

5.8 Heat Transfer from a Fluid Flowing in a Tube

We have incorporated heat exchangers in many of our previous process designs, which we represented by the generic unit shown in Figure 5.22.

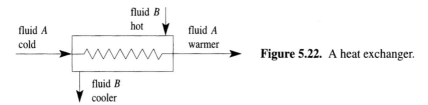

Figure 5.22. A heat exchanger.

Heat exchangers are commonly of the "shell and tube" design. One liquid passes through a bank of parallel tubes. The other liquid flows around the tubes countercurrent to the flow in the tubes contained by the outer shell, as diagrammed in Figure 5.23.

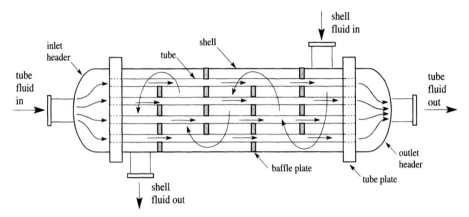

Figure 5.23. Schematic of a shell and tube heat exchanger. The tube fluid enters the inlet header and is distributed into tubes. Typical heat exchangers have hundreds of tubes. The shell fluid flows inside a cylindrical shell, around the outside of the tubes.

An important consideration in designing heat exchangers is the rate of heat transfer to the fluid inside the tubes. Clearly this will depend on the flow rate of the fluid. If the fluid passes through too quickly, it will not absorb much heat. Therefore we expect that dimensional analysis of a heat exchanger includes the parameters of flow in a tube, summarized in Table 5.8c, In addition, we have new parameters associated with heat transfer, listed in Table 5.11. The dimensions of the heat-transfer parameters are less obvious than those in Table 5.8. It is useful to include an extra column for the units of these parameters.

Table 5.11. The parameters of heat transfer for fluid flow through a pipe

Parameter	Symbol	Units (SI)	Dimensions
rate of heat transfer	q	joules/sec	ML^2/T^3
temperature difference	$T_{\text{fluid}} - T_{\text{pipe}}$	°C	Θ
heat capacity	C_p	joules/(kg·°C)	$L^2/T^2\Theta$
thermal conductivity	k	joules/(sec ·m·°C)	$ML/T^3\Theta$

We would not expect a first-year engineer to be able to devise the list in Table 5.11. However, we hope a first-year engineer can appreciate the relevance of these parameters. The temperature difference is the driving force for heat transfer, just like the density difference was the driving force for a sphere to fall (or rise) in a fluid. The heat capacity is literally the fluid's capacity to absorb heat; the temperature rise is caused by the absorption of energy. The thermal conductivity is the transmittance of energy, or the inverse of the resistance to heat transfer; a high conductivity means a low resistance to heat transfer. Similarly, the resistance to a sphere moving through a fluid is the viscosity. The rate of heat transfer, q, is usually combined with the temperature difference and the surface area of the tube (calculated from the pipe diameter and length) to yield a *heat transfer coefficient, h*:

$$h = \frac{q}{(T_{\text{fluid}} - T_{\text{pipe}})(\pi \, dl)} \; [=] \; \frac{ML^2/T^3}{\Theta L^2} \; [=] \; \frac{M}{\Theta T^3}. \tag{5.133}$$

With these parameters one can form four dimensionless groups, three of which are

$$\frac{dv\rho}{\mu} = \text{Re} = Reynolds \; number = \frac{inertial \; force}{viscous \; force}, \tag{5.134}$$

$$\frac{h}{C_P v_\rho} = \text{St} = Stanton \; number = \frac{heat \; transferred}{thermal \; heat \; capacity}, \tag{5.135}$$

$$\frac{C_P \mu}{k} = \text{Pr} = Prandtl \; number = \frac{momentum}{heat \; flux}. \tag{5.136}$$

A fourth dimensionless group includes the pressure drop, ΔP. However, experiments would reveal that this dimensionless group was irrelevant, just as the pendulum angle was revealed to be irrelevant.

The Reynolds number characterizes the dynamics of the fluid flow. The Stanton number characterizes the dynamics of the heat flow. The Prandtl number characterizes the substance being heated as it flows. The Prandtl number varies from substance to substance and varies with temperature; for water, $\text{Pr} = 7.7$ at $15°C$ and 1.5 at $100°C$.

Given three dimensionless groups, one can plot, for example, St (heat transfer dynamics) versus Re (fluid flow dynamics) as a function of Pr (fluid type), as shown in Figure 5.24. The lines through the data sets in Figure 5.24 are fits to equations of the form $\text{St} = c \cdot \text{Re}^{-0.2}$. The proportionality c varies with fluid and with Pr. Note that as Pr increases, the data lie lower on the plot. This suggests that c varies systematically with Pr. Indeed, Figure 5.25 shows that a plot of $\text{St}(\text{Re})^{0.2} (= c)$ as a function of Pr is a straight line. The line drawn in Figure 5.25 is $\text{St}(\text{Re})^{0.2} = 0.023 \cdot \text{Pr}^{-0.6}$.

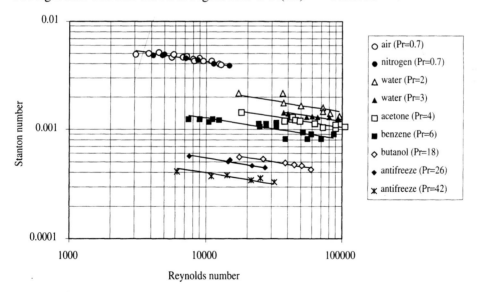

Figure 5.24. St versus Re for various fluids, each of which corresponds to a different value of Pr. (Adapted from Brown *et al.*, 1950, Chapter 29.)

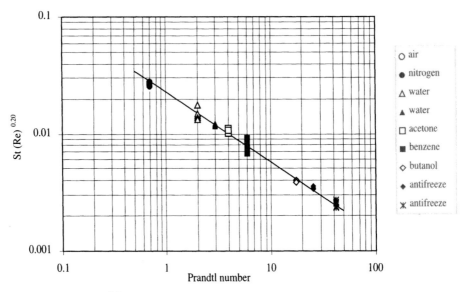

Figure 5.25. $St(Re)^{0.2}$ as a function of Pr. (Adapted from Brown *et al.*, 1950, Chapter 29.)

There is substantial scatter in Figures 5.24 and 5.25. This is common in correlations between dimensionless numbers that describe complex phenomena. Some scatter can be attributed to errors in the measurements, but most scatter indicates that there are one or more parameters missing. Most important, note that predictions from such correlations will not be precise. Whereas the mass balances in Chapter 3 were quite accurate, a prediction from a correlation such as in Figure 5.25 has an accuracy of about ±50%.

5.9 Absorbers and Distillation Columns

A powerful application of dynamic similarity is the analysis of complex systems, systems impractical to model analytically. Examples include the process units we encountered in Chapter 4, absorbers and distillation columns. As you might expect, engineers have developed many design tools for absorbers and distillation columns based on correlations of dimensionless groups. In this section we delve into the inner workings of multistage countercurrent units and examine typical design tools for predicting their behavior.

In the previous chapter we found that an effective design for absorbers and distillation columns is a series of equilibrium stages, as shown in Figure 4.70. Liquid flows down the column and vapor permeates up the column. At each stage the liquid and vapor mix and components transfer from one phase to the other to approach equilibrium. For efficient performance, we want equilibrium to be approached rapidly at each stage. To improve the rate of mass transfer between the liquid and vapor phases, one can increase the amount of interface between the liquid and vapor phases. A common design uses a series of sieve plates, linked by downcomers, as shown in Figure 5.26.

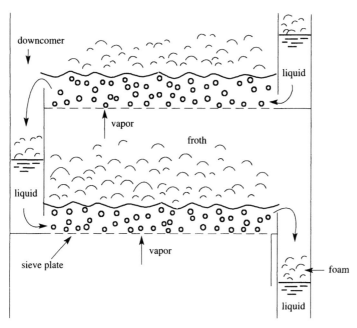

Figure 5.26. Schematic distillation column with sieve plates. Liquid flows from a downcomer and across the sieve plate, into the next downcomer. Vapor flows through the holes in the sieve plate, creating a froth above the liquid.

Liquid passes down a slot and spills onto a sieve plate. Vapor bubbles through holes in the sieve plate and creates a froth, which increases the amount of liquid–vapor interface. The amount of frothing can be increased by adding bubble caps to the sieve holes, shown in Figure 5.27. Bubble caps also decrease leakage of preequilibrated

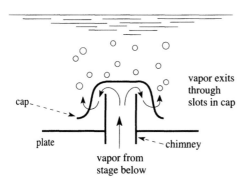

Figure 5.27. Diagram of a bubble cap. (Adapted from King, C. J. 1971. *Separation Processes*, Fig. 4-11.)

liquid to the plate below. The vapor phase equilibrates with the foam and rises to the next sieve plate. The froth spills over into the downcomer and settles into a bubble-free liquid.

Recall that when we designed distillation columns with the graphical McCabe–Thiele method, we specified the *relative* flow rates of liquid and vapor to obtain the operating lines. What diameter of column is needed to accommodate the *absolute* flow rates? If the column is too narrow (*i.e.*, the sieve plate area is too small), the liquid will pass over the sieve plate too quickly and not equilibrate with the vapor. If the column is too wide, the liquid will not cover the tray completely and the vapor will blow through without equilibrating.

Figure 5.28 shows a map of the possible behaviors for a distillation column. The abscissa is a dimensionless group formed from the product of the reflux ratio (L/V)

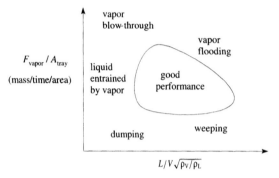

Figure 5.28. Operating regimes for a distillation column. (Adapted from King, C. J. 1971. *Separation Processes*, Fig. 12-8.)

and the square root of the vapor–liquid density ratio. The square root of the vapor–liquid density ratio is about 0.04 at 1 atm. The ratio decreases as the pressure in the column is decreased. The ordinate is the vapor flow rate (mass/time) divided by the area of the tray. With the map in Figure 5.28 one can predict the effect of different tray diameters with constant reflux ratio (a vertical path on the map) or changing the reflux ratio at constant vapor flow rate (a horizontal path). *Weeping* occurs when there is insufficient vapor flow through the sieve holes; the liquid "weeps" through to sieve holes. At lower reflux ratios liquid *dumps* onto the lower tray at low vapor flow and vapor *blows through* the liquid with insufficient contact at high vapor flow. At intermediate vapor flow and low L/V, liquid is entrained by the vapor and carried up to the next tray. Of course, the borders on the map are fuzzy; changes from one behavior to another are gradual, not abrupt.

A design tool for predicting vapor flooding in a distillation column is shown in Figure 5.29. Note that the correlation is only pseudo-dimensionless; the abscissa is dimensionless, but the ordinate and the variable for the family of curves (tray spacing) are not dimensionless. The abscissa is a product of two dimensionless groups, L/G and ρ_G/ρ_L. The L and G in Figure 5.29 are analogous to the L and V we used in Chapter 4. However, L and G have units of lb/hr/ft^2 whereas L and V have units of moles/sec. Here ρ_L and ρ_G are the densities of the liquid and vapor, respectively. The

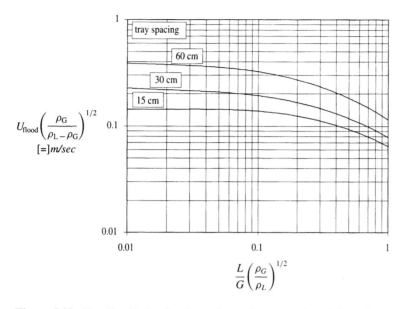

$$U_{\text{flood}}\left(\frac{\rho_G}{\rho_L - \rho_G}\right)^{1/2}$$
$$[=] m/sec$$

$$\frac{L}{G}\left(\frac{\rho_G}{\rho_L}\right)^{1/2}$$

Figure 5.29. Flooding limits for sieve plate columns. (Adapted from Perry R. H., and Green, D. 1984. *Perry's Chemical Engineers' Handbook*, 6th ed., Fig. 18-10.)

ordinate has units of ft/sec, owing to the parameter U_{flood}, the nominal velocity of the gas at which the column begins to flood. U_{flood} can be expressed in terms of the vapor mass flow rate, the vapor density, and the column diameter as follows:

$$vapor\ nominal\ velocity = \frac{vapor\ mass\ flow\ rate}{(vapor\ density)(column\ cross - sectional\ area)}, \tag{5.137}$$

$$U_{\text{flood}} = \frac{F_{\text{vapor}}}{\rho_V \pi r^2} \tag{5.138}$$

$$[=] \frac{\frac{M}{T}}{\left(\frac{M}{L^3}\right)L^2} [=] \frac{L}{T}. \tag{5.139}$$

Also, the parameter designating the family of curves in Figure 5.29 (tray spacing) is not dimensionless. What additional experiments might one perform to transform Figure 5.27 into a universal plot with all parameters dimensionless? Or, in other words, how might you make the parameters U_{flood} and "tray spacing" dimensionless? "Tray spacing" has dimensions of length. What other specifications of the column involve length? How about the column height? Column height would be a poor choice because it is proportional to the tray spacing; if one doubles the tray spacing, the column is twice as tall. A better choice would be the column diameter. How might one convert nominal vapor velocity to a dimensionless group? Hint: What famous dimensionless group involves a fluid velocity? The Reynolds number, of course. We need a characteristic length for the Reynolds number. Similar to flow through a pipe, the column diameter would be a good choice. So the dimensionless group for the vapor velocity might be

$$\mathrm{Re} = \frac{d U_{\text{flood}} \rho_V}{\mu}. \tag{5.140}$$

Another design tool allows one to determine the consequences of operating a column near flooding conditions. An important parameter is the fractional entrainment – the amount of liquid entrained in the gas phase and carried to the plate above. Ideally, no liquid should be carried to the plate above. If one defines a dimensionless group for flooding to be

$$\% \text{ of flood} \equiv 100 \times \frac{\text{number of trays flooded}}{\text{number of trays}}, \tag{5.141}$$

and defines entrainment as the dimensionless quantity,

$$\text{fractional entrainment} \equiv \frac{\text{moles entrained}}{\text{moles in liquid flow}}, \tag{5.142}$$

experimental studies yield the useful design tool shown in Figure 5.30.

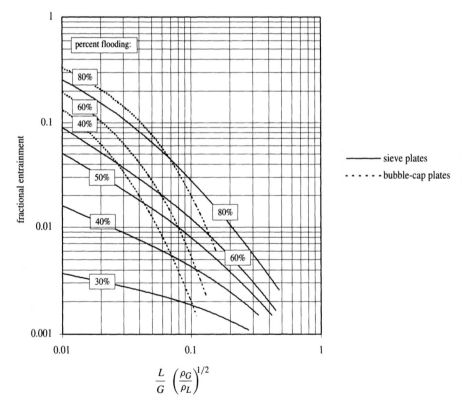

Figure 5.30. Liquid entrainment in a distillation column. (Adapted from Perry R. H., and Green, D. 1984. *Perry's Chemical Engineers' Handbook*, 6th ed., Fig. 18-22.)

There are many more design tools for absorbers and distillation columns. These may be found in the text by King (1971) and in *Perry's Chemical Engineers' Handbook*.

5.10 The Compressibility of a Gas

We have applied dimensional analysis to two process units: heat exchangers and distillation columns. Dimensional analysis can also be applied to fundamental systems, such as a gas of a pure substance. Given some basic properties of a gas, we would like to be able to predict the molar volume at a given temperature and pressure.

In Chapter 3 we cited the ideal gas law as an example of a constrained law. If one is modeling a gas at high temperature and low pressure, the ideal gas law is valid. To extend beyond the limits of the ideal gas law, one could measure the properties of a specific gas and develop methods of graphical analysis similar to those introduced in Chapter 4. A qualitative plot of the behavior outside the ideal limits was shown in Figure 4.1. What if one wants to model a gas outside the ideal-gas region but does not have access to experimental data? Can the behavior of all gases be scaled to a universal curve?

Dimensional analysis provides the basis for a universal curve for gases. As always, we begin by listing the parameters of a gas.

Table 5.12a. The parameters of a gas

Parameter	Symbol	Dimensions
pressure	P	M/LT^2
volume	V	L^3
temperature	T	Θ
gas constant	R	$ML^2/T^2\Theta$

We need to add parameters that distinguish our nonideal gas. Let's examine two causes of nonideality. The equation for an ideal gas predicts that as the temperature approaches zero and the pressure approaches infinity, the volume of the gas approaches zero. But a gas is composed of molecules that have a finite (albeit small) volume. At the limits of low temperature and high pressure, the volume goes to the volume of the liquid (or solid), not zero. Thus at low temperature and high pressure, $V_{real} > V_{ideal}$.

The equation for an ideal gas predicts that at low temperature, as the pressure goes to zero, the volume becomes very large. But because there is a mutual attraction between molecules, one observes that the volume is less than predicted, $V_{real} < V_{ideal}$.

So we could add to Table 5.12a a parameter for the molar volume of the liquid and some interaction force between the gas molecules. The molar volume is a convenient parameter, but the intermolecular force is not. Consider a different approach. Rather than use the *causes* of nonideality, let's examine the *effects* of nonideality. For example, rather than try to calculate intermolecular forces, one might use the boiling point of the liquid. Given two gases with equal molecular weight, the gas with the stronger intermolecular forces will boil at a higher temperature. However, experiments would reveal that the boiling point at a standard pressure is not a good predictor for the nonideal behavior of a gas.

What other properties characterize a gas? Recall the generic phase diagram for a pure substance, Figure 5.31. What "landmarks" do all substances have in common? Every substance has a triple point – the temperature and pressure at which all three

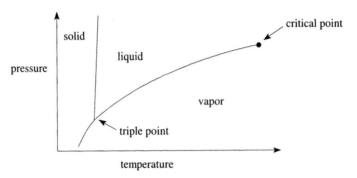

Figure 5.31. Phase diagram for a pure substance.

phases coexist. We could use the pressure and temperature of the triple point as the distinguishing features of our gas. However, like the boiling point at a standard pressure, the triple point is not a good predictor for the nonideal behavior of a gas. The other landmark common to all substances is a critical point – the point at which liquid and gas become a single phase. Experimental studies reveal that the critical temperature and pressure are good indicators of the properties of a gas. We add the critical temperature and pressure to our table of gas parameters.

Table 5.12b. The parameters of a gas

Parameter	Symbol	Dimensions
pressure	P	M/LT^2
volume	V	L^3
temperature	T	Θ
gas constant	R	$ML^2/T^2\Theta$
critical temperature	T_c	Θ
critical pressure	P_c	M/LT^2

You would not have been able to predict that critical temperature and pressure were appropriate for scaling the nonideal behavior of a gas. Until experiments were performed (and you had completed a course in molecular thermodynamics), any other distinguishing features of a gas – boiling point at 1 atm, the triple point, molar volume, etc. – were reasonable choices. And how can we know that the list of parameters in Table 5.12b is sufficient? Again, we would not know until we measured and analyzed data for many gases. If all the data lie on the same plot, the list is sufficient.

We now derive the dimensionless groups. As before we begin by writing a general expression for the dimensionless group, Π:

$$\Pi = P^a V^b T^c R^d T_c^e P_c^f \tag{5.143}$$

$$[=] \left(\frac{M}{LT^2}\right)^a (L^3)^b \Theta^c \left(\frac{ML^2}{T^2\Theta}\right)^d \Theta^e \left(\frac{M}{LT^2}\right)^f \tag{5.144}$$

$$[=] M^{a+d+f} L^{-a+3b+2d-f} T^{-2a-2d-2f} \Theta^{c-d+e}. \tag{5.145}$$

We then set all exponents equal to zero to satisfy the requirement that Π is dimensionless:

Mass, M: $\quad a + d + f = 0,$ $\hfill (5.146)$

Length, L: $\quad -a + 3b + 2d - f = 0,$ $\hfill (5.147)$

Time, T: $\quad -2a - 2d - 2f = 0,$ $\hfill (5.148)$

Temperature, Θ: $\quad c - d + e = 0.$ $\hfill (5.149)$

How many Π groups are there? The Buckingham Π Theorem tells us 6 variables -4 equations $= 2\Pi$ groups. But there is a problem with the 4 equations. This problem would become apparent when you attempted to solve the equations by successive substitution, for example. The equations for mass and time are not independent; multiply Eq. (5.146) by -2 and it becomes identical to Eq. (5.148). This is because M appears only in ratio with T^2. So there are actually only 3 *independent* equations, and thus there are $6 - 3 = 3$ dimensionless groups.

There are many valid choices for the three core variables. However, the standard dimensionless groups used to describe a gas are

$$reduced\ temperature \equiv T_r \equiv \frac{T}{T_c}, \tag{5.150}$$

$$reduced\ pressure \equiv P_r \equiv \frac{P}{P_c}, \tag{5.151}$$

and

$$compressibility \equiv Z \equiv \frac{PV}{RT}. \tag{5.152}$$

The compressibility is a useful indicator of the ideality of a gas. It is conventional to plot Z versus P_r as a function of T_r. Let's predict qualitatively the general appearance of the family of curves on such a plot.

If a gas is ideal, $PV = nRT$. For $n = 1$ we can rewrite the ideal gas law in a dimensionless form,

$$1 = \frac{PV}{RT} = compressibility. \tag{5.153}$$

Thus at very high temperature, we expect $Z = 1$, as shown in Figure 5.32.

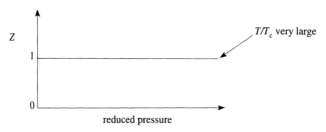

Figure 5.32. Compressibility as a function of pressure at a high temperature.

As discussed earlier, we expect nonideal behavior at low temperature. At high pressures, because of the finite molar volume, $V_{real} > V_{ideal}$ and thus $Z > 1$. At low pressures, the intermolecular interactions cause $V_{real} < V_{ideal}$ and thus $Z < 1$. The predicted behavior of Z is shown on the phase maps in Figures 5.33 and 5.34. Thus we would expect the isotherm for low temperature to have the general shape given in Figure 5.35.

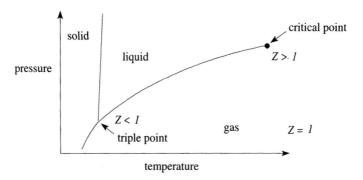

Figure 5.33. Compressibility as a function of pressure and temperature.

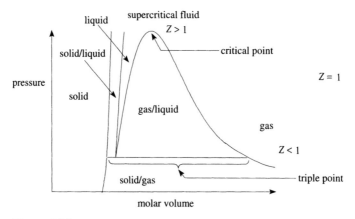

Figure 5.34. Compressibility as a function of pressure and molar volume.

The universal plots of Z versus P_r as a function of T_r are shown for moderate pressures in Figure 5.36a and for high pressures in Figure 5.36b.

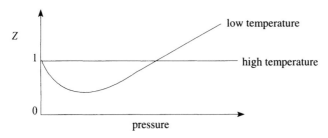

Figure 5.35. Compressibility as a function of pressure at two temperatures.

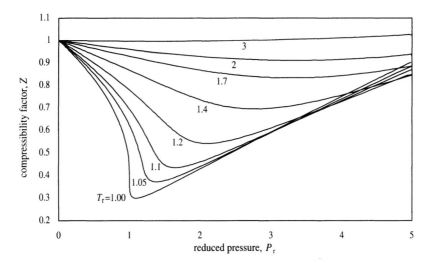

Figure 5.36a. Z versus P_r as a function of T_r for moderate pressures and low temperatures. Curves generated with van der Waals equation.

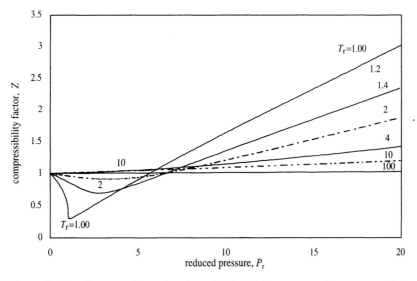

Figure 5.36b. Z versus P_r as a function of T_r for high pressures. Curves generated with van der Waals equation.

285

5.11 Summary

Dimensional analysis has the following advantages:

- *It reveals the relative scaling of parameters.* We found that if the length of the pendulum arm doubles, the period decreases by the square root of 2. If the diameter of a sphere doubles, its terminal velocity increases by a factor of 2, for low Reynolds number.
- *It optimizes an experimental study.* In some cases, dimensional analysis reveals that certain parameters are not pertinent. The period of a pendulum is independent of mass, as is the velocity of walking.
- *It optimizes analysis.* For each system we studied – the pendulum, walking, and spheres moving through fluids – a complex phenomenon described by several variables was reduced to a function of a few dimensionless groupings of the variables. In the analysis of walking the parameters h, l, g, and m reduced to $s/l = f(v^2/gl)$. In the analysis of the terminal velocity of spheres, the six parameters v, D, g, buoyancy, ρ_{fluid}, and μ reduced to a relation between three Π groups,

$$\frac{v^2}{Dg} = f\left(\frac{\rho_{\text{sph}} - \rho_{\text{fluid}}}{\rho_{\text{fluid}}}, \frac{\rho_{\text{fluid}} v D}{\mu}\right). \tag{5.64}$$

- *It reveals the different characteristics of a phenomenon.* For quadrupeds, the style of running changes from a trot to a gallop at a Froude number of about 2.55. The fluid flow pattern around a sphere has three distinct regions – laminar, turbulent, and intermediate – characterized by the Reynolds number.
- *It is expedient.* There are hundreds of phenomena in chemical engineering. When the equation describing one of these phenomena is expressed in a dimensionless form, the equation reduces to one of (about) eight differential equations. (You will encounter differential equations in a mathematics course in your sophomore year.) The advantages of this are obvious: The equation has already been solved and dimensional similarity reveals physical similarity. For example, heat transfer by conduction is mathematically similar to mass transfer by diffusion.

REFERENCES

Alexander, R. McN. 1989. *Dynamics of Dinosaurs and Other Extinct Giants*, Columbia Univ. Press, New York.

Alexander, R. McN. 1992. "How Dinosaurs Ran," *Scientific American*, April, pp. 4–10.

Batchelor, G. K. 1967. *Introduction to Fluid Mechanics*, Cambridge Univ. Press, London.

Bird, R. B., Stewart, W. E., and Lightfoot, E. N. 1960. *Transport Phenomena*, Wiley, New York.

Brown, G. G., Foust, A. S., Katz, D. L., Schneiderwind, R., White, R. R., Wood, W. P., Brown, G. M., Brownell, L. E., Martin, J. J., Williams, G. B., Banchero, J. T., and York, J. L. 1950. *Unit Operations*, Wiley, New York.

King, C. J. 1971. *Separation Processes*, McGraw-Hill, New York.

Perry, R. H., and Chilton, C. H. 1973. *Chemical Engineers' Handbook*, 5th ed., McGraw-Hill, New York.

Perry, R. H., and Green, D. 1984. *Perry's Chemical Engineers' Handbook*, 6th ed., McGraw-Hill, New York.

Zlokarnik, M. 1991. *Dimensional Analysis and Scale-Up in Chemical Engineering*, Springer-Verlag, New York.

EXERCISES

Units and Dimensions

5.1 Consider the following table:

Unit	Base dimensions	Type of unit	Converted to SI units
second	T	base	1 s
gram	M	multiple	1×10^{-3} kg
liter	L^3	derived	1×10^{-3} m^3

Construct a similar table for the following units: acre, Btu, carat, parsec, kilobar, and centipoise.

5.2 Show that the following equations are dimensionally consistent.

(A) Einstein's mass–energy equation,

$$E = mc^2,$$

where E is energy, m is mass, and c is the velocity of light.

(B) Newton's law of gravitation,

$$F = G \frac{m_1 m_2}{d^2},$$

where F is the force between two masses m_1 and m_2 separated by a distance d, and G is the universal gravitational constant.

(C) Einstein's equation for the molar heat capacity of a solid, C_V (in joules/(mol·K)),

$$C_V = 3R \left(\frac{h\upsilon}{kT} \right)^2 \frac{\exp(h\upsilon/kT)}{(\exp(h\upsilon/kT) - 1)^2},$$

where R is the gas constant, h is the Planck constant, υ is a frequency, k is the Boltzmann constant, and T is the temperature.

(D) The Clausius–Clapeyron equation,

$$\frac{\Delta H_{\text{vap}}}{T(V_{\text{gas}} - V_{\text{liquid}})} = \frac{dP}{dT},$$

which predicts the change in vapor pressure, P, caused by a change in temperature, T. ΔH_{vap} is the heat of vaporization (in joules/mole) and V_{gas} and V_{liquid} are the molar volumes of the gas and liquid phases, respectively. **Hint:** Use the definition of a derivative,

$$\frac{dy}{dx} = \lim_{x_1 \to x_2} \frac{y_2 - y_1}{x_2 - x_1},$$

to determine the dimensions of a derivative.

5.3 Show that the following equations are dimensionally consistent.

(A) The ideal gas law,

$$PV = nRT,$$

where P is pressure, V is volume, n is the number of moles, R is the gas constant, and T is temperature.

(B) The Hagen–Poiseuille (pronounced Pwah-zø-yah) law for the volumetric flow rate, Q, of a fluid with viscosity μ through a pipe of radius r,

$$Q = \frac{\pi(\Delta P)r^4}{8\mu l}.$$

The pressure drop over a pipe length l is ΔP.

(C) The energy of an ideal gas,

$$\rho C_V \left(\frac{dT}{dt}\right) = k\left(\frac{d^2 T}{dx^2}\right) - P\left(\frac{dv}{dx}\right),$$

where ρ is the density (in mol/m^3), C_V is the heat capacity at constant volume (in joules/(mol·K)), dT/dt is the derivative of temperature with respect to time, k is the thermal conductivity, $d^2 T/dx^2$ is the second derivative of temperature with respect to distance, P is pressure, and dv/dx is the derivative of velocity with respect to distance. (See the hint in Exercise 5.2(D).)

Deriving Dimensionless Groups

5.4 The mean velocity of a molecule in a gas, v, is determined by the mass of the molecule, m, the temperature of the gas, T, and the Boltzmann constant, $k = 1.38 \times 10^{-23}$ joules/K.

(A) List the dimensions of each of these four quantities.

(B) Combine the four quantities v, m, T, and k to form a dimensionless group.

5.5 Three dimensionless groups were derived to describe fluid flow through a pipe. Core variables ΔP, l, and μ were chosen and we obtained the Euler number, the reduced length, and the Reynolds number.

(A) Derive the dimensionless groups that correspond to the core variables v, ρ, and l.

(B) Show that the groups derived in (A) are dimensionless.

(C) Show that the Euler number and the Reynolds number can be expressed in terms of the groups derived in (A).

5.6 We wish to determine the rate at which a liquid drains from a large tank with a small hole in the bottom. This system is described by the parameters in the table below. Find the dimensionless group(s) that describe(s) this system.

Parameter	Symbol
height of fluid	h
diameter of hole	d
flow rate (gal/min)	Q
fluid density	ρ
fluid viscosity	μ
gravitational acceleration	g

Derive a (set of) dimensionless group(s) that characterizes this system.

(This exercise appeared on an exam. It was estimated that it could be completed in 10 minutes.)

5.7 The Prandtl number, Pr, is a dimensionless number used to characterize heat transfer. The Prandtl number is comprised of three quantities: heat capacity (C_p), viscosity (μ), and thermal conductivity (k). Given that the exponent of C_p is 1, derive the exponents of μ and k.

5.8 The Peclet number is a dimensionless group that appears in the analysis of heat transfer via forced convection. The Peclet number is comprised of five quantities: a characteristic length (l), the fluid velocity (v), heat capacity (C_p), density (ρ), and thermal conductivity (k). Given that the exponent of ρ is 1, derive the Peclet Number.

(This exercise appeared on an exam. It was estimated that it could be completed in 15 minutes.)

5.9 Heat transfer by forced convection is characterized by the following parameters:

Parameter	Symbol
pipe diameter	d
fluid velocity	v
fluid density	ρ
fluid heat capacity	C_P
fluid viscosity	μ
fluid thermal conductivity	k

Derive a (set of) dimensionless group(s) that characterizes this system.

5.10 The Lewis number is a dimensionless group composed of four parameters: fluid density (ρ), fluid heat capacity (C_P), molecular diffusivity (D), and fluid thermal conductivity (k).

(A) A typical thermal conductivity is $k = 0.5$ joules/(m sec K). What are the dimensions of k?

(B) Heat capacity, C_P, is the proportionality between temperature change and energy change for a substance of mass m,

$$\Delta E = m C_p (\Delta T).$$

What are the dimensions of C_P?

(C) Diffusivity, D, appears in Fick's second law, where t is time, x is distance, and C is concentration $[=]$ mol/volume:

$$\frac{dC}{dt} = D\frac{d^2C}{dx^2}.$$

What are the dimensions of D?

(D) Derive an expression for the Lewis number using k as the core variable.

(This exercise appeared on an exam. It was estimated that it could be completed in 15 minutes.)

5.11 The molecules in a fluid move constantly. On average, a molecule will diffuse a distance d in a time period t. Assume that molecular diffusion may be described by the following quantities: distance (d), time (t), mass (m), and diffusion coefficient $(D\,[=]\,L^2/T)$.

(A) Derive the dimensionless group(s) of molecular diffusion.

(B) The distance diffused is proportional to what power of t? That is, for $d \propto t^x$, what is x?

5.12 The rate of unsteady-state heat conduction into a solid sphere is characterized by the following parameters:

Parameter	Symbol
sphere diameter	d
time elapsed	t
sphere density	ρ
heat capacity	C_P
thermal conductivity	k

Derive a (set of) dimensionless group(s) to characterize this system.

(This exercise appeared on an exam. It was estimated that it could be completed in 15 minutes.)

Analyzing Graphical Data

5.13 A fluid of viscosity μ and density ρ flows at velocity v through a rough pipe of diameter d when pressure drop ΔP is applied over a length l. With all other parameters held constant, ΔP is varied and v is measured.

ΔP $(10^5$ Pa$)$	v (m/s)
1	1
4	2
9	3
16	4
100	10

(A) Derive a (set of) dimensionless group(s) that characterizes this system.

Refer to Figure 5.21 for parts (B) and (C).

(B) Is the flow in the pipe laminar, unstable, or turbulent? Explain your conclusion.

(C) How does ΔP depend on μ in this regime? That is, if we write $\Delta P \propto \mu^n$, what is n?

(This exercise appeared on an exam. It was estimated that it could be completed in 15 minutes.)

5.14 A fluid is pumped through a pipe that contains an orifice plate. The diameter of the orifice is half the diameter of the pipe, D_1.

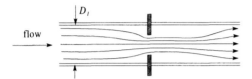

The flow rate through the pipe with the orifice $Q_{orifice}$ is measured as a function of the Reynolds number. $Q_{orifice}$ is divided by the flow rate through the pipe without the orifice, Q_0, and is plotted below. Note that the data are plotted on a semilog coordinate system not a log–log system. Obtain an expression for the flow rate through the orifice (*i.e.*, $Q_{orifice} = ?$) for Re < 100. Check your expression by substituting at least two points in the range Re < 100.

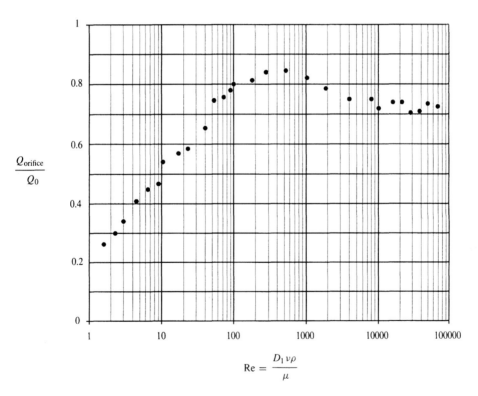

(Figure adapted from Perry, R. H., and Chilton, C. H. 1973. *Chemical Engineers' Handbook*, 5th ed., McGraw-Hill, New York, Fig. 5-18, pp. 5–13. This exercise appeared on an exam. It was estimated that it could be completed in 15 minutes.)

5.15 As discussed in Section 5.8, heat transfer to a fluid flowing through a tube may be described by three dimensionless groups: the Reynolds number (Re), the Prandtl number (Pr), and the Stanton number (St). Use the line drawn through the data in Figure 5.25 to derive an equation that relates Re, Pr, and St. Specifically, derive an equation of the form St = ...

(This exercise appeared on an exam. It was estimated that it could be completed in 15 minutes.)

5.16 Heat transfer from a solid sphere moving through a fluid is characterized by three dimensionless numbers: the Nusselt number (Nu), the Reynolds number (Re), and the Prandtl number (Pr), defined as

$$\mathrm{Nu} = \frac{hd}{k},$$

$$\mathrm{Re} = \frac{dv\rho}{\mu},$$

and

$$\mathrm{Pr} = \frac{C_P\mu}{k},$$

where $h \equiv$ heat transfer coefficient, $\rho \equiv$ fluid density, $v \equiv$ sphere velocity, $k \equiv$ fluid thermal conductivity, $\mu \equiv$ viscosity, $d \equiv$ sphere diameter, and $C_P \equiv$ fluid heat capacity.

Calculate h for a solid sphere of diameter = 3.7 mm moving through water at velocity = 0.29 m/sec.

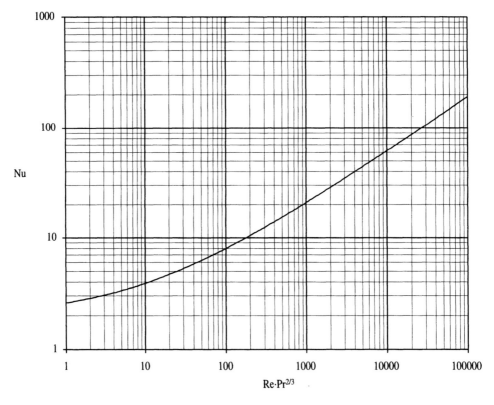

(Figure adapted from Bird, R. B., Stewart, W. E., and Lightfoot, E. N. 1960. *Transport Phenomena*, Wiley, New York, Fig. 13.3-2, p. 409. This exercise appeared on an exam. It was estimated that it could be completed in 20 minutes.)

5.17 A cold metal sphere is warmed as it passes through a hot fluid. As stated in the preceding exercise, this phenomenon is described by three dimensionless groups: the Reynolds number (Re), the Nusselt number (Nu), and the Prandtl number (Pr). The desired values for the parameters in this exercise are as follows: $h = 2{,}500$ joules/(m^2 sec K), $v = 0.17$ m/sec, $k = 0.59$ joules/(m sec K) , $\mu = 1.0 \times 10^{-3}$ Pa sec , $\rho = 1{,}000.$ kg/m^3, and $C_P = 4{,}170$ joules/(kg K).

Use the graph in the preceding exercise to determine the diameter of the sphere needed to obtain these parameters. Note that because d appears in both Nu and Re, this is not a straightforward task. There are several approaches; one is outlined in steps (A), (B), and (C) below. You are not obligated to follow this approach. You may devise your own.

(A) Start with the definition for Nu and derive an equation of the form $d = f_1(\text{Nu})$.

(B) Start with the definitions for Re and Pr and derive an equation of the form $d = f_2(\text{RePr}^{2/3})$.

(C) Plot the function $f_1(\text{Nu}) = f_2(\text{RePr}^{2/3})$ on the graph in the preceding exercise to find d.

(This exercise appeared on an exam. It was estimated that it could be completed in 25 minutes.)

5.18 Hot air is used to pop popcorn and separate popped corn from corn kernels, as follows: An upward flow of hot air levitates a corn kernel; the kernel's terminal velocity equals the velocity of the hot air.

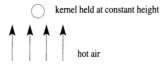
kernel held at constant height

hot air

After the kernel pops, the popped corn rises with the hot air.

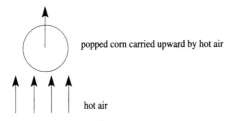
popped corn carried upward by hot air

hot air

(A) Use the plot below to calculate the velocity of the hot air. Note that velocity appears in both Re and the friction factor. The kernel is approximately spherical of diameter 5 mm and mass 0.07 g. The popped corn is also approximately spherical and has diameter 3 cm and mass 0.05 g. The hot air has density 0.9 kg/m^3 and viscosity 3×10^{-5} Pa·sec.

(B) Calculate the upward velocity of popped corn.

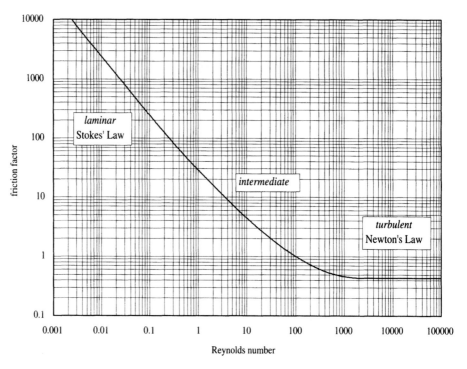

(Figure adapted from Bird, R. B., Stewart, W. E., and Lightfoot, E. N. 1960. *Transport Phenomena*, Fig. 6.3-1. This exercise appeared on an exam. It was estimated that it could be completed in 25 minutes.)

Dynamic Scaling

5.19 The flow through an artificial kidney is described by three dimensionless groups: Re, Eu, and the reduced length. Use these groups to design a model for flow through an artificial kidney. In this hypothetical artificial kidney, blood (density = 1,000. kg/m^3 and viscosity = 1.0×10^{-2} Pa·s) is pumped through 10. m of tubing of diameter 0.0010 m. The average velocity of the blood is 0.010 m/s. To determine the pressure drop in the artificial kidney, we want to measure the pressure drop in a model using glycerin (density = 1,260. kg/m^3, viscosity = 1.49 Pa·s) flowing through a tube of diameter 0.0050 m.

(A) What length of tube do we need for the model system?

(B) What should be the average velocity of the glycerin in our model system?

(C) Assume that we measure a pressure drop of 2.25×10^7 Pa in our model system. What pressure drop do we expect for blood flowing through 10. m of tube in our artificial kidney?

The walls of the tubing in the artificial kidney are very thin to enhance the removal of wastes from the blood. Consequently, the maximum pressure drop must be less than 1.0×10^4 Pa. Given this limit on the pressure drop, it is clear that our artificial kidney cannot be a single tube 10 m long. Rather, it must be many short tubes in parallel such that the total length is 10 m. You may assume that the pressure changes linearly with tube length.

(D) What is the maximum length of one of the many short tubes in parallel?

Warning: Blood flow in a kidney is more complex than described here. Do not use these parameters to design an artificial kidney.

5.20 Waves formed behind a vessel traveling on the surface of a fluid account for much of the frictional loss. This phenomenon is described by two dimensionless groups: the Froude number and the Reynolds number. The quantity D in both dimensionless groups is the depth of the vessel below the surface of the fluid. We wish to study the waves behind a real battleship by using a model battleship.

Quantities for a real battleship:

fluid density: $\rho = 1{,}025 \text{ kg/m}^3$
velocity: $v = 12. \text{ m/s}$
depth: $D = 10. \text{ m}$
viscosity: $\mu = 1.2 \times 10^{-3} \text{ Pa·s}$

(A) The model battleship is scaled such that its depth below the fluid surface is 0.62 m. At what velocity should we move the model through the fluid?

(B) We need to choose the fluid in our model. Assume the model fluid will have a typical density of $1.0 \times 10^3 \text{ kg/m}^3$. Calculate the desired viscosity of the fluid in the model system.

(This exercise appeared on an exam. It was estimated that it could be completed in 20 minutes.)

5.21 A cold metal sphere is warmed as it passes through a hot fluid. This phenomenon is described by three dimensionless groups: the Reynolds number (Re), the Nusselt number (Nu), and the Prandtl number (Pr), defined in Exercise 5.16. We wish to analyze a microscopic process with a macroscopic model. In the microscopic process a copper sphere of diameter 1.0×10^{-4} m moves through water at a velocity of 1.2×10^{-3} m/sec. In our model we will use a copper sphere of diameter 0.10 m in water.

(A) At what velocity should our model sphere move through the water?

(B) We measure a heat transfer coefficient of 2,500 joules/(m^2 sec K) with our model. What is the heat transfer coefficient in the microscopic system?

(This exercise appeared on an exam. It was estimated that it could be completed in 25 minutes.)

5.22 A meteor of mass m and density ρ_m with velocity v strikes the Earth and creates a crater of volume V. The parameters that describe this event are listed below.

	Parameter	Symbol
the meteor	mass	m
	density	ρ_m
	velocity	v
the crater	volume	V
	soil density	ρ_s
constants	gravitational acceleration	g

A set of three dimensionless groups describing this event are

$$\frac{\rho_m}{\rho_s}, \quad \frac{\rho_s V}{m}, \quad \text{and} \quad \frac{g}{v^2}\left(\frac{m}{\rho_m}\right)^{1/3}.$$

(See Schmidt, R., and Housen, K. 1995. "Problem Solving with Dimensional Analysis," *The Industrial Physicist*, Vol. 1, pp. 21–24.)

We wish to model the impact of an enormous meteor ($m = 5.2 \times 10^8$ kg), a round rock of diameter 100. m with velocity 1.1 km/sec (2,500 mph). Our model will use a metal ball with twice the density of the meteor and with velocity 0.26 km/sec. We will perform our experiment in a large centrifuge, so gravitational acceleration is replaced by centrifugal acceleration, which is 500. $\times g_{earth} = 4.9$ km/sec^2. The density of the soil in our model is double the density of the soil at the predicted point of impact.

What is the mass of the "meteor" in our model?

(This exercise appeared on an exam. It was estimated that it could be completed in 15 minutes.)

5.23 Engineers at your chemical company have proposed a process to produce a new chemical. The key unit in the process is unprecedented. It is also very large, very expensive, and the phenomena inside the unit are very complex. You need to know how this unit will perform under a given set of operating conditions, such as the composition of the input, or the temperature.

List a sequence of steps to model this unique unit. The first step is given below. **Do not** execute the steps; just list the steps.

Step 1. Apply dimensional analysis to obtain the Π group(s).

Data Analysis on Spreadsheets

Dimensional analysis involves four steps. First, one derives a set of dimensionless groups. Next, one performs experiments on a convenient system. Third, one analyzes the data. Finally, one applies dynamic similarity to predict the behavior of the system of interest.

An engineer will often pose two fundamental questions when performing the third step, the analysis of the data. First, an engineer may want to know the *certainty* that the data exhibit a specific relationship. For example, experiments on walking will yield a collection of Froude numbers and reduced strides. What is the certainty that reduced stride versus Froude number data lie on a straight line on a log–log plot? That is, what is the certainty that $s/l \propto \text{Fr}^m$? If the certainty is low, perhaps a different functional form is needed.

An engineer's second question is similar, but there is a subtle, important difference. Given a specific functional relationship, what is the *certainty* of the fitted parameters in the equation? For example, *given* that data for walking are represented by the functional form $s/l \propto \text{Fr}^m$, what is the certainty that m is a specific value? Are the confidence limits on $m \pm 0.01$, or ± 0.1, or ± 1, for example?

This tutorial illustrates the proficiency of spreadsheets for reducing experimental data, plotting the reduced data, and analyzing trends. Like the tutorial on spreadsheets for mass balances, this tutorial assumes a basic knowledge of spreadsheets, such as how to use a mouse to select a cell and how to copy and paste. However, the tutorial on spreadsheets for mass balances is not a prerequisite.

Spreadsheet Example 1   Experimental data for the terminal velocities of spheres in water, vegetable oil, and glycerin are presented in three tables below and on the following page. The data are a subset of the experimental data measured by Cornell first-year engineers to obtain a general correlation between the Reynolds number,

$$\mathrm{Re} = \frac{Dv\rho_{\mathrm{fluid}}}{\mu},$$ (5.127)

and the friction factor,

$$\mathit{friction\,factor} = \frac{4}{3}\frac{\rho_{\mathrm{sph}} - \rho_{\mathrm{fluid}}}{\rho_{\mathrm{fluid}}}\frac{1}{\mathrm{Fr}}.$$ (5.70)

Use these data to prepare a plot of the friction factor versus the Reynolds number.

Spheres moving through water (viscosity = 0.0010 Pa·s, density = 1,000 kg/m³)

Sphere	Diam (mm)	Mass (g)	Height (m)	Time (s)
nylon	3.18	0.019	0.50	5.53
nylon	6.35	0.150	0.50	3.90
nylon	6.35	0.150	0.51	3.80
lucite	3.18	0.021	0.51	4.60
teflon	12.70	2.480	0.51	0.75
glass	2.80	0.038	0.50	1.27
aluminum	6.35	0.361	0.51	0.80
steel	3.18	0.133	0.50	0.76
steel	6.35	1.064	0.51	0.45
steel	9.53	3.572	0.51	0.40
steel	9.53	3.572	0.51	0.35
steel	12.70	8.486	0.51	0.40

Spheres moving through vegetable oil (viscosity = 0.05 Pa·s, density = 930 kg/m³)

Sphere	Diam (mm)	Mass (g)	Height (m)	Time (s)
nylon	3.18	0.019	0.64	34.72
nylon	3.18	0.019	0.64	32.80
lucite	6.35	0.172	0.64	9.49
teflon	3.18	0.038	0.64	7.47
aluminum	6.35	0.361	0.51	2.23
steel	1.59	0.016	0.64	5.05
steel	1.59	0.016	0.64	4.99
steel	1.59	0.016	0.64	4.99
steel	3.18	0.133	0.64	1.94
steel	3.18	0.133	0.51	1.79
steel	6.35	1.064	0.51	1.17
steel	9.53	3.572	0.51	0.70

Spheres moving through glycerin (viscosity $= 1.2$ Pa·s
(absorbed moisture from air), density $= 1,250$ kg/m³)

Sphere	Diam (mm)	Mass (g)	Height (m)	Time (s)
polypropylene	3.18	0.014	0.09	41.76
polypropylene	6.35	0.113	0.223	45
nylon	3.18	0.019	0.05	58.79
nylon	3.18	0.019	0.05	64.94
glass	2.80	0.038	0.11	22.56
glass	2.80	0.038	0.11	21.30
aluminum	6.35	0.361	0.23	7.91
steel	1.59	0.016	0.23	33.78
steel	1.59	0.016	0.223	30
steel	3.18	0.133	0.11	3.47
steel	3.18	0.133	0.18	4.98
steel	6.35	1.064	0.23	1.69
steel	6.35	1.064	0.28	1.82

Solution to Spreadsheet Example 1   A spreadsheet solution to this exercise is shown in Figure 5.37. The steps below will lead you through the process of creating this spreadsheet. The method used here is not unique, nor is it the most efficient. If you are skilled at spreadsheets you are encouraged to pursue a different approach to produce a column of Reynolds numbers and a column of friction factors.

1. Enter the title. *Select* cell A1, the cell in column A and row 1; use the mouse to position the cursor in the cell and click the mouse. The border around the cell A1 should be highlighted. Cell A1 is now the *active cell.* Type "*Example 1. Terminal velocity of a sphere in a fluid*" into cell A1. Although only a portion of the title will be visible in cell A1, the entire text is displayed in the *formula bar*, above the worksheet and below the tool bar. After you hit *enter,* the entire text appears in the spreadsheet, because the adjacent cells are empty.

2. Make a table of the three fluids and their properties. Type "fluid" in cell A3, then type the names of the three fluids – "water," "vegetable oil," and "glycerin" – in cells A4 through A6. You may adjust the width of column A by positioning the cursor at the line between the column headings A and B, at the top of the worksheet. The cursor will change to a vertical bar with a horizontal double arrow. Drag the cursor to change the column width.

   Add the properties of the fluids in columns B and C. In cell B2 type "viscosity" and in cell B3 type its units, "Pa·s." Type the fluid viscosities below this column heading. Similarly, type "density" in cell C2, the units "kg/m^3" in cell C3, and the fluid densities below the heading. The in-line caret (^) is the symbol for exponentiation in most spreadsheets.

3. Enter the data for spheres in water. Begin by entering column headings, starting in cell A8 for "sphere" and ending with "t (s)" in cell E8. Now enter the data into rows 9 through 20. The tedium and error associated with entering data can be reduced by copying and pasting. For example, after typing "nylon" into cell A9,

	A	B	C	D	E	F	G	H	I	J	K	L
1	Example 1. Terminal velocity of a sphere in a fluid											
2		viscosity	density									
3	fluid	Pa s	kg/m^3		constants							
4	water	0.0010	1000		pi	3.1416						
5	vegetable oil	0.05	930		g	9.8						
6	glycerin	1.2	1250									
7							velocity	sphere	buoyancy	Reynolds	Froude	friction
8	sphere	dia (mm)	m (g)	ht (m)	t (s)	fluid	m/s	density	kg/m^3	number	number	factor
9	nylon	3.18	0.019	0.5	5.53	water	0.0904	1128	128	288	0.26	0.65
10	nylon	6.35	0.15	0.5	3.9	water	0.1282	1119	119	814	0.26	0.60
11	nylon	6.35	0.15	0.51	3.8	water	0.1342	1119	119	852	0.29	0.55
12	lucite	3.18	0.021	0.51	4.6	water	0.1109	1247	247	353	0.39	0.84
13	teflon	12.7	2.48	0.51	0.75	water	0.6800	2312	1312	8636	3.72	0.47
14	glass	2.8	0.038	0.5	1.27	water	0.3937	3306	2306	1102	5.65	0.54
15	aluminum	6.35	0.361	0.51	0.8	water	0.6375	2693	1693	4048	6.53	0.35
16	steel	3.18	0.133	0.5	0.76	water	0.6579	7899	6899	2092	13.89	0.66
17	steel	6.35	1.064	0.51	0.45	water	1.1333	7936	6936	7197	20.64	0.45
18	steel	9.53	3.572	0.51	0.4	water	1.2750	7882	6882	12151	17.41	0.53
19	steel	9.53	3.572	0.51	0.35	water	1.4571	7882	6882	13887	22.73	0.40
20	steel	12.7	8.486	0.51	0.4	water	1.2750	7912	6912	16193	13.06	0.71
21	nylon	3.18	0.019	0.64	34.72	oil	0.0184	1128	198	1.09	0.01	26.09
22	nylon	3.18	0.019	0.64	32.8	oil	0.0195	1128	198	1.15	0.01	23.29
23	lucite	6.35	0.172	0.64	9.49	oil	0.0674	1283	353	7.97	0.07	6.92
24	teflon	3.18	0.038	0.64	7.47	oil	0.0857	2257	1327	5.07	0.24	8.08
25	aluminum	6.35	0.361	0.51	2.23	oil	0.2287	2693	1763	27.01	0.84	3.01
26	steel	1.59	0.016	0.64	5.05	oil	0.1267	7602	6672	3.75	1.03	9.28
27	steel	1.59	0.016	0.64	4.99	oil	0.1283	7602	6672	3.79	1.06	9.06
28	steel	1.59	0.016	0.64	4.99	oil	0.1283	7602	6672	3.79	1.06	9.06
29	steel	3.18	0.133	0.64	1.94	oil	0.3299	7899	6969	19.51	3.49	2.86
30	steel	3.18	0.133	0.51	1.79	oil	0.2849	7899	6969	16.85	2.60	3.84
31	steel	6.35	1.064	0.51	1.17	oil	0.4359	7936	7006	51.48	3.05	3.29
32	steel	9.53	3.572	0.51	0.7	oil	0.7286	7882	6952	129.1	5.68	1.75

Figure 5.37a. Spreadsheet solution to example 1.

	A	B	C	D	E	F	G	H	I	J	K	L
33	polypropylene	3.18	0.014	0.09	41.76	glycerin	0.0022	831	419	0.0071	0.00015	2995
34	polypropylene	6.35	0.113	0.223	45	glycerin	0.0050	843	407	0.0328	0.00039	1100
35	nylon	3.18	0.019	0.05	58.79	glycerin	0.0009	1128	122	0.0028	0.00002	5587
36	nylon	3.18	0.019	0.05	64.94	glycerin	0.0008	1128	122	0.0026	0.00002	6817
37	glass	2.8	0.038	0.11	22.56	glycerin	0.0049	3306	2056	0.0142	0.00087	2531
38	glass	2.8	0.038	0.11	21.3	glycerin	0.0052	3306	2056	0.0151	0.00097	2256
39	aluminum	6.35	0.361	0.23	7.91	glycerin	0.0291	2693	1443	0.1923	0.01359	113
40	steel	1.59	0.016	0.23	33.78	glycerin	0.0068	7602	6352	0.0113	0.00298	2277
41	steel	1.59	0.016	0.223	30	glycerin	0.0074	7602	6352	0.0123	0.00355	1911
42	steel	3.18	0.133	0.11	3.47	glycerin	0.0317	7899	6649	0.1050	0.03225	220
43	steel	3.18	0.133	0.18	4.98	glycerin	0.0361	7899	6649	0.1197	0.04192	169.2
44	steel	6.35	1.064	0.23	1.69	glycerin	0.1361	7936	6686	0.9002	0.29763	24.0
45	steel	6.35	1.064	0.28	1.82	glycerin	0.1538	7936	6686	1.0176	0.38034	18.8

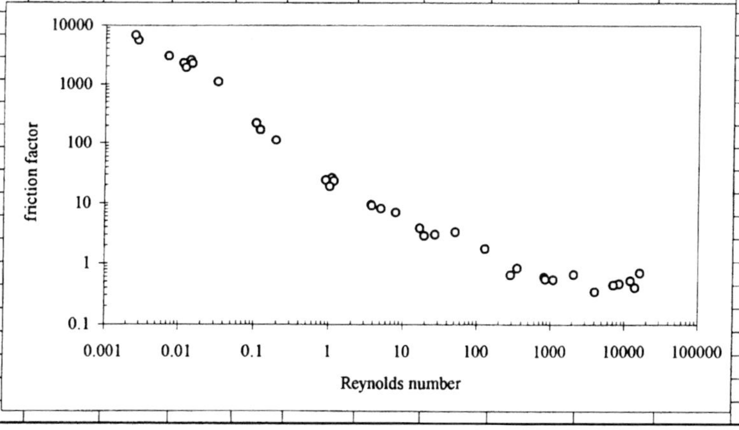

Figure 5.37b. Spreadsheet solution to example 1, (*continued*).

copy cell A9, select cells A10 through A11, and then paste. To select a group of cells, such as A10 through A11, first select cell A10; then drag down to cell A11 while depressing the clicker. Or, first select cell A10; then select the final cell in the group by holding down the *shift* key when clicking. Later, copy "steel" from cell A16 and paste into cells A17 through A20.

4. It will be useful to list the fluid in each row. Enter the column heading "fluid" in cell F8. Type "water" into cell F9; then copy and paste into cells F10 through F20.

5. Enter the data for spheres in vegetable oil and glycerin. Follow steps (3) and (4) to complete rows 21 through 45, columns A through F.

6. Our goal is to reduce the data to the Reynolds number and the friction factor. We will begin by calculating the values of two parameters that constitute these dimensionless groups – the terminal velocity and the buoyancy. We must take care to calculate all parameters *in mks units*.

   Velocity is (distance traveled)/(time elapsed), assuming a uniform velocity. Distance and time are each in mks units, so we don't need any conversion factors. Enter the formula for the first velocity by typing "= D9/E9" into cell G9. The number 0.0904... should appear in cell G9 after you hit *enter*. The number of significant figures will depend on the format and width of the cell.

   As was illustrated in the spreadsheet example in Chapter 3, it is easier to create a formula by selecting cells rather than typing cell addresses. For example, the formula for the first velocity can be created by first typing "=" into cell G9. This notifies the spreadsheet you are entering a formula. Then select cell D9 with the mouse. The characters "D9" will appear in cell G9 and the formula bar. Type the symbol for multiplication, "/". Now select cell E9. This completes the formula. Hit *enter*.

   Calculate all the velocities. Copy the contents of cell G9 and paste into cells G10 through G45. Note that the cell addresses in the formula are incremented automatically. For example, the formula in G10 is "= D10/E10" and so on down the column.

7. To calculate the buoyancy, it is prudent to first calculate the sphere density, in mks units. Type the heading "sphere" in cell H7 and "density" in cell H8. Density is mass/volume, and the volume of a sphere is $\pi d^3/6$. We could simplify this formula by calculating the constant $\pi/6$. However, this formula provides an excuse to introduce absolute cell references. Start a list of constants. Type the heading "constants" in cell E3 and type "pi" in cell E4. Now enter a value for π in cell F4. Five significant figures is more than enough, given the accuracy of the data; type 3.1416 in cell F4. Now create the formula for density in cell H9. Enter "= (C9/1000)/(F4*(B9/1000)^3/6)" in cell H9. Note that both mass and diameter must be converted to mks units: grams to kilograms and millimeters to meters. The first pair of parentheses are only for clarity. The same result is obtained without the first pair of parentheses.

   Paste the formula for sphere density into the cells below. But wait – the reference to the cell that contains π, cell F4, is incremented to F5, F6, and so on. We want the reference to cell F4 to remain unchanged down the column. We must indicate an *absolute cell reference* by replacing F4 with F4. Now you can paste the formula from cell H9 into cells H10 through H45. Because only the row is incremented as we paste down a column, we actually only needed to indicate that the row was absolute. Replacing F4 with F$4 is sufficient.

Computing the sphere density as a separate quantity allows one to scan for errors – errors in the formula for density and errors in the data. Check the nylon spheres – all should have about the same density, about $1,120 \text{ kg/m}^3$. Teflon should be about $2,300 \text{ kg/m}^3$, steel about $7,700 \text{ kg/m}^3$, polypropylene about 840 kg/m^3, and so on.

8. Calculate the buoyancy – the difference between the sphere density and the fluid density – in column I. Create formulas that use the densities for water, oil, and glycerin in cells C4, C5, and C6, respectively. If you used the absolute references correctly, the buoyancy of a given material in a given fluid should be constant. Also, a given material, such as steel, should be least buoyant in oil and most buoyant in glycerin.

You probably noticed something suspicious about some of the buoyancies – some are negative. This was also evident to the students who measured the data. Nylon and polypropylene spheres float on glycerin. Negative buoyancies represent spheres *rising* in glycerin. Spheres rising at their terminal velocity lie on the same correlation as spheres falling at their terminal velocity.

However, negative buoyancies will wreak havoc with the friction factor, which must be a positive number. Let's guarantee that all the buoyancies are positive. A simple solution is to calculate the absolute value. Enter the formula "=ABS(H9-C4)" in cell I9 and paste down the column, taking care to reset the absolute reference to the densities of oil and water.

9. We can now calculate the dimensionless groups. Start with the Reynolds number. Apply your spreadsheet skills to create a formula for Re in cell J9, taking care that all parameters are in mks units. Paste the formula down column J. Remember to use absolute references for ρ_{fluid} and μ, and remember to change the absolute references when the fluid changes. Compare your results with the spreadsheet printed in Figure 5.37.

10. We now calculate the Froude number,

$$\text{Fr} = \frac{v^2}{Dg}, \tag{5.58}$$

as an intermediate to the friction factor. The gravitational acceleration on Earth is $9.8 \text{ m}^2/\text{s}$. We add g to the list of constants. Type "g" in cell E5 and "9.8" in cell F5. Calculate Fr in column K and check your results.

11. Finally, we calculate the friction factor. Calculate the friction factor in column L and check your results. Did you remember to use an absolute reference for ρ_{fluid} and change the absolute reference when the fluid changed?

12. Prepare a log–log plot of the data. First, select the data to be plotted. The x values are in column J, rows 9 through 45. The y values are in column L, rows 9 through 45. Because the columns are not adjacent, selecting both at once requires special attention. Dragging from cell J9 to cell L45 would select column K as well. First, select cells J9 through J45. Then hold the *Ctrl* key while selecting cells L9 through L45.

The details of plotting vary with software type. The procedure for Microsoft® Excel™ begins with clicking on the *ChartWizard*© icon in the *tool bar*, which changes the cursor from an arrow to cross-hairs. Position the cross-hairs where you would like one corner of your plot, drag the cross-hairs to the position of the opposite corner, and release. This sets the size of your graph. The size is easily changed later, by dragging any corner or border.

ChartWizard© now leads you through five steps, which are self-explanatory. At the second step, choose a *XY (Scatter) chart* and choose the log–log option, designated by nonuniform spacing of the gridlines in both the *x* and *y* directions. After step 5 a plot appears, albeit an esthetically unpleasing plot. Explore various formatting options by clicking on various features of the plot – the axes, the body, the gridlines – to produce a plot of publication quality, as shown in the example.

Spreadsheet Example 2

Determine the equation for a straight line through the data in the preceding example for Re < 2. Evaluate the certainty that the data lie on a straight line. Given that a straight line will be fit to the data, evaluate the certainty of the fitted parameters, the slope and the intercept.

Solution to Spreadsheet Example 2   Begin by culling the data for Re < 2. One could do this point by point since there are only 37 data points to inspect. But this would be tedious for larger data sets. Plus, selecting point by point lacks style. Instead, we will arrange the data in ascending Re, and then truncate the data for Re > 2.

1. Open a new worksheet. Select cell A1 and enter the title "*Example 2. Terminal velocity of a sphere in Stokes regime.*" Now, copy the column of Reynolds numbers from the previous worksheet into column A of the new worksheet. Copy the heading as well, but do not copy the formulas in each cell, as follows. Change windows to the first worksheet, select cells J7 through J45, and copy. Change windows to the new worksheet and select cell A2. Pull down the *edit* menu and choose paste *special*; specify that only the *values* are pasted. Check that the formulas were not pasted by selecting a cell in column A. Only its value should appear in the *formula bar* – no formula.

   Paste the friction factors from the first worksheet to column B of the new worksheet. You may now close the first worksheet.

2. Sort the data. Select the data in columns A and B. Pull down the *Data* menu and choose *Sort* to open a dialog box. Specify an *ascending* sort on column A. Because you selected both columns A and B, the Re and f pairs are retained when the data are sorted.

3. Delete the data for Re > 2. If all is well to this point, you should find that the data in row 19 and below have Re > 2. Select these data (both columns A and B) and delete, by pulling down the *Edit* menu and choosing *Clear* (specify *All* in the dialog box). The first two columns of your spreadsheet should resemble the example in Figure 5.38, give or take a few insignificant figures.

   At this point one could plot the data and choose a canned routine to fit a straight line to the data. Admittedly, this approach has style. However, canned procedures

	A	B	C	D	E	F	G	H	I	J
1	Example 2. Terminal velocity of a sphere in Stokes' regime									
2	Reynolds	friction					log(Re)*			
3	number	factor	log(Re)	log(f)	[log(Re)]^2	[log(f)]^2	log(f)	f (calc'd)	deviat'n^2	
4	0.0026	6817.3	-2.593	3.834	6.726	14.697	-9.942	9071.1	0.015	
5	0.0028	5587.2	-2.550	3.747	6.503	14.041	-9.556	8221.6	0.028	large deviation
6	0.0071	2995.3	-2.146	3.476	4.607	12.086	-7.462	3280.1	0.002	
7	0.0113	2277.3	-1.948	3.357	3.794	11.272	-6.540	2087.6	0.001	
8	0.0123	1910.7	-1.910	3.281	3.647	10.766	-6.266	1914.2	0.000	
9	0.0142	2531.3	-1.847	3.403	3.412	11.583	-6.286	1659.9	0.034	large deviation
10	0.0151	2256.4	-1.822	3.353	3.320	11.245	-6.110	1568.3	0.025	large deviation
11	0.0328	1100.5	-1.484	3.042	2.203	9.251	-4.515	727.3	0.032	large deviation
12	0.1050	219.9	-0.979	2.342	0.958	5.486	-2.293	230.1	0.000	
13	0.1197	169.2	-0.922	2.228	0.850	4.966	-2.054	202.2	0.006	
14	0.1923	113.3	-0.716	2.054	0.513	4.219	-1.471	126.5	0.002	
15	0.9002	24.0	-0.046	1.380	0.002	1.903	-0.063	27.5	0.004	
16	1.0176	18.8	0.008	1.273	0.000	1.621	0.010	24.4	0.013	
17	1.0903	26.1	0.038	1.417	0.001	2.006	0.053	22.8	0.003	
18	1.1541	23.3	0.062	1.367	0.004	1.869	0.085	21.5	0.001	
19		sum:	-18.86	39.56	36.54	117.01	-62.41		0.17	
20										
21	points:	n	15							
22	slope:	m	-0.988							
23	intercept:	b	1.395							
24	uncertainty in y		0.113							
25	uncertainty in slope		0.032							
26	uncert'nty in intercept		0.049							

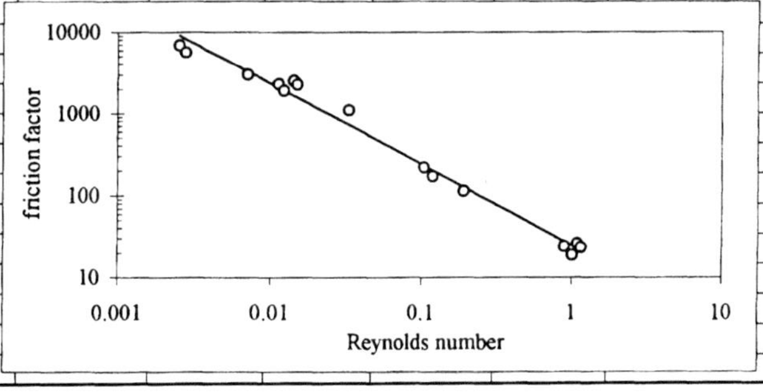

Figure 5.38. Spreadsheet solution to example 2.

often omit details of the fit. Is there a subset of data that consistently lies above the curve? Perhaps all the data for spheres in oil? Maybe our value for the viscosity for oil is too low? More important, not all canned routines provide the information necessary to evaluate the certainty of the model and the certainty of the fitted parameters.

We now take up the method of least squares for fitting a straight line to a set of data.

Linear Regression and the Method of Least Squares. We seek the best fit of a straight line,

$$y = mx + b,$$

to a set of n data points. The slope and intercept that minimize the sum of the squared deviations between each data point and the line are

$$slope = m = \frac{n \sum xy - \sum x \sum y}{n \sum x^2 - (\sum x)^2},$$

$$intercept = b = \frac{\sum y - m \sum x}{n}.$$

The certainty that the data are represented by a straight line can be evaluated by calculating the uncertainty in the values of y, the sum of the squared deviations between the experimental data and the calculated data,

$$\text{uncertainty in } y \equiv \sigma_y \approx \left[\frac{\sum (y - mx - b)^2}{n - 2} \right]^{1/2}.$$

Given that a straight line will be used, we can estimate the uncertainty in the slope and intercept with the following equations:

$$\text{uncertainty in the slope} \equiv \sigma_m \approx \left[\frac{n\sigma_y^2}{n \sum x^2 - (\sum x)^2} \right]^{1/2},$$

$$\text{uncertainty in the intercept} \equiv \sigma_b \approx \left[\frac{\sigma_y^2 \sum x^2}{n \sum x^2 - (\sum x)^2} \right]^{1/2}.$$

We now use a spreadsheet to calculate the least-squares fit and evaluate the uncertainty of the calculated values and the uncertainty of the fitted parameters.

4. First we must remember that because the data are linear on log–log paper, x is $\log_{10} \text{Re}$, not Re, and y is $\log_{10} f$, not f. We create two new columns: $\log_{10} \text{Re}$ and $\log_{10} f$. Type the label "log(Re)" in cell C3 and enter the formula "$= \log 10(A4)$" in cell C4. Paste the formula from cell C4 into cells C5 through C18. Type "log f" in cell D3 and copy the formula from cell C4 into cell D4. The formula in cell D4 should read "$= \log10(B4)$". Complete column D.

5. Compute various summations, $\sum x$, $\sum x^2$, $\sum y$, $\sum y^2$, and $\sum xy$. Spreadsheets provide functions to calculate summations. Begin with $\sum x$. In cell C19 enter the formula "$= \text{SUM(C4:C18)}$." To calculate $\sum y$ copy cell C19 and paste into cell D19. To remind us that row 19 contains sums, type "sum:" in cell B19. In columns E, F, and G, compute $\sum x^2$, $\sum y^2$, and $\sum xy$. Finally, calculate the sums of these columns by copying cell C19 and pasting into cells E19, F19, and G19. Compare your spreadsheet through column G with the spreadsheet example in Figure 5.38.

6. Compute the slope and intercept. Type the labels "n," "m," and "b" in cells B21, B22, and B23. Enter the number of data points (15) in cell C21. Use the formulas above to calculate the slope and intercept in cells C22 and C23. You should obtain -0.988 for the slope and 1.395 for the intercept.

7. Use the slope and intercept to calculate values for the friction factor, given the values for the Reynolds numbers in column A. As was discussed in the text, the equation

$$\log_{10} f = m[\log_{10} \text{Re}] + b$$

can be rearranged to

$$f = 10^b \, \text{Re}^m.$$

Use the above formula and calculate friction factors in cells H4 through H18. Did you remember to use absolute references for the slope and intercept?

8. Calculate squared deviations between the experimental values for y and the calculated values for y. In cell I4 enter the formula "$= (D4 - \log10(H4))^{\wedge}2$." Remember – we must calculate the differences between base-10 logarithms of friction factors, not differences between friction factors. Paste the formula from cell I4 down the column. In cell I19 calculate the sum of cells I4 through I18.

9. Calculate the uncertainty in y, the logarithm of the friction factor. In cell C24 enter the formula "$= \text{SQRT}(I19/(C21-2))$." We obtain an uncertainty of 0.113, which means the typical deviation between the logarithm of the experimental friction factor and the logarithm of the calculated friction factor is 0.113. The difference between logarithms of two numbers is equivalent to the logarithm of the ratio of the two numbers. In other words, the ratio of two numbers is 10 raised to the difference in logarithms. In this case, the ratio is $10^{0.113} = 1.30$. Thus a typical calculated friction factor will deviate by as much as a factor of 1.30; the calculated friction factor will typically be between 30% larger and 30% smaller than the experimental friction factor.

Does an uncertainty in y of 0.113 justify a fit to a straight line? The uncertainty *alone* is insufficient to judge. We need to compare the uncertainty of $\pm 30\%$ to the confidence limit of the experimental data. If the experimental uncertainty is $\pm 40\%$, a straight line is justified. If the experimental uncertainty is $\pm 10\%$, a straight line is questionable. We would try different functional forms until the calculated uncertainty was comparable to the experimental uncertainty. Similarly, if the experimental uncertainty is $\pm 40\%$, there is no justification for replacing our straight line with, for example, a quadratic function even though the calculated uncertainty might be improved to $\pm 20\%$.

Estimating the experimental uncertainty lies beyond the intended range of this book. It would require details about the experiment not pertinent to this tutorial and would require an explanation of the method of propagation of errors.

10. Calculate the uncertainty in the fitted parameters – the slope and the intercept. In cell C25 enter the formula "$= \text{SQRT}(C21*C24*C24/(C21*E19 - C19*C19))$." This should yield a value of 0.032. The slope is thus -0.988 ± 0.032. When reporting uncertainties, only one significant figure is meaningful, unless the first digit is 1. Similarly, it is not meaningful to report the fitted parameters to decimal places beyond the uncertainty. Thus the slope is -0.99 ± 0.03. We calculate the uncertainty of the intercept to be 0.049. Thus the intercept is 1.40 ± 0.05. Stokes's law predicts a slope of -1 and an intercept of 1.398 ($10^{1.398} = 25$). Our fitted parameters and their uncertainties are consistent with Stokes's law.

11. Plot the experimental data (columns A and B) and the calculated data (columns A and I). Plot the data as points and the calculated data as a line (no points). You are invited to prepare a plot larger than presented here. We endeavored to fit the example onto one page of this textbook.

12. Examine the results. The largest deviations are for the points with Re = 0.0028, 0.0142, 0.0151, and 0.0328. Reopen the spreadsheet from the first example and locate these data points. The points correspond to the data in rows 35, 37, 38, and 34, respectively. All data are from experiments with glycerin, which is a difficult fluid to work with because it is sticky and gooey. But the problem is not the value

for the viscosity of glycerin, because some glycerin points lie above the line and some lie below. Two of the data points are for glass spheres – that's suspicious. The densities of these glass spheres are 3,300 kg/m^3, which is abnormally high. Silica glass typically has a density of 2,200 kg/m^3. The mass or the diameter might be wrong. The mass was measured on a calibrated electronic balance, so the mass is probably correct. Sphere diameters (2.8 mm) were measured with calipers with accuracy of ± 0.1 mm. What diameter would yield the typical glass density of 2,200 kg/m^3? 3.2 mm. An error of 0.4 mm is conceivable.

We can use the spreadsheets to investigate further. Start with the first spreadsheet and save a copy under a new file name. Now change the diameters of the glass spheres in cells B37 and B38 to 3.2 mm. Go to the second spreadsheet and save a copy under a new file name. Paste the "adjusted" Reynolds numbers and friction factors for the glass spheres into the second spreadsheet. Note uncertainty in y decreases to 0.086 from 0.113. However, the fitted slope and intercept deviate further from the Stokes's law parameters: The slope changes to -0.97 ± 0.02 from -0.99 ± 0.03 (theoretical value is 1) and the intercept changes to 1.39 ± 0.04 from 1.40 ± 0.05 (theoretical value is 1.398).

Do we leave the glass sphere diameters at 3.2 mm, because the data fit the correlation better? Absolutely not! Delete both files with the "adjusted" data. "Adjusting" data is fraud. Such scientific misconduct would warrant professional sanctions. "Adjusting" data goes against the fundamental goal of science – to seek the truth. Can we delete the data for the glass spheres, because we suspect they are faulty? Absolutely not! It is also fraud to discard data because it doesn't agree with your pet theory or (as has been known to happen) because it doesn't agree with your mentor's pet theory.

Searching for the source of experimental deviation can suggest new measurements. In this case, we suspect it is difficult to accurately measure the diameter of small spheres. What do you think *you* would find if *you* remeasured the glass spheres? Of course, it would be hard to avoid finding 3.2 mm, because we are prejudiced. Rather you should ask someone else to measure the glass spheres, someone not biased toward any particular answer. And, this person should remeasure the diameters of *all* the small spheres. If the method of measuring small spheres is suspect, it is suspect for all small spheres, not just those with the largest deviations from the correlation. Or perhaps another method of determining a sphere's diameter is needed? Perhaps one could place several identical small spheres in a graduated cylinder with water, measure the volume displaced, and then calculate the diameter?

It is important to recognize potential sources of scientific fraud, especially self-deception. These topics are discussed in the engaging book *Betrayers of the Truth*, by William Broad and Nicholas Wade (Simon and Schuster, New York, 1982).

Now, use the skills you acquired in the preceding tutorial to complete the following exercise.

5.24 Analyze the relation between a quadruped's body mass and the circumferences of its major leg bones, given by the data below.

(A) Prepare a plot of body mass (y axis) versus the sum of the humerus and femur bone circumferences (x axis). Simply adding the bone circumferences may seem naive. If we knew the dominant stress on the bones – compressive, bending, or torsional– we would combine the bone circumferences differently. For example, if compressive forces dominated, because the strain caused by compressive forces goes as the cross-sectional area, we would add the squares of the circumferences, then take the square root. However, because the humerus and femur circumferences are comparable, the method of combining the two has little effect. The larger effect is the differences in body structures.

Bone circumferences and body masses of quadrupedal mammals

Mammal	Leg bone circumference (mm)		Mass (kg)
	Humerus	Femur	
meadow mouse	4.9	5.5	0.047
guinea pig	10	15	0.385
gray squirrel	10	13	0.399
opossum	27	23	3.92
gray fox	28	26	4.20
raccoon	30	28	4.82
nutria	21	28	4.84
bobcat	31	32	5.82
porcupine	30	34	7.20
otter	32	28	9.68
coyote	36	36	12.7
cloud leopard	45	41	13.5
duiker	31	46	13.9
yellow baboon	55	57	28.6
cheetah	67	69	38.0
cougar	62	60	44.0
wolf	62	62	48.1
bushbuck	56	62	50.9
impala	65	69	60.5
warthog	83	72	90.5
nyala	99	97	135.
lion	104	94	144.
black bear	98	94	218.
grizzly bear	124	107	256.
blue wildebeest	115	100	257.
Cape Mountain zebra	132	143	262.
kudu	140	135	301.
Burchell's zebra	129	147	378.
polar bear	158	135	448.
giraffe	192	173	710.
bison	192	168	1,179.
hippopotamus	209	208	1,950.
elephant ("Jumbo")	459	413	5,897.

(Data from Anderson, J. F., Hall-Martin, A., and Russell, D. A. 1985. "Long-Bone Circumference and Weight in Mammals, Birds, and Dinosaurs," *Journal of Zoology, London (A)*, **207**: 53–61.)

(B) Determine the *proportionality constant* and exponent *m* in a relation of the form

$$(body\ mass) = (proportionality\ constant)$$
$$\times\ (humerus + femur\ bone\ circumference)^m$$

from a least-squares fit to the data. Calculate uncertainties for the proportionality constant and the exponent *m*.

You should find that *m* is surprisingly close to 3. This suggests that animals *do* scale linearly, contrary to the argument at the beginning of this chapter. What gives? The key issue is the *stress* on the bone, not the *strength* of the bone. Strain doesn't necessarily scale with body mass. The stress, or force, on a bone is proportional to body mass *times acceleration*. Earth's gravitational field causes acceleration, whether an animal is standing, walking, galloping, or jumping. So if bears hopped around as squirrels do, bears would surely break bones. Similarly, if elephants jumped about as terriers do, elephants would surely break bones. Again, one must consider the *dynamics* when scaling animate systems.

(C) Use the relation you determined in (B) to draw a line through the data and calculate the individual deviations between the actual body masses and the predicted body masses. Critically analyze the fit. Is this a universal correlation? Are there any types of mammals that always lie above or below the curve? Catlike mammals? Deerlike mammals? Bearlike mammals? Are small mammals consistently above or below the curve? Large mammals? Medium mammals? Do the data deviate from a straight line, suggesting a different functional form is needed?

(D) Use the correlation developed in (B) to estimate the masses of the dinosaurs in the table below. Check one of your estimates with ours: We estimate the apatosaurus mass to be 39,000 kg, to two significant figures.

Bone circumferences and body masses of quadrupedal dinosaurs

Dinosaur	Leg bone circumference (mm)		Estimated Mass (kg)
	Humerus	Femur	
styracosaurus	288	370	4,080
diplodocus	320	405	10,560
brachiosaurus	654	730	78,260
apatosaurus	629	845	27,870
			32,420

(Data from Anderson *et al.* (1985) and Colbert, E. H. 1962. "The Weights of Dinosaurs," *American Museum Novitates,* **2076**: 1–16. Weights were estimated from the volumes of models, assuming a density of 900 kg/m^3. More recent studies assume densities comparable to water, 1,000 kg/m^3 (see Alexander, R. McN. 1983. "Posture and Gait of Dinosaurs," *Zoological Journal of the Linnean Society,* **83**: 1–25), which would increase the masses in the table above by 11%.)

(E) As opposed to the mammal masses, the dinosaur masses tabulated above are not experimental data. There are no data for the masses of live dinosaurs. Colbert molded dinosaur models, measured the models' volumes, and estimated dinosaur masses by assuming a dinosaur density and lung volume. Based on your predictions from dynamic similarity, which of Colbert's models was overly stocky or overly thin? That is, which of your estimates deviate excessively from Colbert's estimates? (An excessive deviation is one that exceeds the typical deviations you observed for mammals.)

(F) (*optional*) Increase the range of the correlation by adding data for insects. Insects have an exoskeleton, so the leg "bone" circumference is just the leg circumference. True, insect legs are fleshy inside, but so are mammal bones. And as you will learn in statics, the strength of a cylindrical member depends little on the material in the center.

Properties of some compounds at 1 atm

Quantity	Symbol	Water	Acetone	1-Butanol
molecular wt (amu)		18	58	74
density (kg m^{-3})	ρ	1,000.	790.	947
viscosity (Pa · s)	μ	1.0×10^{-3}	—	—
melting point (K)	T_{melt}	273.	178.	183.
boiling point (K)	T_{vap}	373.	329.	390.
heat capacity (joules kg^{-1} K^{-1})	C_P	4,190 (liq) 1,860 (gas)	2,170 (liq) 1,290 (gas)	1,850 (gas)
heat of melting (joules mol^{-1})	ΔH_{melt}	6.01×10^3	5.72×10^3	9.37×10^3
heat of vaporization (joules mol^{-1})	ΔH_{vap}	4.07×10^4	2.91×10^4	4.31×10^4
thermal conductivity (joules m^{-1} sec^{-1} K^{-1})	k	0.59	—	—

6

Transient Processes

I N CHAPTER 3 WE ANALYZED PROCESSES by developing models based on fundamental laws, such as conservation principles. We restricted our analyses to systems at steady state, namely where none of the variables changed with time. This simplified the mathematical modeling since we dealt only with algebraic equations; such models, as we saw, are extremely useful for analyzing steady-state systems.

Many chemical processes, however, do not operate at steady state. A chemical reaction performed one batch at a time, such as the experiments you have done in your secondary school and first-year chemistry classes, is an example of an unsteady process. Thus we need design tools for processes that are inherently time dependent. Furthermore, even when designing a steady-state process, time-dependent problems occur: For example, how does one *start* one of the elegant processes we designed in Chapter 2? A member of the faculty at a leading U.S. university tells an interesting anecdote. When he was a design engineer employed at a major oil company he designed a continuous reactor that was constructed but could not be started. Furthermore, this happened twice! The hard-earned lesson is that *a thorough understanding of transients is required to start, stop, and control a continuous process.*

A nonsteady process, or transient process, is one where at least one of the process variables changes with time. Modeling a transient process usually involves a differential equation. Since your calculus course has now prepared you to deal with simple differential equations, you can now appreciate the engineering aspects of transient phenomena.

For the last chapter of this book we are going to apply the three methods presented in the previous chapters: mathematical modeling, graphical analysis, and dimensional analysis. We will begin with mathematical modeling based upon fundamental and constitutive laws: the fundamental laws are chiefly conservation of mass, energy, and momentum. Momentum is a vector quantity and the mathematics that describe it are largely beyond the scope of this text. Mass and energy are scalar quantities (at least in the absence of special and general relativistic effects) and are the fundamental dependent variables around which we build our conservation laws. For the moment we will separate problems in which mass is conserved from problems in which energy is conserved: Proper modeling of energy conservation requires engineering thermodynamics. In most real situations models for conservation of both mass and energy are developed simultaneously, leading to more complicated mathematics, which we will defer to later courses.

The fundamentals used in the construction of our transient mathematical models derive from nineteenth-century physics, such as the conservation of mass and energy. The antiquity of the physics may lead you to suspect that our task of analysis is trivial. If our goal was an expression for the mass or energy, then our task may indeed be easy. But more often we seek an expression for a measurable quantity that characterizes the state of the system. Examples of such characterizing variables are the height of a fluid in a tank, or the temperature in a reactor, or the pressure drop across a porous bed of catalyst. The difficulty in analyzing a system comes when we must express a fundamental variable – such as mass or energy – in terms of a characterizing variable. Thermodynamics provides succinct and pedagogically tractable relations between physical observables, such as temperature and pressure, and fundamental quantities, such as energy and mass. And as before, the choice of system boundaries can greatly simplify an analysis. Similarly, for transient processes we will find that the appropriate choice of time boundaries can greatly simplify an analysis.

We begin with a formal statement of a conservation principle:

> The rate of change of a quantity within a system equals the rate at which the quantity enters that system plus the rate at which that quantity is generated within the system minus the rate at which that quantity leaves the system and minus the rate at which that quantity is consumed within the system.

A convenient distillation of the above statement is

> The rate of accumulation equals input plus generation minus output minus consumption.

The previous chapters on mass conservation dealt with systems at steady state, namely when accumulation equaled zero. We now generalize mass conservation to include accumulation; we will discuss the generation and consumption terms later in the chapter.

6.1 The Basics – A Surge Tank

A surge tank is a simple system rich in transient phenomena and important engineering lessons. In its typical application, a surge tank may be used to average transients in a flow stream; the flow into the tank may surge and ebb, but the flow out of the tank will be stable. We will use the production of citric acid as a case study in the use of surge tanks, and in the suppression of transients in process design. The process flowsheet in Figure 6.1 shows the overall design for citric acid production; a fermentor is followed by an absorber that absorbs citric acid from the aqueous phase into the organic phase, thereby separating the product from the reaction mixture.

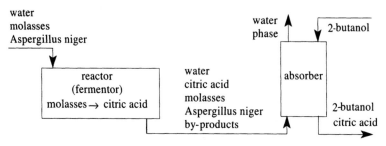

Figure 6.1. Citric acid produced in a fermentor and extracted in an absorber.

The process is more complicated than shown in Figure 6.1, because the flow out of the fermentor is incompatible with the flow required by the absorber. The fermentor produces citric acid in batches, whereas the absorber operates best as a continuous device. A surge tank allows us to connect batch fermentation with continuous separations, as shown in Figure 6.2.

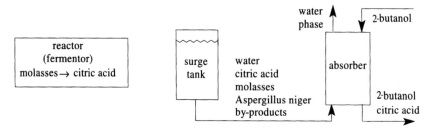

Figure 6.2. Batch fermentation takes place while the surge tank provides a steady flow to the absorber.

Upon completion of the fermentation process, the fermentor will empty its contents into the surge tank. The fermentor will then be refilled, and the process begins again (Figure 6.3).

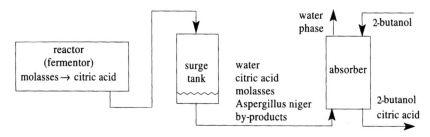

Figure 6.3. Fermentor empties its contents into the surge tank.

Figure 6.4 shows three graphs depicting the stream flow rates out of the fermentor and into the surge tank, the stream flow rates into the absorber, and the level of fluid in

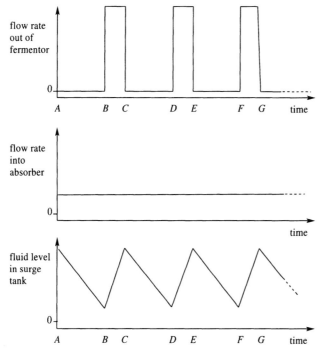

Figure 6.4. Stream flow rates out of the fermentor and into the surge tank (top), the stream flow rates into the absorber (middle), and fluid level in the surge tank (bottom).

the surge tank. Fermentation occurs during time A to time B; there is no flow out of the fermentor and the fluid level in the surge tank falls. The fermentor is emptied into the surge tank during time B to time C, and the level in the surge tank rises. The process repeats, refilling the fermentor and conducting fermentation from time C to time D, and draining the fermentor from D to E. The surge tank accepts the large transients arising from batch production and provides the steady flow required by the absorber. Of course, the surge tank must never overflow or empty completely. We also see the time scales of the transients that the surge tank must suppress. We now have a design problem: What specifications does the surge tank need to perform this function?

Let's consider just the surge tank, and apply some of the mathematical analysis tools we developed in Chapter 3. Consider a surge tank with volumetric flow rate in Q_{in} and volumetric flow rate out Q_{out}, as shown in Figure 6.5.

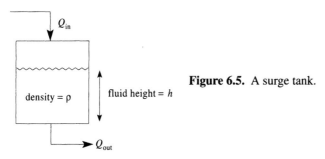

Figure 6.5. A surge tank.

Let us first analyze the simple scenario in which Q_{in} and Q_{out} are pumped at constant, but not necessarily equal, rates. Further assume that the cross-sectional area of the surge tank, A_{tank}, is constant along the height of the tank. Let us apply conservation principles to this surge tank to model its behavior. Specifically, we wish to model the height of the fluid as a function of time. Is there a fundamental law that concerns the fluid height, such as "conservation of height?" No. What is conserved? Mass is conserved. We must begin with mass and express in terms of height. To introduce the dimension of length, one can express the total mass in terms of the density, ρ, and the volume, V:

$$mass \,[=] \left(\frac{mass}{volume} \right) volume \,[=] (density)(volume), \tag{6.1}$$

$$M_{fluid} = \rho_{fluid} V_{fluid}. \tag{6.2}$$

Equation (6.2) requires uniform density throughout the tank, which is valid for many liquid systems; we could always add a stirring mechanism if we were concerned about density gradients. Because the cross-sectional area of the tank is constant, the volume of the fluid is the product of the fluid height, h, and A_{tank}:

$$V = A_{tank} h, \tag{6.3}$$

$$M_{fluid} = \rho_{fluid} A_{tank} h. \tag{6.4}$$

We can use Eq. (6.4) to translate an expression in terms of mass into an expression in terms of height.

We begin with the conservation of mass – *rate of accumulation equals input plus generation minus output minus consumption* – and translate into mathematics. What are the appropriate system boundaries? Answer: the tank. We associate the generation and consumption terms of our formal conservation statement with the reactants and products of chemical reactions. In the absence of any reactions, therefore, both the generation and consumption terms are zero. Like the total mass in the system, we express the mass flow rates F_{in} and F_{out} in terms of density and volumetric flow rates:

$$rate \, of \, mass \, flow \, in \,[=] \frac{mass}{time} \,[=] \left(\frac{mass}{volume} \right) \left(\frac{volume}{time} \right)$$

$$= (density \, of \, incoming \, fluid) \times (volumetric \, flow \, rate \, in), \tag{6.5}$$

$$F_{in} = \rho_{in} Q_{in}. \tag{6.6}$$

Similarly we have

$$F_{out} = \rho_{out} Q_{out}. \tag{6.7}$$

The accumulation is the rate at which total mass in the system changes with time,

$$rate \, of \, accumulation = \frac{d M_{tank}}{dt}, \tag{6.8}$$

which can be rewritten in terms of h using Eq. (6.4),

$$\frac{dM_{\text{tank}}}{dt} = \frac{d}{dt}(\rho_{\text{fluid}} A_{\text{tank}} h) = \rho_{\text{fluid}} A_{\text{tank}} \frac{dh}{dt}. \tag{6.9}$$

We can combine the terms in the conservation statement

$$\frac{dM_{\text{tank}}}{dt} = F_{\text{in}} - F_{\text{out}} = \rho_{\text{in}} Q_{\text{in}} - \rho_{\text{out}} Q_{\text{out}} \tag{6.10}$$

and arrive at the equation

$$\rho_{\text{fluid}} A_{\text{tank}} \frac{dh}{dt} = \rho_{\text{in}} Q_{\text{in}} - \rho_{\text{out}} Q_{\text{out}}. \tag{6.11}$$

We assume further that the density of the fluid entering the tank is equal to that exiting the tank, and equal to that in the tank,

$$\rho_{\text{fluid}} = \rho_{\text{in}} = \rho_{\text{out}}, \tag{6.12}$$

which simplifies Eq. (6.11) to

$$A_{\text{tank}} \frac{dh}{dt} = Q_{\text{in}} - Q_{\text{out}}. \tag{6.13}$$

We now solve Eq. (6.13). Its solution is simplified by the condition that Q_{in} and Q_{out} are constant, which allows one to move all variables dependent on h to one side of the equation and all variables dependent on t to the other side. (The constants may reside on either side.) Thus

$$A_{\text{tank}} \, dh = (Q_{\text{in}} - Q_{\text{out}}) \, dt. \tag{6.14}$$

The differential Eq. (6.13) is thus *separable* and as such is easily solved by integration:

$$A_{\text{tank}} \int dh = (Q_{\text{in}} - Q_{\text{out}}) \int dt, \tag{6.15}$$

$$A_{\text{tank}} h + (a \, constant) = (Q_{\text{in}} - Q_{\text{out}})t + (another \, constant). \tag{6.16}$$

Combining the constants and solving for h yields

$$h = \frac{1}{A_{\text{tank}}}[(Q_{\text{in}} - Q_{\text{out}})t + (a \, constant)]. \tag{6.17}$$

This formula for the fluid height as a function of time contains an unknown constant; can we eliminate this from our analysis? Let's exploit the fact that our analysis tells us that height is linearly proportional to time; Equation (6.17) defines a straight line for which we know the slope, but not the intercept. To uniquely determine the line on a plot of h versus t we must specify one point on the line, and thus we need to know h at a given time to determine the constant in Eq. (6.17). This datum is an example of an *initial condition*. The solution of differential equations containing a derivative with respect to time may be solved completely only when an initial condition is specified. Similarly, differential equations that contain a derivative with respect to position may be solved completely only if the conditions of the system are specified at a geometric boundary – hence the term *boundary condition*. For the surge tank we will specify a condition at a specific time; we arbitrarily designate time such that at $t = 0$ the height of the fluid in the tank h is equal to h_0. Substituting into Eq. (6.17) we have

$$h_0 = \frac{1}{A_{\text{tank}}}[(Q_{\text{in}} - Q_{\text{out}}) \times 0 + (a\ constant)], \tag{6.18}$$

$$A_{\text{tank}}h_0 = a\ constant. \tag{6.19}$$

Thus

$$h = \frac{1}{A_{\text{tank}}}(Q_{\text{in}} - Q_{\text{out}})t + h_0. \tag{6.20}$$

A more efficient juncture to introduce the initial condition(s) is before integration, Eq. (6.15). We create two definite integrals using the initial condition $h = h_0$ at $t = 0$ for the lower limits of each integral,

$$A_{\text{tank}} \int_{h_0} dh = (Q_{\text{in}} - Q_{\text{out}}) \int_0 dt, \tag{6.21}$$

and the unknown h at some later t as the upper limits,

$$A_{\text{tank}} \int_{h_0}^{h} dh = (Q_{\text{in}} - Q_{\text{out}}) \int_0^{t} dt. \tag{6.22}$$

Equation (6.22) represents an evolution from a certain height h_0 at time $= 0$ to a different h at a later time. Paying careful attention to the limits of integration we have

$$A_{\text{tank}}(h - h_0) = (Q_{\text{in}} - Q_{\text{out}})(t - 0), \tag{6.23}$$

which yields the same result as Eq. (6.20). Thus the height of fluid in the tank increases linearly with time if $Q_{\text{in}} - Q_{\text{out}} > 0$, decreases linearly with time if $Q_{\text{in}} - Q_{\text{out}} < 0$, and remains constant if $Q_{\text{in}} = Q_{\text{out}}$. These three mathematical conditions make intuitive sense: When $Q_{\text{in}} - Q_{\text{out}} > 0$ the rate of accumulation is positive and the surge tank fills up with fluid; when $Q_{\text{in}} - Q_{\text{out}} < 0$ the rate of accumulation is negative and the surge tank empties. The situation when $Q_{\text{in}} = Q_{\text{out}}$ is exactly at steady state, and the fluid height in the surge tank remains constant.

A surge tank drained by gravity poses a slightly more challenging problem. The flow rate of fluid through an orifice at the bottom is a function of the height of water in the tank, as shown in Figure 6.6. Qualitatively, one expects the flow rate to decrease as the fluid height decreases. Thus a model for a draining surge tank will include the relationship between the height of fluid in the tank and the flow rate through the hole in the bottom.

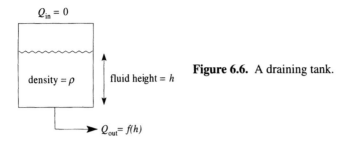

$Q_{in} = 0$

density $= \rho$ | fluid height $= h$

$Q_{out} = f(h)$

Figure 6.6. A draining tank.

Dimensional analysis may be used to obtain the relationship between h and Q_{out}. Using the techniques presented in Chapter 5, we first list the relevant parameters in Table 6.1.

Table 6.1. The parameters of a draining tank

Parameter	Variable	Dimensions
fluid height	h	L
volumetric flow rate	Q_{out}	L^3/T
orifice cross-sectional area	$A_{orifice}$	L^2
gravity	g	L/T^2

Upon arranging the exponents of the three key dimensions (mass, time, length) into algebraic equations and choosing the height h and Q_{out} as core variables we derive two dimensionless groups:

$$\Pi_1 = \frac{h^2}{A_{orifice}} \quad \text{and} \quad \Pi_2 = \frac{Q_{out}^2}{g A_{orifice}^{5/2}}. \tag{6.24}$$

Thus we know there exists a functional relationship of the form

$$\Pi_1 = f(\Pi_2), \tag{6.25}$$

$$\frac{h^2}{A_{orifice}} = f\left(\frac{Q_{out}^2}{g A_{orifice}^{5/2}}\right). \tag{6.26}$$

One then goes to the laboratory and measures Q_{out} as a function of $A_{orifice}$ and h. A plot of the data reveals a simple relationship:

$$Q_{out} = 0.6 A_{orifice}(gh)^{1/2}. \tag{6.27}$$

This is the so-called *orifice equation*, a constitutive law that relates h and Q_{out}.

Using this expression in our mass balance equation, along with the fact that $Q_{in} = 0$, we have

$$\frac{dh}{dt} = \frac{1}{A_{tank}}(Q_{in} - Q_{out}) \quad \text{with} \quad Q_{in} = 0, \tag{6.13}$$

$$\frac{dh}{dt} = -\frac{1}{A_{tank}}(0.6 A_{orifice}(gh)^{1/2}). \tag{6.28}$$

Although this differential equation is more complicated than Eq. (6.13), it is still separable. Upon separation of variables we have

$$h^{-1/2} \, dh = -0.6 \frac{A_{\text{orifice}}}{A_{\text{tank}}} g^{1/2} \, dt \tag{6.29}$$

and we can integrate both sides:

$$\int_{h_0}^{h} h^{-1/2} \, dh = -0.6 \frac{A_{\text{orifice}}}{A_{\text{tank}}} g^{1/2} \int_{0}^{t} dt. \tag{6.30}$$

Equation (6.30) uses the same initial conditions as before. We have moved the constants out of the integral over time. Evaluating the integrals yields

$$2\left(h^{1/2} - h_0^{1/2}\right) = -0.6 \frac{A_{\text{orifice}}}{A_{\text{tank}}} g^{1/2}(t - 0), \tag{6.31}$$

$$h = \left[h_0^{1/2} - 0.3 \frac{A_{\text{orifice}}}{A_{\text{tank}}} g^{1/2} t\right]^2. \tag{6.32}$$

Equation (6.32) may be generalized using dimensional consistency and scaling. Rather than pose a problem specific to this tank, one may pose a problem applicable to all draining tanks. How does one eliminate the information specific to this problem to solve a more general problem?

As we discussed in Chapter 5, systems may be rendered dynamically similar by converting to a dimensionless form. A step toward solving the more general problem is to recognize that the ratio h/h_0 is dimensionless and represents the fraction of water remaining in the tank. We define

$$x \equiv \textit{fraction of water remaining in tank} = \frac{h}{h_0}, \tag{6.33}$$

and it follows that

$$\frac{dx}{dh} = \frac{d}{dh}\left(\frac{h}{h_0}\right) = \frac{1}{h_0}. \tag{6.34}$$

Thus

$$dh = h_0 \, dx. \tag{6.35}$$

We also introduce a dimensionless time. This is not as straightforward as the dimensionless height because no other variable has dimensions of time. The gravitational constant g has dimensions of L/T^2, and thus $(h_0/g)^{1/2}$ has dimensions of time. We thus define

$$\tau \equiv \textit{reduced time} = \left(\frac{g}{h_0}\right)^{1/2} t, \tag{6.36}$$

and it follows that

$$\frac{d\tau}{dt} = \frac{d}{dt}\left(\frac{g}{h_0}\right)^{1/2}t = \left(\frac{g}{h_0}\right)^{1/2}. \tag{6.37}$$

Thus

$$dt = \left(\frac{h_0}{g}\right)^{1/2}d\tau. \tag{6.38}$$

Finally we define a dimensionless ratio of the relative cross-sectional areas of the tank and the orifice:

$$\alpha \equiv \textit{relative areas of tank and orifice} = \frac{A_{\text{orifice}}}{A_{\text{tank}}}. \tag{6.39}$$

Returning to our mass balance expression,

$$\frac{dh}{dt} = -\frac{1}{A_{\text{tank}}}(0.6A_{\text{orifice}}(gh)^{1/2}), \tag{6.28}$$

we substitute the dimensionless variables from Eqs. (6.33), (6.35), (6.38), and (6.39) to yield

$$\frac{h_0\,dx}{\left(\frac{h_0}{g}\right)^{1/2}d\tau} = -0.6\alpha(gh_0 x)^{1/2}, \tag{6.40}$$

which simplifies to

$$\frac{dx}{d\tau} = -0.6\alpha x^{1/2}. \tag{6.41}$$

Again we separate variables, which gives

$$x^{-1/2}\,dx = -0.6\alpha\,d\tau. \tag{6.42}$$

We now integrate, and apply the initial condition $h = h_0$ at $t = 0$, which in dimensionless quantities corresponds to $x = 1$ at $\tau = 0$:

$$\int_1^x x^{-1/2}\,dx = -0.6\alpha \int_0^\tau d\tau. \tag{6.43}$$

Solving the integrals and substituting the limits gives

$$2(x^{1/2} - 1^{1/2}) = -0.6\alpha(\tau - 0), \tag{6.44}$$

which can be rearranged to obtain

$$x = (1 - 0.3\alpha\tau)^2. \tag{6.45}$$

.cts that the fraction of water remaining in *any similar tank will*
. Notice also that when posed in a dimensionless form the equation is
..nd thus is easier to manipulate mathematically.

We return now to the design problem illustrated in Figure 6.4. If we choose a gravity-drained surge tank before our absorber, then the flow rate into the absorber will not be constant: Equation (6.27) tells us it will vary as the square root of fluid level. Thus we must use a pump, or some other controlling device, after the surge tank to guarantee constant absorber input.

Any mathematical model is subject to the limits imposed by its founding assumptions. For example, when we invoked the orifice equation we implicitly assumed that the fluid height in the tank preserved the physics of the draining process. When the tank is nearly empty the physics will be different – a whirlpool of air will form in the outlet, for example. Thus at long times the orifice equation, and subsequently Eq. (6.45), will not be valid. Similarly, a tank with a closed top, with variable cross-sectional area (such as a funnel), or density gradients in the fluid requires different mathematical models.

This example of the draining tank was adapted from the textbook by Russell and Denn (1972). You are encouraged to study the other analyses of dynamic systems in their textbook.

6.2   Residence Times and Sewage Treatment

Consider a mathematical model for a more complex physical situation where the fluid streams contain solutes. This situation is common in industrial practice, for example in the treatment of residential sewage.

A common method for treating sewage exposes residential sewage to bacteria that digest the organic material into CO_2 and water. Consider the system diagrammed in Figure 6.7. In this simplified process, sewage is pumped into a well-mixed aeration tank and $[\text{bacteria}]_{\text{tank}} = [\text{bacteria}]_{\text{treated}} = 8$ g bacteria/gal sewage. For the settling

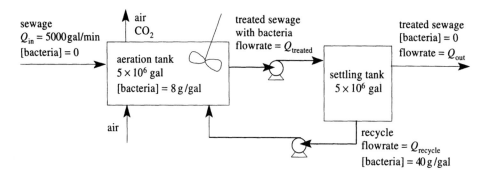

Figure 6.7. A process for treating residential sewage.

tank, the bacteria concentrations in the outlet streams differ from the average concentration in the tank. That's the purpose of the settling tank – to recycle bacteria. In this example the *design* was based on a steady-state analysis; the *operation*, however, will be plagued with transients.

Let's analyze the steady-state operation to determine the volumetric flow rates $Q_{treated}$ and $Q_{recycle}$. We will assume that (1) the sewage density is independent of the bacteria concentration or organic matter concentration, (2) the process is at steady state, (3) the aeration tank is well mixed, (4) the flow rate of the inlet air is equal to the flow rate of the outlet air plus CO_2, and (5) the growth rate of the bacteria is zero (this is controlled by the flow rate of air to the aeration tank).

Given that there are two unknowns, $Q_{treated}$ and $Q_{recycle}$, you should suspect that you need two independent equations. A mass balance on the entire system is trivial, namely that 5,000 gallons per minute enter and leave the treatment facility. We recognize, however, that in addition to the overall mass balance, we can apply the conservation of mass principle to the *components* that make up the fluid streams. Thus we can write a mass balance for the bacteria as well as for the total mass.

Let's choose the system boundaries to be the aeration tank because we have the most information about this unit. We continue to use the symbol F for the rate of total mass input and Q for volumetric flow rates; applying the conservation of mass to the aeration tank at steady state yields

$$F_{total, in} + F_{total, recycle} = F_{total, treated}, \tag{6.46}$$

$$Q_{in}\rho_{in} + Q_{recycle}\rho_{recycle} = Q_{treated}\rho_{treated}. \tag{6.47}$$

Because we assumed that the densities are equal and we are given that $Q_{in} = 5,000$ gal/min,

$$5,000 + Q_{recycle} = Q_{treated}. \tag{6.48}$$

A steady-state bacteria balance around the aeration tank yields

$$F_{bacteria, in} + F_{bacteria, recycle} = F_{bacteria, treated}, \tag{6.49}$$

$$Q_{in}[bacteria]_{in} + Q_{recycle}[bacteria]_{recycle} = Q_{treated}[bacteria]_{treated}, \tag{6.50}$$

$$0 + Q_{recycle}(40) = Q_{treated}(8), \tag{6.51}$$

$$5Q_{recycle} = Q_{treated}. \tag{6.52}$$

Equations (6.48) and (6.52) can be solved by substitution to yield

$$Q_{recycle} = 1,250 \text{ gal/min} \quad \text{and} \quad Q_{treated} = 6,250 \text{ gal/min}. \tag{6.53}$$

Now imagine that the pump on the recycle stream fails. As the managing engineer for this process, you must take action. For example, you must predict how long before the sewage treated by the process is unsafe to release. That is, when will the bacteria

...eration tank decrease to a level that the organic matter is not ...ted?

...st priority is to control the transients. What problem should be addressed Without the recycle, the level in the aeration tank will fall and the level in the settling tank will rise. Your first objective should be to maintain the levels in each tank. How does one drain the aeration tank more slowly and fill the settling tank more slowly? The best solution is to throttle the pump for the $Q_{treated}$ stream down to 5,000 gallons per minute from 6,250 gallons per minute.

We expect the bacteria concentration in the aeration tank to fall with a concomitant rise in the level of sewage in the discharge, as sketched in Figure 6.8. Of course, neither change will necessarily be linear with time. Figure 6.8 is only a qualitative sketch.

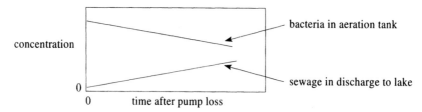

Figure 6.8. Qualitative expectations for the concentrations of sewage and bacteria after the failure of the recycling pump.

We need to know how long we can continue to treat the sewage. Thus we need to know the transient behavior of the bacteria concentration in the aeration tank and the sewage concentration in the settling tank. After consulting with various bacteriologists and treatment experts, we find that a minimum level of 4 g bacteria per gallon in the aeration tank is necessary to assure safe levels of sewage in the discharge.

Now we have a transient problem with regards to the bacteria concentration in the treatment tank. Clearly we need to rewrite our mass conservation expression for bacteria, this time including the accumulation term. We thus have

$$\frac{d}{dt}([\text{bacteria}]_{\text{tank}} V) = [\text{bacteria}]_{\text{in}} Q_{\text{in}} - [\text{bacteria}]_{\text{treated}} Q_{\text{treated}}, \quad (6.54)$$

where V is the volume of the aeration tank. Recall that the aeration tank is well mixed, so $[\text{bacteria}]_{\text{tank}} = [\text{bacteria}]_{\text{treated}}$. Since density is constant, and $[\text{bacteria}]_{\text{in}} = 0$, and because the volume is stabilized by throttling the Q_{treated} pump, we have

$$\frac{d}{dt}[\text{bacteria}] = -\left(\frac{Q_{\text{treated}}}{V}\right)[\text{bacteria}]. \quad (6.55)$$

Upon separating the variables dependent on concentration from those dependent on time we get

$$\frac{d[\text{bacteria}]}{[\text{bacteria}]} = -\left(\frac{Q_{\text{treated}}}{V}\right) dt. \quad (6.56)$$

Just as with the case of the draining tank, we now must incorporate the *initial conditions*. Let us define $t = 0$ as the time the pump failed. At $t = 0$ the bacteria concentration in the tank was 8 g bacteria/gal sewage. We are interested in calculating the time t_{bad} when the bacteria concentration falls to 4 g bacteria/gal sewage. We can now set the limits and integrate Eq. (6.56):

$$\int_8^4 \frac{d[\text{bacteria}]}{[\text{bacteria}]} = -\left(\frac{Q_{\text{treated}}}{V}\right) \int_0^{t_{bad}} dt. \tag{6.57}$$

Upon integrating and substituting the limits, we obtain

$$\ln(4) - \ln(8) = -\frac{Q_{\text{treated}}}{V}(t_{bad} - 0), \tag{6.58}$$

$$t_{bad} = -\frac{V}{Q_{\text{treated}}} \ln \frac{4}{8} = 700 \text{ min.} \tag{6.59}$$

Analysis of this sewage treatment process is interesting for three reasons. First, the choice of the system boundaries requires some judgment (in this case, the unit for which you have the most data). Second, it represents an actual problem that a BS chemical engineer related to us. Third, it demonstrates that operating a process requires an understanding of both transient and steady-state modeling.

Equations (6.55)–(6.59) contain an interesting ratio of quantities, the flow rate out of the tank divided by the fluid volume in the tank, Q_{treated}/V. It has dimensions of reciprocal time:

$$\frac{Q_{\text{treated}}}{V} = \frac{volumetric\ flow\ rate}{fluid\ volume} [=] \frac{L^3/T}{L^3} [=] \frac{1}{T}. \tag{6.60}$$

The inverse of this ratio indicates the average period of time a molecule resides in the tank and is called the *residence time* of water in the tank:

$$\frac{fluid\ volume}{volumetric\ flow\ rate} \equiv residence\ time \equiv \tau. \tag{6.61}$$

Returning to Eq. (6.58) we can write a general equation that relates the concentration, C, in terms of the initial concentration, C_0,

$$\ln\left(\frac{C}{C_0}\right) = -\frac{1}{\tau}t. \qquad \bullet \tag{6.62}$$

Solving for the concentration, we get

$$C = C_0 \exp(-t/\tau). \tag{6.63}$$

Equation (6.63) tells us that in all problems of this type the concentration of solute will fall to zero *exponentially*. The time scale of the exponential decrease is governed by the residence time of the system. Figure 6.9 shows the bacteria concentration in the aeration tank as a function of time.

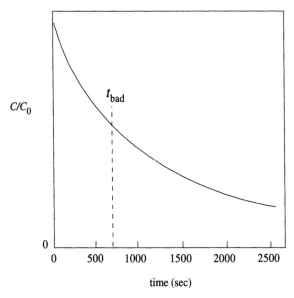

Figure 6.9. The decay of concentration as a function of time.

The exponential term on the right-hand side of (6.63) is a number that multiplies the initial concentration C_0. Consider the magnitude of the exponential for various increments of the ratio t/τ. The data in Table 6.2 suggest that to an accuracy of about 5%, a period of about three residence times must pass for the bacteria concentration to decay to zero.

Table 6.2. Exponential decay

t/τ	$\exp(-t/\tau)$
0.1	0.90
0.5	0.61
1	0.37
2	0.14
3	0.05
5	0.007

It is useful to compare the residence time to the dimensionless time introduced in the draining tank problem. For the draining tank the quantity $(h_0/g)^{1/2}$ has dimensions of time and also forms a characteristic residence time. In this case the characteristic time reflects the time for the tank to drain. Indeed, if you use Eq. (6.44) and set x equal to zero, one finds that the tank empties in a time $(h_0/g)^{1/2}/0.3\alpha$.

Let us generalize from our study of the sewage treatment plant, the surge tank, and the draining tank. Characteristic time scales describe the dynamics of chemical engineering systems. In the case of well-stirred tanks, the characteristic time is the *residence* time, given by the ratio of the tank volume to the characteristic output (or input) flow rate. For draining tanks we found that the characteristic time is the time to drain the tank. The effects of transients on tanks can be estimated by comparing the time scale of the perturbation to the characteristic time; input fluctuations rapid with respect to the characteristic time have little effect. Input fluctuations having time scales long with respect to the characteristic time affect the output of the tank; thus the tank serves no purpose. A surge tank's residence time must be longer than time scales of the anticipated fluctuations. This is a key design criterion for a transient process.

The relationship among what goes in, what comes out, and what fluctuates within a system is called *process control*. Chemical engineering students are required to study this subject in some detail, usually as a course and laboratory in the latter half of the undergraduate curriculum. Prerequisite is a thorough understanding of chemical processes, as well as second-year college calculus and differential equations.

6.3 Rate Constants – Modeling Atmospheric Chemistry

Anthropogenic activities have prompted many scientists and engineers to look at the effects of chemicals on the composition of the Earth's upper atmosphere. Of particular concern is the role certain chemicals play in creating or destroying stratospheric ozone, O_3. Ozone in the stratosphere is important to life on Earth because it absorbs solar radiation in the *ultraviolet* (*uv*) portion of the electromagnetic spectrum.[1] This high-energy portion of the solar spectrum is capable of indiscriminately breaking chemical bonds and thus can be harmful to plant and animal life on Earth; degradation of this layer is of international concern.

One occasionally hears that man-made ozone should be used to replace the ozone destroyed by chemical reaction with man-made pollutants. This viewpoint (that is, using chemical engineering on a global scale to remedy a pollution problem) is naive. Approximately 10^{13} watts of solar radiation are used in the production of Earth's natural ozone layer; this is three times the total power produced by humans. In other words, replacing 10% of the stratospheric ozone would take about one third of all the power plants on Earth! Assuming one could solve the power requirement, one must also engineer a means of delivering the ozone to the stratosphere, with the constraint that ozone spontaneously explodes at low concentrations. The present international strategy is to ban production of the chemicals that react with ozone in the upper atmosphere.

It is difficult to conduct experiments in the upper atmosphere, and thus our understanding of transport and chemistry in the stratosphere relies on modeling. These models use sophisticated computer algorithms that include hundreds of chemical reactions, diffusion of different species throughout the atmosphere, and fluctuating energy input from the solar radiation. To use these models we need to know the rate that chemicals are generated and consumed in the atmosphere (see Figure 6.10).

---

[1] Ozone in the troposphere, down where we live, is harmful to human health and is to be avoided.

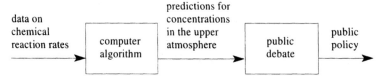

Figure 6.10. A process to convert technical data into public policy.

An important chemical reaction is the decomposition of nitrous oxide into oxygen and nitrogen:

$$2N_2O \rightarrow 2N_2 + O_2. \tag{6.64}$$

Nitrous oxide contributes to ozone depletion through photochemical reactions in the stratosphere. It reacts with an oxygen atom to form NO; N_2O is the chief source of NO in the stratosphere through the reaction

$$N_2O + O \rightarrow 2NO. \tag{6.65}$$

NO catalyzes the removal of ozone in the upper stratosphere through the cycle in reactions (6.66) and (6.67), the net result of which is reaction (6.68):

$$NO + O_3 \rightarrow NO_2 + O_2 \tag{6.66}$$

$$\underline{NO_2 + O \rightarrow NO + O_2} \tag{6.67}$$

$$\text{net: } O + O_3 \rightarrow 2O_2 \tag{6.68}$$

Nitrous oxides are formed naturally by decomposition of nitrogen compounds in the soil. N_2O is also a man-made chemical used as a propellant for food spray cans and as an anesthetic. It can also be a by-product of pollution control devices installed on smokestacks at power plants. In any case, man-made nitrous oxides formed in the lower part of the atmosphere diffuse to the upper atmosphere where they contribute to ozone depletion. Knowledge of the rate of reaction (6.64) would help us model how long nitrous oxide survives in the atmosphere.

Our goal is to design a chemical reactor that measures the rate of reaction (6.64). Our chemical reactor will be a simple "batch" reactor (*i.e.*, one that is ostensibly simple because it has no input or output as shown in Figure 6.11). For our purposes, it is a constant-volume, constant-temperature vessel. We presume that if we measure the concentration of chemicals in the reactor as a function of time we will be able to write a characteristic rate law for the rate of N_2O decomposition. Usually one would measure the pressure in the reactor to calculate the N_2O concentration in the reactor.

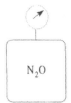

Figure 6.11. A batch reactor for studying the decomposition of N_2O.

However, we will simplify our analysis by assuming we have a spectrometer that can measure the N_2O concentration in the reactor.

For this reactor we will need model equations that describe the concentration of nitrous oxide with time. How will this system evolve with time? Clearly the N_2O concentration will fall and the concentrations of both N_2 and O_2 will rise such that the concentration of N_2 is double the concentration of O_2. (This latter fact comes from the stoichiometry of the reaction.) Because we have assumed reaction (6.64) is irreversible, ultimately the concentration of N_2O will fall to zero; we expect, then, something like Figure 6.12.

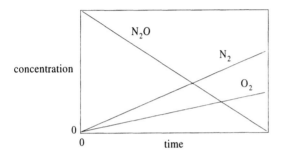

Figure 6.12. Qualitative expectations for the concentrations of N_2O, N_2, and O_2 as a function of time.

Our model equations will come from the principle of conservation of mass. We will apply the statement

rate of accumulation equals input plus generation minus output minus consumption

to this particular physical situation.

The total mass balance is trivial. With no input, output, generation, or consumption there is no accumulation. What we seek, however, is a conservation statement for the nitrous oxide inside the reactor. Thus we choose the reactor as the system boundaries and consider only the mass of nitrous oxide. There is neither input nor output of N_2O. The chemical reaction given by Eq. (6.64) tells us, however, that nitrous oxide is *consumed* by the reaction to produce nitrogen and oxygen. Thus we have

$input = 0,$

$output = 0,$

$generation = 0,$

consumption = an experession derived from reaction (6.64).

What is this expression for consumption? In other words, how do we articulate mathematically consumption by chemical reaction? We first define a variable for the rate a substance is consumed by chemical reaction. We seek an intensive variable, that is, a variable that does not depend on the size of the reactor. This will allow us to use the

327

rate in systems of different scale. We write

rate at which N_2O *is consumed by chemical reaction*

$$= (reaction\ rate) \times (reactor\ volume) = rV\ [=]\ \frac{moles}{time} \qquad (6.69)$$

and thus

$$r\ [=]\ \frac{moles}{(volume)(time)}. \qquad (6.70)$$

The rate of accumulation of nitrous oxide is the rate at which the amount of N_2O changes with time, which in mathematical terms is

$$rate\ of\ accumulation\ of\ N_2O = \frac{d}{dt}(moles\ of\ N_2O\ in\ the\ reactor) \qquad (6.71)$$

$$= \frac{d}{dt}([N_2O]V) \qquad (6.72)$$

$$= V\frac{d}{dt}[N_2O]. \qquad (6.73)$$

Our mass conservation expression for N_2O in the reactor may now be written:

$$V\frac{d}{dt}[N_2O] = 0 + 0 - 0 - rV, \qquad (6.74)$$

$$\frac{d}{dt}[N_2O] = -r. \qquad (6.75)$$

The differential equation (6.75) appears trivial until we realize that the reaction rate r depends on the N_2O concentration. In other words, we expect that the reaction rate of N_2O will be higher at higher N_2O concentrations, as shown qualitatively in Figure 6.13. We must now determine how the reaction rate depends on the N_2O concentration.

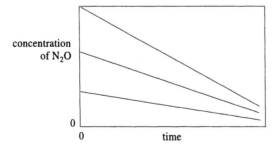

Figure 6.13. Qualitative expectation for the amount of N_2O as a function of time for different initial concentrations. The rate of decay, which is the slope of the line, is faster for higher initial concentrations.

Modern chemical physics uses principles of quantum mechanics and statistical mechanics to model reaction rates in terms of concentrations and temperature. Reactions in gases are caused by collisions between molecules. One can use statistics to argue that the reaction rate of N_2O is proportional to the rate of N_2O–N_2O collisions. The rate of N_2O–N_2O collisions is proportional to the square of the N_2O concentration. We *assume* that a nitrous oxide molecule must collide with another nitrous oxide molecule to react, so that

$$r = k[N_2O]^2, \tag{6.76}$$

where k is the *rate constant*. Although k is called a "constant" it is not a universal constant. It will almost certainly depend on temperature.

We use Eq. (6.76) to substitute for r in Eq. (6.75) and obtain

$$\frac{d}{dt}[N_2O] = -k[N_2O]^2. \tag{6.77}$$

Again we find that the differential equation may be solved by separation of variables:

$$\frac{d[N_2O]}{[N_2O]^2} = -k\,dt. \tag{6.78}$$

Now we implement an initial condition. At time $t = 0$ we designate the concentration of nitrous oxide to be $[N_2O]_0$, and at a time t later the concentration is $[N_2O]$. We add these limits to the integrals formed from Eq. (6.78):

$$\int_{[N_2O]_0}^{[N_2O]} \frac{d[N_2O]}{[N_2O]^2} = -k \int_0^t dt. \tag{6.79}$$

Evaluating the integrals we obtain

$$-\left(\frac{1}{[N_2O]} - \frac{1}{[N_2O]_0}\right) = -k(t - 0), \tag{6.80}$$

which we solve for $[N_2O]$:

$$[N_2O] = \frac{[N_2O]_0}{1 + kt[N_2O]_0}. \tag{6.81}$$

In a typical experiment, we would charge the batch reactor with N_2O and then measure the time required for the N_2O concentration to drop by 50%, the half-life, $t_{1/2}$. Half-life data for N_2O at 1,015 K are shown in Table 6.3.

Table 6.3. Half-lives for the decomposition of N_2O at 1,015 K

$[N_2O]_0$ (mol/L)	$t_{1/2}$ (sec)
0.135	1,060
0.286	500
0.416	344
0.683	209

Equation (6.81) may be used to derive a formula for $t_{1/2}$, the time for 50% conversion of nitrous oxide. When 50% of the nitrous oxide is reacted, then $[N_2O] = 0.5\,[N_2O]_0$, and (6.81) yields

$$t_{1/2} = \frac{1}{k[N_2O]_0}. \tag{6.82}$$

Thus a plot of $t_{1/2}$ versus $1/[N_2O]_0$ should yield a straight line with the slope of $1/k$. The data from Table 6.3 are plotted in Figure 6.14.

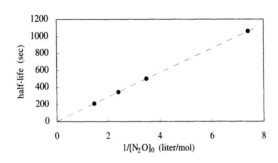

Figure 6.14. The half-life for the decomposition of N_2O as a function of the reciprocal of the concentration.

The straight line is obtained from a best fit of a straight line to the four data points. Because the fit to a straight line is good, our assumptions for the rate equation, Eq. (6.76), are justified. Using the slope of the line in Figure 6.10 (143 mol·s/L), and Eq. (6.82), we calculate the rate constant for reaction (6.64):

$$k = \frac{1}{\text{slope}} = \frac{1}{143 \, \text{mol·s/L}} = 0.007 \, \text{L/mol·s}. \tag{6.83}$$

Our batch reactor may be used to generate rate constants for a variety of temperatures and pressures. These laboratory-derived rate constants can subsequently be used in quantitative models for atmospheric chemistry. In other words, if we choose the upper atmosphere for the system boundary, then we may write a conservation of mass statement for nitrous oxide that looks like:

rate of accumulation equals input plus generation minus output minus consumption,

where at least one of the consumption terms is $rV = k[N_2O]^2$ and $k = 0.007$ L/mol·s at 1,015 K.

6.4 Optimization – Batch Reactors

Batch reactors are rarely used in the chemical industry to produce bulk commodity chemicals. In specialty chemical applications, however, batch reactors may be used because it is relatively easy and inexpensive to set up a synthesis on an occasional basis. The citric acid production described earlier is another example: Fermentation is inherently a batch process. Whereas the N_2O decomposition illustrates how batch reactors may be used to measure reaction rates, commercial applications of batch reactors often involve more complicated chemical systems, such as enzymatic reactions or reactions taking place in the presence of multiple phases. We now consider operating issues associated with producing a chemical in a batch reactor where the chemical is an intermediate in a series of reactions.

Suppose we wish to dehydrogenate n-butane, $CH_3CH_2CH_2CH_3$, to produce 1-butene, $CH_2=CHCH_2CH_3$, in a batch reactor:

$$CH_3CH_2CH_2CH_3 \rightarrow CH_2=CHCH_2CH_3 + H_2. \tag{6.84}$$

Suppose further that we want to avoid dehydrogenation of 1-butene to butadiene:

$$CH_2=CHCH_2CH_3 \rightarrow CH_2=CHCH=CH_2 + H_2. \tag{6.85}$$

Reactions (6.84) and (6.85) may each be assigned rates:

$$r_1 \equiv rate\ at\ which\ CH_3CH_2CH_2CH_3\ is\ consumed\ [=] \frac{moles}{(volume)(time)}, \tag{6.86}$$

$$r_2 \equiv rate\ at\ which\ CH_2=CHCH_2CH_3\ is\ consumed\ [=] \frac{moles}{(volume)(time)}. \tag{6.87}$$

The principle of conservation of mass may now be applied to the batch reactor; we apply

rate of accumulation equals input plus generation minus output minus consumption

to each component. For $CH_3CH_2CH_2CH_3$, the five terms in the above expression are

$$accumulation = \frac{d}{dt}[CH_3CH_2CH_2CH_3]V,$$

$$generation = 0,$$

$$input = 0, \tag{6.88}$$

$$output = 0,$$

$$consumption = -r_1 V.$$

Because volume is constant in our batch reactor, we combine the data in (6.88) to arrive at

$$\frac{d}{dt}[CH_3CH_2CH_2CH_3] = -r_1. \tag{6.89}$$

Applying mass conservation to each component we obtain three additional equations:

$$\frac{d}{dt}[CH_2{=}CHCH_2CH_3] = r_1 - r_2, \tag{6.90}$$

$$\frac{d}{dt}[H_2] = r_1 + r_2, \tag{6.91}$$

$$\frac{d}{dt}[CH_2{=}CHCH{=}CH_2] = r_2. \tag{6.92}$$

We now postulate rate expressions for r_1 and r_2. In the previous example we assumed the rate expression varied as the concentration of reactant squared; this is a *second-order* rate expression because the exponent 2 appears in the rate expression. For this case, however, we note that chemists have previously found that each decomposition is proportional to the amount of reactant. That is, reactions (6.84) and (6.85) are *first order*. The expressions for the rates r_1 and r_2 are thus given by

$$r_1 = k_1[CH_3CH_2CH_2CH_3], \tag{6.93}$$

$$r_2 = k_2[CH_2{=}CHCH_2CH_3]. \tag{6.94}$$

From Eqs. (6.89)–(6.92) we construct the differential equations governing the compositions in our reactor:

$$\frac{d}{dt}[CH_3CH_2CH_2CH_3] = -k_1[CH_3CH_2CH_2CH_3], \tag{6.95}$$

$$\frac{d}{dt}[CH_2{=}CHCH_2CH_3] = k_1[CH_3CH_2CH_2CH_3] - k_2[CH_2{=}CHCH_2CH_3], \tag{6.96}$$

$$\frac{d}{dt}[H_2] = k_1[CH_3CH_2CH_2CH_3] + k_2[CH_2{=}CHCH_2CH_3], \tag{6.97}$$

$$\frac{d}{dt}[CH_2{=}CHCH{=}CH_2] = k_2[CH_2{=}CHCH_2CH_3]. \tag{6.98}$$

The chemical composition in our reactor at any time may be determined by solving the system of differential equations (6.95)–(6.98). Equation (6.95) may be solved by separation and integration. The solution to Eq. (6.95) lies within the grasp of a student who is (by now) well into a first course in calculus. Separating into terms that depend on $[CH_3CH_2CH_2CH_3]$ on the left and terms that depend on time on the right, integrating, and adding limits gives us

$$\int_{[CH_3CH_2CH_2CH_3]_0}^{[CH_3CH_2CH_2CH_3]} \frac{d[CH_3CH_2CH_2CH_3]}{[CH_3CH_2CH_2CH_3]} = -\int_0^t k_1\, dt. \tag{6.99}$$

Performing the integration we get

$$\ln[CH_3CH_2CH_2CH_3] - \ln[CH_3CH_2CH_2CH_3]_0 = -k_1t, \qquad (6.100)$$

or

$$[CH_3CH_2CH_2CH_3] = [CH_3CH_2CH_2CH_3]_0 \exp(-k_1t). \qquad (6.101)$$

The concentration of $CH_3CH_2CH_2CH_3$ thus decreases exponentially with a time constant of $1/k_1$, as shown in Figure 6.15. Recall that this is the same behavior we predicted for the bacteria in the sewage tank, as shown in Figure 6.9.

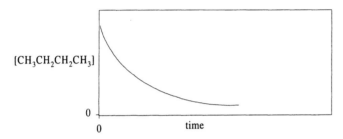

Figure 6.15. A sketch of the concentration of $CH_3CH_2CH_2CH_3$ as a function of time.

We substitute the expression for $[CH_3CH_2CH_2CH_3]$ from Eq. (6.101) into Eq. (6.96) to yield

$$\frac{d}{dt}[CH_2{=}CHCH_2CH_3] = k_1[CH_3CH_2CH_2CH_3]_0 \exp(-k_1t)$$
$$-k_2[CH_2{=}CHCH_2CH_3]. \qquad (6.102)$$

We need to solve Eq. (6.102) to optimize the production of $CH_2{=}CHCH_2CH_3$. However, Eq. (6.102) cannot be separated. The procedure to solve Eq. (6.102) is probably not in your repertoire. Throughout your career as an engineer you will encounter equations that are beyond your ability to solve analytically, even after four semesters of calculus. Indeed, some of the equations you will encounter will not have analytical solutions. Let's explore a method of approximating the time dependence of the concentration of $CH_2{=}CHCH_2CH_3$. We will follow the incremental change in $[CH_2{=}CHCH_2CH_3]$.

Consider the rate expression for $[CH_2{=}CHCH_2CH_3]$. Assuming that there is initially no $CH_2{=}CHCH_2CH_3$ in the reactor, at *short times* the expression (6.102) is approximately

$$\frac{d}{dt}[CH_2{=}CHCH_2CH_3] \approx k_1[CH_3CH_2CH_2CH_3]_0 \exp(-k_1t), \qquad (6.103)$$

which is greater than zero. A derivative greater than zero means the function grows with time; thus at short times $[CH_2{=}CHCH_2CH_3]$ increases, as sketched in Figure 6.16.

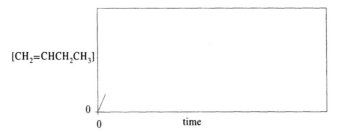

Figure 6.16. A sketch of the concentration of $CH_2{=}CHCH_2CH_3$ as a function of time for very short times.

However, the rate of increase of $[CH_2{=}CHCH_2CH_3]$ decreases because $\exp(-k_1/t)$ decreases with increasing time and the second term in Eq. (6.102) comes into play. This trend continues, as shown in Figure 6.17. Clearly the accuracy of these incremental estimates can be improved by decreasing the time increment.

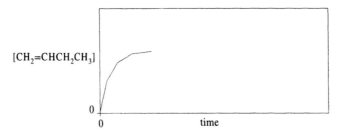

Figure 6.17. A sketch of the concentration of $CH_2{=}CHCH_2CH_3$ as a function of time for short times.

At long times, $CH_3CH_2CH_2CH_3$ is depleted, and thus $[CH_3CH_2CH_2CH_3] \approx 0$; the rate expression for $CH_2{=}CHCH_2CH_3$ becomes

$$\frac{d}{dt}[CH_2{=}CHCH_2CH_3] \approx -k_2[CH_2{=}CHCH_2CH_3], \tag{6.104}$$

which is less than zero. A derivative less than zero means the function decreases with time; thus the rate of decrease of $[CH_2{=}CHCH_2CH_3]$ decreases for longer times, because $[CH_2{=}CHCH_2CH_3]$ decreases. We add the long-time behavior for three increments to yield Figure 6.18. The behavior for intermediate times can be estimated by sketching a smooth curve between the estimates at short and long times.

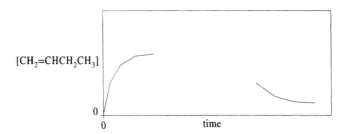

Figure 6.18. A sketch of the concentration of $CH_2{=}CHCH_2CH_3$ as a function of time for short and long times.

These limiting cases clearly suggest that $[CH_2=CHCH_2CH_3]$, the desired product, passes through a maximum. Recall that a function is at a maximum when its derivative equals zero. Thus the maximum $[CH_2=CHCH_2CH_3]$ is obtained by setting Eq. (6.102) equal to zero:

$$0 = k_1[CH_3CH_2CH_2CH_3]_0 \exp(-k_1 t_{max}) - k_2[CH_2=CHCH_2CH_3]_{max}, \quad (6.105)$$

$$[CH_2=CHCH_2CH_3]_{max} = \frac{k_1}{k_2}[CH_3CH_2CH_2CH_3]_0 \exp(-k_1 t_{max}). \quad (6.106)$$

And from Eq. (6.106) one obtains the time at which the maximum of $[CH_2=CHCH_2CH_3]$ occurs:

$$t_{max} = \frac{1}{k_1} \ln\left(\frac{k_1}{k_2}\frac{[CH_3CH_2CH_2CH_3]_0}{[CH_2=CHCH_2CH_3]_{max}}\right). \quad (6.107)$$

Thus if k_1 is large (the desired reaction (6.84) is fast) and k_2 is small (the undesired reaction (6.85) is slow), the maximum concentration of $CH_2=CHCH_2CH_3$ is comparable to the initial concentration of $CH_2=CHCH_2CH_3$, and the maximum is reached in short time. If the converse is true – if k_1 is small and k_2 is large – the maximum concentration of $CH_2=CHCH_2CH_3$ is small.

Finally, we sketch the time behavior of the undesired product, $CH_2=CHCH=CH_2$. From Eq. (6.98) we see that we can estimate $[CH_2=CHCH=CH_2]$ as a function of time from Figure 6.14. First, we see that the rate of increase of $[CH_2=CHCH=CH_2]$ is never negative. At short times, there is little $CH_2=CHCH_2CH_3$ in the reactor and thus the rate of increase of $CH_2=CHCH=CH_2$ is slow. At longer times, $CH_2=CHCH_2CH_3$ increases and the rate of increase of $CH_2=CHCH=CH_2$ increases. This trend is sketched in Figure 6.19.

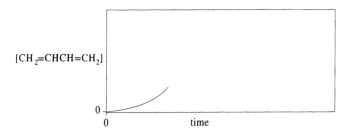

Figure 6.19. A sketch of the concentration of $CH_2=CHCH=CH_2$ as a function of time for short times.

At much longer times, $CH_2=CHCH_2CH_3$ falls to zero and the rate of increase of $CH_2=CHCH=CH_2$ falls to zero. In other words, at long times, $[CH_2=CHCH=CH_2]$ asymptotically reaches a maximum. This is sketched in Figure 6.20.

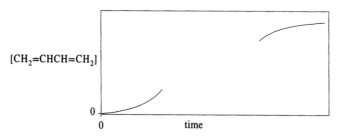

Figure 6.20. A sketch of the concentration of $CH_2=CHCH=CH_2$ as a function of time for short and long times.

Without solving differential Eqs. (6.95)–(6.98) we have predicted the qualitative concentrations of the reactant, intermediate, and undesired product in our batch reactor. Quantitative estimates using time steps done with computers is called *numerical integration*. An introduction to numerical integration is given in the exercises at the end of this chapter.

The species concentrations as a function of time for a typical reaction $A \rightarrow B \rightarrow C$, obtained either by numerical or analytical methods, yield a plot similar to Figure 6.21.

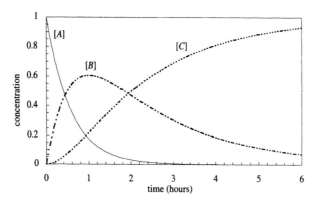

Figure 6.21. Concentrations of reactant A, product C, and intermediate B as a function of time for a series reaction conducted in a batch reactor.

There is clearly a time that the amount of compound B reaches a maximum, about one hour in this example. A chemical plant for the continuous production of compound B, then, would contain a large reactor that was charged with reactants, run for one hour, and then emptied into a surge tank, which feeds a separator. The reactor is subsequently recharged, and the process repeated.

6.5 Continuous Production – Plug Flow Reactors

Batch production is not compatible with the *steady-state* processes we designed in Chapter 2. We could insert a surge tank after a batch reactor, as was shown for citric acid production in Section 6.1. Instead, we could use a *plug flow reactor (PFR)* to produce compound *B* continuously and produce the same maximum concentration of *B* as the batch reactor. The plug flow reactor is so named because a *plug* of fluid moves along the reactor *without mixing with the plug in front of it or behind it*: It is a pipe. We will describe how it works using our understanding of the batch reactor.

Imagine a batch reactor on a conveyor belt, as shown in Figure 6.22. The reactor is filled with reactants and placed on the conveyor. The conveyor's speed is adjusted so that the batch reactor reaches the end just as the concentration of *B* reaches its maximum. As the batch reactor falls off the end of the conveyor, we catch it and empty its contents into our separations process. The reactor is then recharged and placed at the beginning of the conveyor.

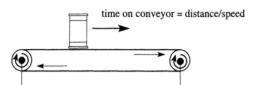

Figure 6.22. A batch reactor on a conveyor belt.

The flow from this conveyor reactor can be smoothed by adding more reactors as shown in Figure 6.23. If the conveyor is loaded with many, tiny reactors, the output is nearly continuous.

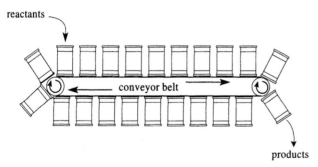

Figure 6.23. A continuous series of batch reactors on a conveyor belt.

The last step toward visualizing a plug flow reactor is to imagine that the batch reactors become infinitesimally small and infinitely numerous. The conveyor is replaced by a pipe, and the conveyor belt is replaced by a solvent that contains the reactants. Each "plug" of liquid moving down the pipe is the equivalent of a batch reactor on the conveyor. The plugs are infinitesimally thin and thus infinitely numerous. Such a plug flow reactor is shown in Figure 6.24. The time each plug spends in the pipe, or residence time, is the pipe volume divided by the volumetric flow rate. The volumetric flow rate is adjusted so the time in the pipe maximizes the amount of

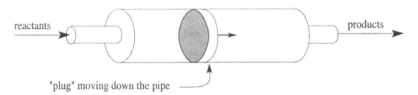

reactants

products

"plug" moving down the pipe

Figure 6.24. The plug flow reactor.

intermediate B. Or, given a desired volumetric flow rate, the volume of the plug flow reactor is designed to achieve the desired residence time.

Chemical engineering students take a course entitled *reaction engineering* during their junior year; this course teaches both analytical and graphical methods for designing plug flow reactors for complex reaction kinetics and heterogeneous physical systems.

6.6 Multiple Steady States – Catalytic Converters

An automobile catalytic converter is also a *continuous reactor* that expedites three different chemical reactions. First, it catalyzes the reduction of NO (with the concomitant oxidation of CO):

$$NO + CO \rightarrow \frac{1}{2}N_2 + CO_2. \tag{6.108}$$

The CO in excess from this reaction is oxidized via

$$CO + \frac{1}{2}O_2 \rightarrow CO_2 \tag{6.109}$$

as are hydrocarbons via

$$hydrocarbons + O_2 \rightarrow CO_2 + H_2O. \tag{6.110}$$

The three-way catalyst for these reactions is a formulation of metal particles supported on a metal-oxide surface. In this section we will study the design and operation of a reactor to conduct reaction (6.108).

Reactions at catalytic surfaces are characterized by unusual rate expressions. Unlike simple reactions in liquids, where the reaction rate increases with reactant concentration, the rate for a surface reaction often depends on reactant concentration in the manner shown in Figure 6.25. The features in Figure 6.25 arise from the mechanism of the surface-catalyzed reaction; for CO to react, both CO and the coreactant NO must be adsorbed on the surface of the solid catalyst. At low [CO], increasing [CO] increases the amount of CO on the surface and thus increases the rate of reaction of CO. But above a critical concentration of CO, too much CO adsorbs on the surface and it excludes the adsorption of NO. Thus the rate of reaction decreases at high [CO].

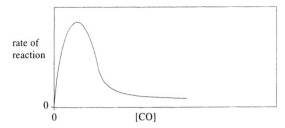

Figure 6.25. Rate of a surface-catalyzed reaction as a function of reactant concentration.

The reaction comprises a series of elementary steps involving CO, NO, and adsorption sites, designated S:

$$CO + S \leftrightarrow CO\!-\!S, \tag{6.111}$$

$$NO + S \leftrightarrow NO\!-\!S, \tag{6.112}$$

$$CO\!-\!S + NO\!-\!S \rightarrow \frac{1}{2}N_2 + CO_2 + 2S. \tag{6.113}$$

The double arrows indicate reversible reactions. Because these reversible adsorption/desorption reactions (6.111) and (6.112) are much faster than the surface reaction (6.113), the reversible reactions are at equilibrium. The mechanism of reactions (6.111)–(6.113) leads to the rate expression

$$\textit{rate of reaction} = \frac{k_1[CO][NO]}{(1 + K_1[CO] + K_2[NO])^2}, \tag{6.114}$$

where k_1 is the rate constant of the surface reaction, and K_1 and K_2 are the equilibrium constants for the adsorption/desorption reactions. A qualitative plot of Eq. (6.114) is given in Figure 6.25.

As noted in the previous section, most chemical reactors are *continuous*, unlike the batch reactor. There are several types of continuous reactors, including the PFR. Another common industrial reactor is the Continuous Stirred-Tank Reactor (CSTR) diagrammed in Figure 6.26. Because the contents of a CSTR are *well stirred*, the composition of the effluent is the same as the composition at every point inside the reactor. Uniform composition is a key assumption in the analysis of a CSTR.

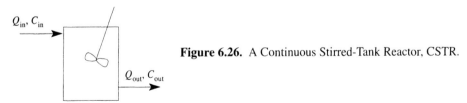

Figure 6.26. A Continuous Stirred-Tank Reactor, CSTR.

Let us use a CSTR to catalytically reduce NO with CO, reaction (6.108). The catalytic converter in your car is *not* a CSTR; it is a PFR. Our analysis will suggest one reason *why* it is not a CSTR.

We apply the mass conservation principle to the CO in the reactor. First list the five terms in the time-dependent mass balance on CO:

$$input = [CO]_{in} Q_{in},$$

$$output = [CO]_{out} Q_{out},$$

$$generation = 0,$$

(6.115)

$$consumption = rV,$$

$$accumulation = \frac{d}{dt}[CO]_{reactor} V.$$

Combining these terms our mass balance for CO is given by

$$\frac{d}{dt}[CO]_{reactor} V = [CO]_{in} \dot{Q}_{in} - [CO]_{out} Q_{out} - rV.$$

(6.116)

At *steady state* the derivative with respect to time equals zero and $Q_{in} = Q_{out} \equiv Q$, which yields

$$[CO]_{in} - [CO]_{out} = r\frac{V}{Q}.$$

(6.117)

Recall that the ratio V/Q is the residence time for a molecule in the reactor; thus we write

$$[CO]_{in} - [CO]_{out} = r\tau.$$

(6.118)

Thus the difference in concentration between inlet and outlet equals the rate of reaction times the residence time. This makes sense. If the rate of reaction is increased, for example by increasing the temperature, Eq. (6.118) predicts that the difference between inlet and outlet concentrations should increase. If the reactor is enlarged, the residence time increases and the difference between inlet and outlet concentrations should increase.

Equation (6.114) provides a mathematical expression for the rate of reaction, r, in terms of the concentration of [CO]. But rather than solve this nonlinear algebraic equation analytically (*i.e.*, derive an expression $[CO]_{out} = \ldots$) we will solve Eq. (6.118) graphically, as follows. First, we plot the left side of Eq. (6.118) as a function of $[CO]_{reactor}$ in Figure 6.27. Recall that $[CO]_{reactor} = [CO]_{out}$ for a CSTR.

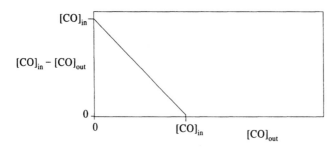

Figure 6.27. A plot of $[CO]_{in} - [CO]_{out}$ versus $[CO]_{out}$.

Now we superimpose on Figure 6.27 the right side of Eq. (6.118) as a function of [CO]. This curve is obtained by multiplying the rate shown in Figure 6.25 by the residence time. The superposition of the two curves shown in Figure 6.28 generates a *locus* of points that satisfy Eq. (6.118). Figure 6.28 reveals that there are *three* steady-state operating points for our CSTR. Note that we can change these operating points,

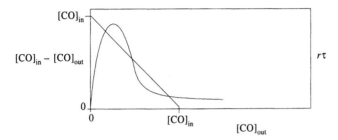

Figure 6.28. A plot of $r\tau$ versus $[CO]_{out}$.

for example, by changing the size of the reactor (design) or the flow rate through the reactor (operation). For example, increasing the reactor size increases τ, which moves the rate curve up, which changes the intersections. Similarly, decreasing the flow rate increases τ, which moves the rate curve up, which changes the intersections.

Not all the steady-state operating points derived from Figure 6.28 can be obtained in practice, however. Some of the steady-state points are stable and some are unstable. If the reactor is perturbed slightly from steady state and it returns to the same steady state, the operating point is *stable*. Similarly, if the reactor is perturbed slightly from steady state and it then migrates further from steady state, the operating point is *unstable*. To examine the stability of a steady-state point, we must examine the time dependence of the system, as given in Eq. (6.116),

$$\frac{d}{dt}[CO]_{reactor}V = [CO]_{in}Q_{in} - [CO]_{out}Q_{out} - rV. \tag{6.116}$$

Assume the reactor volume is constant and the flow rate in equals the flow rate out. Divide each side of Eq. (6.116) by the flow rate to obtain

$$\tau\frac{d}{dt}[CO]_{reactor} = ([CO]_{in} - CO]_{out}) - r\tau. \tag{6.119}$$

Assume now that we are operating at the first steady-state point, the intersection of the two curves in Figure 6.28 at the smallest concentration of CO. Suppose, for some reason, that [CO] falls below this steady-state point. Will [CO] continue to decrease, that is, is $d[CO]/dt$ less than zero? Equation (6.119) tells us that the time dependence of [CO] is given by the difference between the two curves in Figure 6.23. In this case, the difference is positive – the line for $[CO]_{in} - [CO]_{out}$ is above the line for $r\tau$ – and thus $d[CO]/dt$ is positive; [CO] increases back to steady state. Similarly, suppose that [CO] rises above the first steady-state point. The difference between the curves is negative in this region and thus $d[CO]/dt$ is negative; [CO] decreases back to steady state. The first steady-state operating point is stable.

The same analysis reveals that the second steady-state operating point is unstable. If for some reason [CO] falls below the second steady-state point it continues to fall until it reaches the first steady-state point. Why? Because the difference between the line for $[CO]_{in} - [CO]_{out}$ and the line for $r\tau$ is negative. Similarly, if [CO] momentarily rises above the second steady-state point, it continues to rise until it reaches the third steady-state operating point. An analysis of the third operating point reveals that it is stable.

We conclude with an interesting question on the operation of our CSTR. Recall that we wish to react as much CO as possible. When we start our reactor, $[CO]_{out}$ is equal to $[CO]_{in}$. The reaction starts and [CO] decreases, but only to the third operating point. Whenever [CO] decreases below the third steady-state point, our stability analysis predicts that [CO] will return to the third steady-state point. How does one reach the desirable operating point at the lower [CO]?

The steady-state operation of this CSTR predicted acceptable operation. However, start-up is going to be tricky, if not impossible. Perhaps a CSTR is a poor choice for this reaction. Note that this design problem would not be detected by analyzing the steady-state operation alone. The transient behavior must also be analyzed.

The analysis of multiple steady states in reactors is described in more detail in the text by Denn (1986), from which this example was adapted. We encourage you to investigate this textbook on process modeling.

REFERENCES

Denn, M. M. 1986. *Process Modeling*, Pitman Publishing Co., Marshfield, MA.
Russell, T. W. F., and Denn, M. M. 1972. *Introduction to Chemical Engineering Analysis*, Wiley, New York.

EXERCISES

6.1 A round lake of diameter d is fed by a river and drained by seepage. The flow rate of the river is equal to the rate of seepage and thus the water level in the lake is constant. Coincidentally, the level in the lake matches the height of a wide spillway at the north end. Should the water level increase, water flows over the spillway into a flood plain.

Suddenly the flow rate of the river increases to 1.5 times its normal rate. The extra water flows over the spillway at a rate determined by the height of the lake above the spillway,

$$Q_{spillway} = \alpha h^{3/2} \text{ (in gal/min)},$$

such that α is a constant determined in part by the width of the spillway.

(A) If the flow rate of the river persists indefinitely at 1.5 times the normal rate, to what level will the lake rise, relative to the height of the spillway? You may assume that the surface area of the lake remains approximately constant as the level rises.

(B) After the lake rises to the steady-state flood level calculated in (A), the flow rate of the river suddenly returns to normal. Derive a formula for the height of the lake as a function of time after the river flow returns to normal.

6.2 A different lake is fed by a river and is depleted by evaporation (no seepage). The rate of evaporation equals the rate of flow from the river. Suddenly a pollutant (compound X) is continuously dumped into the river. The pollutant does not evaporate and thus accumulates in the lake.

(A) Calculate the concentration of pollutant in the lake as a function of time (in kg pollutant/gal water).

> Volume of water in lake $\equiv V_{lake} = 1.8 \times 10^{10}$ gal.
> Flow rate of river $\equiv Q_{river} = 3.1 \times 10^8$ gal/day.
> Concentration of X in river $\equiv [X] = 1.3 \times 10^{-6}$ kg/gal water.

(B) The pollutant decomposes to inert substances at a rate proportional to its concentration in water:

> rate of decomposition of pollutant $= k[X]$.

$[X]$ has units of (kg X)/(gal water) and k is a constant with units (gal water)/day. Calculate the concentration of pollutant in the lake as a function of time (in kg pollutant/gal water) when decomposition is included.

(C) Calculate the steady-state concentration of pollutant in the lake.

6.3 A meteor strikes the Earth and forms a conical crater. The cone is shaped such that at a depth x to the bottom of the crater, the diameter of the crater is $4x$. Thus the volume when filled to a depth x is

$$volume = \frac{4}{3}\pi x^3.$$

Much later a river with volumetric flow rate Q_{river} begins to fill the crater.

(A) Assume that no water leaves the crater by evaporation or by seepage and derive an equation for the depth of water as a function of time.

(B) Assume now that water escapes from the crater lake by evaporation. The rate of evaporation (in gal/min) is proportional to the surface area of the lake:

> rate of evaporation $= kx^2$,

where k is a constant determined by the temperature, humidity, and rate of solar heating. What will be the level of the lake at steady state?

(C) After the lake reaches steady state, the flow to the crater suddenly stops (because the river is diverted to Los Angeles). The water level in the lake drops, owing to evaporation. Derive a formula for the level of the lake as a function of time, starting from the time the river is diverted to Los Angeles.

6.4 A round lake of diameter d is fed by a river and drained by seepage. The flow rate of the river is equal to the rate of seepage and thus the water level in the lake is constant. Should the lake level rise, the water from the lake flows through a triangular notch in a retaining wall and into a flood plain. Normally the water level matches the bottom of the triangular notch.

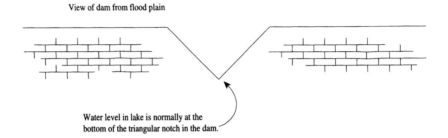

View of dam from flood plain

Water level in lake is normally at the
bottom of the triangular notch in the dam.

Suddenly the flow rate of the river increases to 1.5 times the normal rate. The extra water flows through the triangular notch at a rate determined by the height of the lake above the bottom of the notch,

$$Q_{\text{notch}} = \alpha h^{5/2} \text{ (in gal/min)},$$

such that α is a constant determined by the specifics of the notch.

(A) If the flow rate persists indefinitely at 1.5 times the normal rate, to what level will the lake level rise, relative to the height above the bottom of the notch? You may assume that the surface area of the lake remains approximately constant as the level rises.

(B) After the lake level has risen to the steady-state flood level calculated in (A), the flow rate of the river suddenly returns to normal. Derive a formula for the height of the lake as a function of time after the river flow returns to normal.

6.5 In Section 6.3, the concentration of N_2O in the batch reactor was measured by spectroscopy. Assume that we don't have a spectrometer and instead must use the pressure in the reactor to calculate the concentration of N_2O. The reactor contains initially only N_2O at a concentration $[N_2O]_0$ and the initial pressure is P_0. Derive an expression for $[N_2O]$ in terms of P, P_0, and $[N_2O]_0$.

6.6 Consider the reversible reaction

$$A \leftrightarrow B.$$

A is consumed in the forward reaction at a rate $= k_f[A] [=]$ moles/(time $\times$ volume). A is generated in the backward reaction at a rate $= k_b[B]$.

(A) Write a mass balance on A for a batch reactor that contains both A and B.

(B) Consider a batch reactor that contains initially only A at a concentration $[A]_0$, in moles/liter. Express $[B]$ in terms of $[A]_0$ and $[A]$.

(C) Substitute your expression for $[B]$ into the mass balance and solve for $[A]$. Your expression for $[A]$ should contain only k_f, k_b, $[A]_0$, and t. (You may wish to consult the table of indefinite integrals in Appendix E.)

(D) Check your expression in the limits $t = 0$ and $t = \infty$. At equilibrium, the ratio of concentrations in the batch reactor will be $[A]/[B] = k_b/k_f$.

6.7 A simultaneously reacts to form B and C,

The rate of the reaction $A \rightarrow B$ is $r_1 = k_1[A]$ and the rate of the reaction $A \rightarrow C$ is $r_2 = k_2[A]$. Consider a batch reactor that initially contains only A.

(A) Derive an equation for $[A]$ as a function of time. In other words, derive an equation of the form $[A] = \ldots$ where the right side contains only constants, t, and $[A]_0$.

(B) Derive an equation for $[B]$ as a function of time.

(This exercise appeared on an exam. It was estimated that it could be completed in 20 minutes.)

6.8 The "heat capacity" of my house is 3,300 kJ/°C. That is, 3,300 kJ raises the temperature of my house 1°C. The heater in my house can supply heat at a maximum rate of 5.2×10^4 kJ/hour.

(A) I return from vacation to a cold house. The inside temperature is 5°C (41°F) and the outside temperature is -15°C (5°F). I set the heater at its maximum rate at 8:00 pm. At what time will the temperature in my house be 20°C (68°F)?

(B) Heat escapes from my house by conduction through the walls and roof. The rate of heat loss, q_{loss} in kJ/hour, is proportional to the difference between the inside temperature and the outside temperature:

$$q_{loss} = k(T_{inside} - T_{outside}),$$

where $k = 740$ kJ/(°C hour). Repeat the calculation in (A), but include heat loss by conduction through the walls and roof.

(This exercise appeared on an exam. It was estimated that it could be completed in 25 minutes.)

6.9 The reaction of chemical P to chemical Q releases heat:

$$P \rightarrow Q + 770 \text{ kJ/mole}.$$

Because pure P reacts explosively, the reaction is conducted in a dilute water solution. Consider a batch reactor (no flow in or out) initially charged with 1.0 kg of water and 0.12 mol of $P(= 0.013\,\text{kg}P)$ at 50.°C. Thus $[P]_0 = 0.12$ mol/kg water. The reactor is thermally insulated.

(A) Calculate the temperature in the reactor after P has completely reacted to form Q. You may assume that the heat capacity of the dilute solution is the same as that of water.

(B) Obtain a mathematical expression for the temperature in the reactor as a function of $[P]$.

(C) The rate of the reaction $P \rightarrow Q$ increases as the temperature increases. Under the conditions here the rate is approximately proportional to the temperature:

$$\text{rate of reaction} = \frac{d[P]}{dt} = -\alpha T[P],$$

such that α is a constant. Derive an expression for $[P]$ as a function of time. Note that T is a function of time. Note also that T is a function of $[P]$.

(This exercise appeared on an exam. It was estimated that it could be completed in 30 minutes.)

Numerical Integration of Differential Equations

For simple rate equations, the integrated form can be obtained analytically. For example, the rate equation,

$$\frac{d[A]}{dt} = -k[A], \tag{1}$$

can be separated and integrated to yield

$$[A] = [A]_0 \exp(-kt). \tag{2}$$

Some complex rate laws cannot be integrated analytically (or are too difficult to integrate analytically) and one must appeal to numerical methods. Numerical methods yield $[A]$ at time increments of Δt, beginning with $[A]_0$:

$$[A]_{\Delta t} = [A]_0 + \left(\frac{d[A]}{dt}\right)_0 (\Delta t). \tag{3}$$

That is, Eq. (1) is used to calculate $d[A]/dt$ at $t = 0$; $(d[A]/dt)_0 = -k[A]_0$ and Eq. (3) becomes

$$[A]_{\Delta t} = [A]_0 - k[A]_0(\Delta t). \tag{4}$$

We continue in the same manner to calculate $[A]$ at $t = 2\Delta t$, using $(d[A]/dt)_{\Delta t} = -k[A]_{\Delta t}$ and

$$[A]_{2\Delta t} = [A]_{\Delta t} - k[A]_{\Delta t}(\Delta t). \tag{5}$$

In general, $[A]$ at any time t can be used to calculate $[A]$ at a time $t + \Delta t$ as follows:

$$[A]_{t+\Delta t} = [A]_t + \left(\frac{d[A]}{dt}\right)_t (\Delta t). \tag{6}$$

This algorithm of numerical integration is called the *Euler method*. The error is proportional to Δt. A more detailed discussion of the Euler method, as well as methods with better accuracy (and concomitant higher complexity), can be found in any text on numerical methods. A fine text with many examples in the context of chemical engineering is *Applied Numerical Methods* by B. Carnahan, H. A. Luther, and J. O. Wilkes (Krieger Publ., Melbourne, FL, 1990).

6.10 Integrate numerically rate equation (1) given in the previous page, with $k = 1$ sec^{-1}. Use a spreadsheet (or write a computer program) to produce a table with at least the following six columns:

Column 1. n, the iteration step. This column should begin with 0 and increase in increments of 1.

Column 2. t, the time. This column should begin with 0 and increase in increments of Δt.

Column 3. $[A]_t$, the numerically integrated value for $[A]$. The column should begin with an arbitrary $[A]_0$ of your choosing, and subsequent values should be computed using formula (6).

Column 4. $(d[A]/dt)_t$. This is computed with the formula

$$\left(\frac{d[A]}{dt}\right)_t = -k[A]_t.$$

Column 5. $[A]_t$ obtained from the analytical solution, Eq. (2).

Column 6. The relative error, defined as the difference between the numerical and analytical values for $[A]_t$, divided by the analytical value for $[A]_t$.

Use your spreadsheet (or program) to supply the following information. (You are encouraged to add additional rows to the table.)

$[A]_0$: _____

Analytical value for $[A]_t$ at 4 sec: _____

Δt	Number of steps	$[A]_t$ at 4 sec (numerical)	Relative error (%)
0.333	12		
0.1	40		
0.0333	120		
0.01	400		

6.11 Use numerical integration to obtain $[A]$, $[B]$, and $[C]$ as functions of time for the following reaction:

$$A \xrightarrow{k_a} B \xrightarrow{k_b} C,$$

such that $k_a = 0.8$ sec^{-1} and $k_b = 0.7$ sec^{-1} and for initial concentrations of $[A]_0 = 1.0$, $[B]_0 = [C]_0 = 0.0$. Plot $[A]$, $[B]$, and $[C]$ as a function of time for $t = 0$ to $t = 10$ sec. Assume that each reaction is first order. That is, $r_a = k_a[A]$ and $r_b = k_b[B]$.

Compare the results of your numerical integration to the analytical results:

$$[A] = [A]_0 e^{-k_a t},$$

$$[B] = [A]_0 \frac{k_a}{k_a - k_b}(e^{-k_b t} - e^{-k_a t}),$$

$$[C] = [A]_0 \frac{1}{k_a - k_b}[k_a(1 - e^{-k_b t}) - k_b(1 - e^{-k_a t})].$$

6.12 Use numerical integration to obtain $[A]$, $[B]$, and $[C]$ as functions of time for the following reaction:

$$A \underset{k_{-a}}{\overset{k_a}{\longleftrightarrow}} B \overset{k_b}{\rightarrow} C,$$

such that $k_a = 0.8$ sec^{-1}, $k_{-a} = 1$ sec^{-1}, and $k_b = 0.7$ sec^{-1} and for the initial concentrations of $[A]_0 = 1.0$ and $[B]_0 = [C]_0 = 0.0$. Plot $[A]$, $[B]$, and $[C]$ as functions of time for $t = 0$ to $t = 10$ sec. Assume that each reaction is first order. That is, $r_a = k_a[A]$, $r_{-a} = k_{-a}[B]$, and $r_b = k_b[B]$.

Compare the results of your numerical integration to the analytical results:

$$[A] = [A]_0 \left[\frac{k_a(\lambda_2 - k_b)}{\lambda_2(\lambda_2 - \lambda_3)} e^{-\lambda_2 t} + \frac{k_a(k_b - \lambda_3)}{\lambda_3(\lambda_2 - \lambda_3)} e^{-\lambda_3 t} \right],$$

$$[B] = [A]_0 \left[\frac{-k_a}{\lambda_2 - \lambda_3} e^{-\lambda_2 t} + \frac{k_a}{\lambda_2 - \lambda_3} e^{-\lambda_3 t} \right],$$

$$[C] = [A]_0 \left[\frac{k_a k_b}{\lambda_2 \lambda_3} + \frac{k_a k_b}{\lambda_2(\lambda_2 - \lambda_3)} e^{-\lambda_2 t} - \frac{k_a k_b}{\lambda_3(\lambda_2 - \lambda_3)} e^{-\lambda_3 t} \right],$$

such that

$$\lambda_2 = \frac{1}{2}(k_a + k_{-a} + k_b + [(k_a + k_{-a} + k_b)^2 - 4k_a k_b]^{1/2}),$$

$$\lambda_3 = \frac{1}{2}(k_a + k_{-a} + k_b - [(k_a + k_{-a} + k_b)^2 - 4k_a k_b]^{1/2}).$$

(Reference: Moore, J. W., and Pearson, R. G. 1981. *Kinetics and Mechanism*, 3rd ed., Wiley, New York, p. 313.)

Note that when the reaction progressed from $A \rightarrow B$ to $A \leftrightarrow B \rightarrow C$ in the preceding three exercises the complexity of the analytical solutions increased dramatically. However, the complexity of the numerical calculation increased only modestly.

While working the preceding exercises, you may encounter integrals that may not be in your repertoire. Even after studying the calculus for a year you will encounter integrals you cannot solve. Yes, it's true. When training and/or memory fails you, it is useful to consult a table of integrals. Appendix E contains a brief list of indefinite integrals germane to the preceding exercises. The *CRC Handbook of Chemistry and Physics* has a short table and there are several books with larger compilations. Our favorites are *Tables of Integrals and Other Mathematical Data* by H. B. Dwight (300+ pages) and *Table of Integrals, Series, and Products* by I. S. Gradshteyn and I. M. Ryzhik (1,100+ pages).

APPENDIX A

List of Symbols

A area (m^2)

C_P molar heat capacity at constant pressure (joules/(kg·K) or joules/(mol·K))

C_V molar heat capacity at constant volume (joules/(mol·K))

f a function

$F_{A,i}$ mass flow rate of component A in stream i (kg/s)

g gravitational constant (9.8 m/s^2)

k Boltzmann's constant, or rate constant for a chemical reaction, or thermal conductivity (joules/(s·m·C)), or roughness factor for flow in pipes

L liquid flow rate in a separation unit (mol/s)

$M_{A,i}$ mass of component A in phase i (kg)

N_A Avogadro's constant (6.02 × 10^{23}/mol)

P pressure (Pa, although atm is also used)

q_i rate of heat flow of stream i (joules/s)

Q_i volumetric flow rate of stream i (m^3/s)

R gas constant (8.31 m^3·Pa/(mol·K))

T temperature (°C or K)

t time (s)

v velocity (m/s)

V volume (m^3), or vapor flow rate in a separation unit (mol/s)

x_i mol fraction of i in the liquid phase

y_i mol fraction of i in the vapor phase

Greek Symbols

μ viscosity (Pa·s)

ρ density (kg/m^3)

Dimensions

L   length

M   mass

N   amount

T   time

Θ temperature

Symbols

$\equiv$ is defined as

$[=]$ has dimensions of

$[A]_t$ mass (molar) concentration of A at time t

Dimensionless Groups

Fr	Froude number	Pr	Prandtl number
Nu	Nusselt number	Re	Reynolds number
Pe	Peclet number	Sh	Sherwood number
		St	Stanton number

Units, Conversion Factors, and Physical Constants

B.1   Units

We use the mks (meter–kilogram–second) SI (Système International) system of units in (most) calculations in this text. The base units (see Chapter 5) of the mks system are given in Table B.1.

Table B.1. Base units in the mks system

Dimension	Unit	Symbol
length	meter	m
mass	kilogram	kg
time	second	s
temperature	kelvin or degrees celsius	K or °C
amount	mole	mol

The base units are combined to form derived units, some of which are shown in Table B.2.

Table B.2. Some derived units in the mks system

Quantity	Units	Name	Symbol
area	m^2	—	—
volume	m^3	—	—
velocity	m/s	—	—
acceleration	m/s^2	—	—
momentum	$m \cdot kg/s$	—	—
force	$m \cdot kg/s^2$	newton	N
energy	$m^2 \cdot kg/s^2$	joule	J
power	$m^2 \cdot kg/s^3$	watt	W
pressure	$kg/m \cdot s^2$	pascal	Pa

B.3 Conversion Factors

Area

1 acre	$= 4047 \text{ m}^2$

Force

1 dyne	$\equiv 1 \times 10^{-5} \text{ N}$
1 pound-force (lbf)	$= 4.448 \text{ N}$

Length

1 angstrom (Å)	$\equiv 1 \times 10^{-10} \text{ m}$
1 fathom	$= 1.829 \text{ m}$
1 foot (ft)	$= 0.3048 \text{ m}$
1 inch (in)	$= 0.0254 \text{ m}$
1 light year	$= 9.46 \times 10^{15} \text{ m}$
1 mile	$= 1,609 \text{ m}$
1 yard	$= 0.9144 \text{ m}$

Energy

1 Btu	$= 1054.4 \text{ J}$
1 calorie	$= 4.184 \text{ J}$
1 erg	$\equiv 1 \times 10^{-7} \text{ J}$

Mass

1 ounce (avoirdupois)	$= 2.835 \times 10^{-2} \text{ kg}$
1 ounce (troy)	$= 3.110 \times 10^{-2} \text{ kg}$
1 pound (lb, 16 oz avoirdupois)	$= 0.4536 \text{ kg}$
1 pound (12 oz troy)	$= 0.3732 \text{ kg}$
1 ton (short, 2,000 lb)	$= 907.2 \text{ kg}$

Power

1 horsepower (hp)	$= 745.7 \text{ W}$

Pressure

1 atmosphere (atm)	$= 1.013 \times 10^5 \text{ Pa}$
1 bar	$\equiv 1 \times 10^5 \text{ Pa}$
1 inch of Hg	$= 3,386 \text{ Pa}$
1 foot of water	$= 2,989 \text{ Pa}$
1 pound-force/in^2 (psi)	$= 6,895 \text{ Pa}$
1 torr (1 mm Hg)	$= 133.3 \text{ Pa}$

Temperature

kelvin (K)	$= \text{celsius} + 273.15$
fahrenheit (F)	$= 1.8 \times \text{celsius} + 32$

Time

1 day	$= 86,400 \text{ s}$
1 hour (hr)	$= 3,600 \text{ s}$
1 minute (min)	$= 60 \text{ s}$
1 year (yr)	$= 3.154 \times 10^7 \text{ s}$

Velocity

1 knot	$= 0.5144 \text{ m/s}$
1 mile/hour (mph)	$= 0.4470 \text{ m/s}$

Viscosity

1 centipoise (cp)	$= 1 \times 10^{-3} \text{ Pa·s}$

Volume

1 barrel (oil, 42 gal)	$= 0.1590 \text{ m}^3$
1 barrel (bbl, 31.5 gal)	$= 0.1192 \text{ m}^3$
1 fluid ounce (US)	$= 2.957 \times 10^{-5} \text{ m}^3$
1 gallon (US, liquid)	$= 3.785 \times 10^{-3} \text{ m}^3$
1 liter (L)	$\equiv 1 \times 10^{-3} \text{ m}^3$
1 quart (US, liquid)	$= 9.464 \times 10^{-4} \text{ m}^3$

B.4   Physical Constants

Avogadro constant	N_A	6.022×10^{23} mol^{-1}
Boltzmann constant	$k = R/N_A$	1.381×10^{-23} J/K
gas constant	R	8.314 J/(mol·K)
		8.314 m^3·Pa/(mol·K)
		0.08206 liter·atm/(mol·K)
gravitational acceleration at sea level	g	9.8 m/s^2
Planck constant	h	6.626×10^{-34} J·s
speed of light in vacuum	c	3.0×10^8 m/s

Significant Figures

QUANTITIES are composed of two parts: a number (e.g., 5.31) and a unit (e.g., grams). The number must have the proper amount of significant figures, defined as follows:

> Significant figures – the number of digits from the first nonzero digit on the left to the last nonzero digit on the right.

The amount of significant figures in a number is perhaps determined easiest from the scientific notation for the number. Some examples are listed in Table C.1.

Table C.1. Significant figures

Number	Scientific notation	Significant figures
3.0	3.0×10^0	2
23	2.3×10^1	2
0.0353	3.53×10^{-2}	3
1,000	1×10^3	1
1,000.	1.000×10^3	4
1,000.0	1.0000×10^3	5

The numbers 1,000 and 1,000. each have only 1 significant figure if one applies the definition above. This is wrong; 1,000 has 1 significant figure and 1,000. has 4. We need to augment the definition as follows:

> Significant figures – the number of digits from the first nonzero digit on the left to the last nonzero digit on the right, or to the last digit if there is a decimal point.

How does one determine the significant figures in a number obtained from an arithmetic operation? Follow this algorithm for addition and subtraction:

1. Note the position of the last significant figure in each number being added or subtracted.
2. Note the position of the left-most last significant figure of all the numbers to be added.
3. The last significant figure of the sum is given by the left-most last significant figure.

Consider two examples:

$$\begin{array}{r} 6750 \\ +\quad 10.\mathbf{3} \\ \hline 6760.3 \end{array}$$

The correct sum is 6,760 after truncating to three significant figures.

$$\begin{array}{r} 1.0000 \\ +0.22 \\ \hline 1.2200 \end{array}$$

The correct sum is 1.22 after truncating to three significant figures.

Follow this algorithm for multiplication and division:

1. Determine the amount of significant figures in each multiplicand or divisor.
2. The amount of significant figures in the product is equal to the amount of significant figures in the multiplicand or divisor with the least amount of significant figures.

Example: Convert 87.0 kg water/min to units of gal/hour:

$$\left(\frac{87.0\,\text{kg water}}{\text{min}}\right)\left(\frac{60\,\text{min}}{1\,\text{hour}}\right)\left(\frac{1.000\,\text{g}}{1\,\text{kg}}\right)\left(\frac{1.000\,\text{ml water}}{1.000\,\text{g water}}\right)\left(\frac{1\,\text{gal}}{4.405 \times 10^{-3}\,\text{mL}}\right)$$

$$= 1{,}185.0\,\text{gal/hour}.$$

The units are converted by multiplying by factors of 1. For example, because 60 min = 1 hr,

$$\left(\frac{60\,\text{min}}{1\,\text{hour}}\right) = 1$$

by definition. And each quantity has infinite significant figures, by definition. Another factor of 1 is formed from a physical property specific to this problem, the density of water, for which 1.000 mL water = 1.000 g water, and

$$\left(\frac{1.000\,\text{mL water}}{1.000\,\text{g water}}\right) = 1.$$

It is assumed here that the density of water is known to four significant figures. The multiplicand with the least significant figures is 87.0 kg water/min, which has three. Thus the proper converted quantity is $1{,}180$ gal/hr, or 1.18×10^3 gal/hr.

The last example brings us to the convention for rounding off numbers that end in a 5:

If the digit after the last significant figure is a 5 and the last significant figure is even, round down. If the digit after the last significant figure is a 5 and the last significant figure is odd, round up.

Thus 1,185 rounds off to 1,180 and 1,175 rounds off to 1,180.

Consider another example: The 1995–96 tuition at Cornell was \$10,000.00/semester. Convert this to units of \$/in-class hours. Assume that the average load of a first-semester first-year student at Cornell is 14 credits, as recommended by the enlightened policies of the College of Engineering. Further assume that 14 credits entails about 16 in-class hours per week. Therefore

$$\left(\frac{10,000.00\ \$}{1\ \text{semester}}\right)\left(\frac{1\ \text{semester}}{13\ \text{weeks}}\right)\left(\frac{1\ \text{week}}{16\ \text{in-class hours}}\right) = \$48.08/\text{in-class hour.}$$

Although 10,000.00 has seven significant figures, our assumptions have only two. The answer is thus \$48/in-class hour. A bargain at any price.

Here are examples of conversions that use an incorrect number of significant figures. The August 26, 1991 issue of *Chemical and Engineering News* reported that "In Japan blue roses cost \$78." Why the arbitrary price of \$78? Why not \$75, or \$80? It seems that blue roses cost 10,000 yen, which is a nice round number with one significant figure. When one converts yen to dollars using the exchange rate at that time, one gets

$$\left(\frac{10,000\ \text{yen}}{\text{blue rose}}\right)\left(\frac{\$1.000}{128\ \text{yen}}\right) = \$78/\text{blue rose.}$$

Converting to one significant figure, the correct price is \$80/blue rose.

On average, the temperature of the human body is 37°C. What is the average temperature in Fahrenheit? We use the well-known formula to convert from centigrade to Fahrenheit:

$$37°C\left(\frac{9.00°F}{5.00°C}\right) + 32.00°F = 98.6°F.$$

Converting the answer to two significant figures, we get an average body temperature of 99°F.

Finally, retain extra digits during a calculation and truncate numbers to the proper number of significant figures at the end. Do not truncate at an intermediate stage of your calculation.

REFERENCE

Felder R. M., and Rousseau, R. W. 1986. *Elementary Principles of Chemical Processes*, 2nd ed., Wiley, New York, pp. 19–21.

Graph Paper

EXPERIMENTAL DATA often span many orders of magnitude. An example is the friction factor of a sphere moving through a fluid as a function of the Reynolds number, as discussed in Chapter 5. Some typical data for laminar flow are presented in Table D.1.

Table D.1. Data for a sphere moving through a fluid

Reynolds number	Friction factor
0.9	27
0.33	73
0.12	200
0.074	324
0.023	300
0.0091	2,635
0.0065	3,700
0.0027	8,888
0.0011	21,900

It is useful to analyze experimental data by plotting. However, the plot shown in Figure D.1 is not very useful. The data are crowded near the origin or lie on either the *x* or *y* axes. Figure D.1 would not be useful for interpolating between points or scanning for suspicious data, for example.

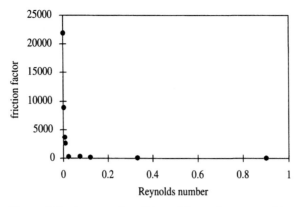

Figure D.1. An x–y plot of Reynolds number versus friction factor.

The utility of the plot is improved by plotting the logarithm of the data, as shown in Figure D.2. Figure D.2 clearly shows a linear correlation, as reinforced by the straight line through the data. Figure D.2 also reveals that the point measured for Re = 0.023 is suspect; it lies below the correlation of the other data.

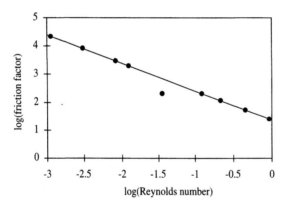

Figure D.2. An x–y plot of $\log_{10}$(Reynolds number) versus $\log_{10}$(friction factor).

Rather than compute the base-10 logarithm of each coordinate, one can plot directly onto *log–log graph paper*, as shown in Figure D.3. The gradations on both the abscissa and ordinate increase logarithmically.

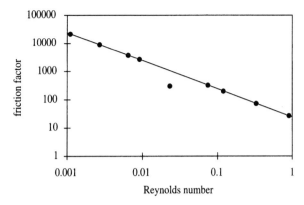

Figure D.3. A log–log plot of Reynolds number versus friction factor.

For clarity, the gridlines have been omitted in Figure D.3. An example of log–log paper with gridlines is shown in Figure D.4. Specifically, this is called 2 cycle × 2 cycle log–log graph paper because there are two powers of ten on each axis. (Figure D.3 is 3 cycle × 5 cycle log–log graph paper.) Note that the gridlines are not evenly spaced. The first gridline moving from the left is 2, not 1.1. The first gridline to the left of 10 is 9. The distance between 1 and 2 is greater than the distance between 9 and 10. Note that 10 is halfway between 1 and 100 because 10^1 is halfway between 10^0 and 10^2. That is, $\log_{10}10^1(=1)$ is halfway between $\log_{10} 10^0(=0)$ and $\log_{10} 10^2(=2)$. Distances

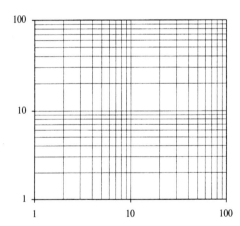

Figure D.4. 2 cycle × 2 cycle log–log graph paper.

on a log scale are proportional to the exponent of ten. The midpoint between 1 and 10 is not 5.5; rather it is the midpoint between 10^0 and 10^1, which is $10^{0.5} = 3.16$. To find the midpoint between two numbers, average their logarithms (base 10) and raise 10 to that power. What is the midpoint between 1 and 2 on the graph? It is not 1.5 but 10 raised to the power of the average of $\log_{10}(1)$ and $\log_{10}(2)$, which is $10^{0.15051} = 1.414 = 2^{0.5}$.

Graph paper with a logarithmic scale on the ordinate and a linear scale on the abscissa is called *semilog graph paper*. An example of 2-cycle semilog paper is shown in Figure D.5. The range on the abscissa in both the log–log graph paper in Figure D.4 and the semilog graph paper in Figure D.5 is 1 to 100. What if your data are in a different range, for example, from 0.001 to 0.1? No problem. Just replace 1 with 0.001, replace 10 with 0.01, and replace 100 with 0.1.

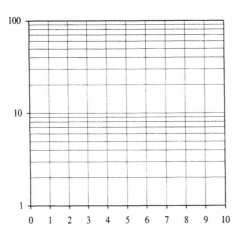

Figure D.5. 2-cycle semilog graph paper.

Log–log plots and semilog plots are useful for determining the functional relationships of data. The human eye is adept at recognizing straight lines but less able to distinguish x^2 from e^x especially if only a portion of the data is available. However $y = x^2$ is a straight line on log–log graph paper, but it is not on x–y or semilog paper. The function $y = e^x$ is a straight line on semilog paper but not on x–y or log–log paper. How does one determine the mathematical equation of a straight line on log–log paper, such as the line shown in Figure D.6?

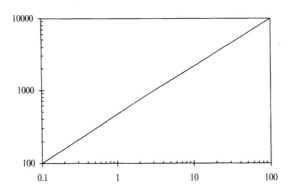

Figure D.6. A straight line on a log–log plot.

A straight line means the functional form is

$$y = mx + b, \tag{D.1}$$

where m is the slope and b is the y-intercept. But be careful – Figure D.6 is not a plot of x versus y but a plot of $\log_{10}(x)$ versus $\log_{10}(y)$. So a straight line on a log–log plot means the functional form is

$$\log_{10}(y) = m \log_{10}(x) + b. \tag{D.2}$$

Let's calculate the slope m of the line in Figure D.6. We will compute the slope from the ratio of rise (the distance traveled in vertically) to the run (the distance traveled horizontally):

$$\text{slope} = \frac{\text{rise}}{\text{run}} = \frac{10{,}000 - 100}{100 - 0.1} = 99.1. \tag{D.3}$$

Correct? No, we didn't use the proper coordinates to calculate the rise and run of the line. Equation (D.3) uses x and y. We need to use $\log_{10}(x)$ and $\log_{10}(y)$. Therefore

$$\text{slope} = \frac{\text{rise}}{\text{run}} = \frac{\log_{10}(10{,}000) - \log_{10}(100)}{\log_{10}(100) - \log_{10}(0.1)} = \frac{\log_{10}(10^4) - \log_{10}(10^2)}{\log_{10}(10^2) - \log_{10}(10^{-1})}$$

$$= \frac{4 - 2}{2 - (-1)} = \frac{2}{3}. \tag{D.4}$$

And what of b, the y-intercept? Again be careful – although the line in Figure D.6 crosses the y axis at $y = 100$, this is not the intercept. Note that the y axis is at $x = 0.1$, not $x = 0$. There is no intercept on a log–log graph because x is never equal to zero. To find b we substitute the slope into Eq. (D.2):

$$\log_{10}(y) = \frac{2}{3} \log_{10}(x) + b \tag{D.5}$$

$$= \log_{10}(x^{2/3}) + b. \tag{D.6}$$

We now raise both sides of Eq. (D.6) to the exponent of 10:

$$10^{\log_{10}(y)} = 10^{[\log_{10}(x^{2/3}) + b]}, \tag{D.7}$$

$$y = 10^{[\log_{10}(x^{2/3})]} \times 10^b \tag{D.8}$$

$$= x^{2/3} \times 10^b. \tag{D.9}$$

To determine b (or actually, to determine 10^b), we substitute a point that lies on the line, such as $x = 0.1$, $y = 100$. This gives

$$10^b = \frac{y}{x^{2/3}} = \frac{100}{0.1^{2/3}} = 464. \tag{D.10}$$

Thus the equation of the straight line in Figure D.6 is

$$y = 464 \, x^{2/3}. \tag{D.11}$$

It is always prudent to check one's solution. Does the line go through another known point, such as $x = 100$, $y = 10{,}000$? Does

$$10{,}000 = 464 \times 100^{2/3}? \tag{D.12}$$

This yields

$$10,000 \approx 9,997. \qquad\qquad\qquad (D.13)$$

It checks. Now, test your skills by determining the equation for the line in Figure D.7.

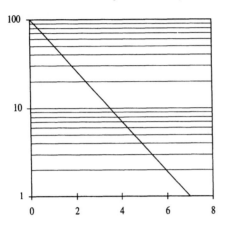

Figure D.7. A straight line on a semilog plot.

(We calculate that the line corresponds to $y = 100(10^{-2x/7})$, or $y = 100e^{-0.66x}$.)

Mathematics, Mechanics, and Thermodynamics

Algebra

$$a^m a^n = a^{m+n} \qquad (ab)^m = a^m b^m$$
$$(a^m)^n = a^{mn} \qquad a^0 = 1$$
$$a^{-m} = \frac{1}{a^m}$$
$$\ln e^x = x \qquad \ln x^n = n \ln x$$
$$\ln a + \ln b = \ln(ab)$$

Quadratic formula: The roots of

$$ax^2 + bx + c = 0$$

are

$$x = \frac{-b \pm \sqrt{b^2 - 4ac}}{2a} \quad \text{for} \quad a \neq 0.$$

Geometry

A = area, C = circumference, V = volume,
r = radius, b = length of base, h = height

Planar geometry:

circle: $\quad A = \pi r^2, C = 2\pi r$
triangle: $\quad A = bh/2$

Solid geometry:

sphere: $\quad V = \dfrac{4}{3}\pi r^3, A = 4\pi r^2$
cylinder: $\quad V = \pi h r^2$
cone: $\quad V = \dfrac{1}{3}\pi r^2 h$

Theorem of Pythagoras: For a right triangle with hypotenuse c and legs a and b, $a^2 + b^2 = c^2$.

Trigonometry

$$\sin^2 \alpha + \cos^2 \alpha = 1$$
$$\sin(\alpha + \beta) = \sin \alpha \cos \beta + \cos \alpha \sin \beta$$
$$\cos(\alpha + \beta) = \cos \alpha \cos \beta + \sin \alpha \sin \beta$$
$$e^{ix} = \cos x + i \sin x$$

The Calculus

Differentiation:

$$\frac{d}{dx} x^n = n x^{n-1}$$

$$\frac{d}{dx} e^{ax} = a e^{ax} \qquad \frac{d}{dx} \ln(x) = \frac{1}{x}$$

Product rule:

$$\frac{d}{dx}[f(x)g(x)] = f(x)\frac{d}{dx}g(x) + g(x)\frac{d}{dx}f(x)$$

Integration:

$$\int x^n dx = \frac{x^{n+1}}{n+1} \qquad \int \frac{1}{x} dx = \ln|x|$$

$$\int (a + bx)\, dx = \frac{1}{2b}(a + bx)^2$$

$$\int \frac{dx}{a + bx} = \frac{1}{b}\ln|a + bx|$$

$$\int \frac{dx}{x(a + bx)} = \frac{-1}{a}\ln\left|\frac{a + bx}{x}\right|$$

$$\int \frac{dx}{(a + bx)^2} = \frac{-1}{b(a + bx)}$$

$$\int \frac{dx}{(a + fx)(c + gx)} = \frac{-1}{ag - cf}\ln\left|\frac{c + gx}{a + fx}\right|$$

Mechanics

Linear momentum:	$p = mv$
Kinetic energy:	$K.E. = \frac{1}{2}mv^2$
Potential energy at Earth's surface:	$P.E. = mgh$

Thermodynamics

Ideal gas law: $PV = nRT$
Heat capacity:

for an ideal gas,

$$C_P = C_V + R$$

for a monatomic ideal gas (He, Ne, Hg),

$$C_P = \frac{5}{2}R$$

for a diatomic ideal gas (O_2, N_2),

$$C_P = \frac{7}{2}R$$

The Greek alphabet

alpha	α	A	nu	ν	N
beta	β	B	xi	ξ	Ξ
gamma	γ	Γ	omicron	o	O
delta	δ	Δ	pi	π	Π
epsilon	ε	E	rho	ρ	P
zeta	ζ	Z	sigma	σ	Σ
eta	η	H	tau	τ	T
theta	θ	Θ	upsilon	υ	Y
iota	ι	I	phi	ϕ, φ	Φ
kappa	κ	K	chi	χ	X
lambda	λ	Λ	psi	ψ	Ψ
mu	μ	M	omega	ω	Ω

Glossary of Chemical Engineering

absorption   the assimilation of a chemical (usually a gas or liquid) *into* a liquid or porous solid.

adsorption   the deposition of a chemical (usually a gas or liquid) *onto* the surface of a solid. Charcoal in an aquarium filter or in a tap water purifier *adsorbs* organic chemicals from water.

azeotrope   a mixture of two or more components such that the composition of the liquid is the same as the vapor. Not all mixtures form an azeotrope. Water and ethanol form an azeotrope at about 95% water.

base units   the units of the base dimensions, which are length, time, mass, temperature, amount, electric charge, and luminous intensity. In the mks SI system, the base units are meter, second, kilogram, kelvin (or celsius), mole, ampere, and candela.

batch reactor   a vessel into which reactants are loaded, induced to react (by increasing the temperature, and/or increasing the pressure, and/or adding a catalyst), then discharged. A batch reactor is not at steady state, although its contents may approach equilibrium. A batch reactor is inherently noncontinuous but can be integrated into a continuous process if preceded by a hold-up tank and followed by a surge tank. A crock pot is a batch reactor.

biochemical engineering   the design and analysis of processes based on biological reactions, for example to produce a drug or destroy a pollutant.

bioengineering   the application of engineering design and analysis to biological systems. The two chief types of bioengineering are biochemical engineering and biotechnology.

biotechnology   the application of science and engineering to organisms, cells, and biomolecular processes. Specific examples are the design and analysis of devices used in biological systems, such as artificial organs or diagnostic equipment.

bubble point   the temperature at which the first bubble of vapor forms in a liquid at a given pressure or the pressure at which the first bubble of vapor forms in a liquid at a given temperature. The bubble point of a pure liquid is the boiling point.

capital cost   the expenditure to purchase and install equipment. The chief capital cost in a chemical process is usually the reactor.

catalyst   a substance that accelerates a chemical reaction, without being consumed or modified. A catalyst increases the rate that a system proceeds to equilibrium but does not change the equilibrium concentrations.

centrifugal pump   a pump that increases the kinetic energy of a fluid by centrifugal force, then converts the kinetic energy to increase the fluid pressure and reduce the fluid velocity.

centrifugation   separation by density quickened by spinning the mixture. The increase in separation rate is determined by the ratio of centripetal acceleration in the spinning mixture to gravitational acceleration. Centrifugation is common for biological separations, such as separating blood into plasma (the liquid portion) and the cells and platelets.

chemical kinetics   study of the mechanism and rate of chemical reactions.

closed system   an isolated system; a system with no mass and/or energy flow across its borders.

condensation   the conversion from vapor to liquid.

conservation law   a statement that the amount of some thing is invariant. Invariant things include energy, momentum, and electric charge. The amount of mass is invariant, unless mass is converted into energy.

constitutive equation   a constrained law that relates a driving force to a flow of mass, energy, or momentum. For example, Fourier's law of heat conduction relates the driving force of a temperature gradient to a flux of energy.

constrained law   A statement governing physical or chemical behavior restricted to certain conditions. The ideal gas law, $PV = nRT$, is a constrained law; it is valid only at low pressure and high temperature.

continuous stirred-tank reactor (CSTR)   a vessel into which reactants are introduced at a constant rate, and effluent is discharged at the same rate. The CSTR's contents are stirred to maintain uniform composition and temperature. To aid stirring, the vessel usually has comparable height and width, like a tank.

The effluent has the same composition and temperature as the vessel's contents. Although more accurately called a continuous flow stirred-tank reactor, this terminology is less common.

core variable   a variable that appears in exactly one dimensionless group describing some phenomenon.

derived unit   a unit created by the product of the base units. The joule, a unit of energy in the mks SI system, is the product of kg, m, and s: 1 joule = 1 $kg \cdot m^2/s^2$.

dew point   the temperature at which the first drop of liquid condenses from a vapor at a given pressure or the pressure at which the first drop of liquid condenses from a vapor at a given temperature. The dew point of a pure liquid is the boiling point.

dimension   a physical quality. Any thing can be completely described by specifying its base dimensions: length, time, mass, temperature, amount, electric charge, and luminous intensity. Any parameter, such as energy or momentum, has dimensions. The dimensions of energy are $mass \cdot length^2/time^2$.

dimensional analysis   the process of combining the parameters that describe a phenomenon (such as mass, velocity, or viscosity) into dimensionless groups (such as the Reynolds number).

displacement pump   a pump that decreases the volume of a chamber that contains the fluid, which increases the pressure of the fluid, forcing it out of the chamber. The chamber is then expanded and refilled. Your heart is a displacement pump.

distillate   the components with lower boiling points, and thus higher volatility. The distillate comes off the top of a distillation column. Also known as *light ends* or *tops*.

distillation   separation of liquid mixtures based on differences in component volatilities.

dynamic similarity   two systems are dynamically similar if the magnitudes of the dimensionless groups describing the systems are equal. A steel sphere falling through molasses may be dynamically similar to a weather balloon rising in air.

elastomer   A polymeric material that returns to its initial shape after deformation. Elastomers are usually formed by cross-linking polymer molecules. Rubber bands, silicone rubbers, and automobile tires are elastomers.

electrolytic reactor   a device for converting chemical energy into electrical energy. Some electrolytic reactors convert H_2 and O_2 into H_2O and generate electric current.

empirical analysis   the study of a phenomenon by measuring behavior to find the correlation between parameters. Empirical analysis does not delve into underlying principles. Also known as *parametric analysis*.

enthalpy   an energy used to characterize open systems. The change in enthalpy is the sum of (1) the change in internal energy upon moving through the system and (2) the work applied to move that substance through the system.

entrainment   liquid droplets carried upward by a vapor bubbled through the liquid.

equation of state   a mathematical expression that relates various thermodynamics properties. The ideal gas law, $PV = nRT$, is an equation of state for a gas.

equilibrium   a system in which the conditions at any position do not change with time, and the conditions are uniform with position. The height of the water in a lake is at steady state; the water height is the same everywhere.

extensive property   a thermodynamic parameter that depends on the amount (extent) of a substance. Mass, volume, and internal energy are extensive

properties. If 1 kg of water is added to 1 kg of water, the total mass is 2 kg.

extraction   a separation that extracts a component from a liquid mixture by dissolving the component into an immiscible solvent.

flash drum   A vessel in which a liquid is vaporized (*flashed*) by suddenly increasing the temperature and/or decreasing the pressure.

flowsheet   a diagram used to represent a chemical process. Process units are represented by simple geometric shapes and interconnecting pipes are represented by lines and arrows.

fluid   a substance capable of flowing; a gas, a liquid, and in some cases, a finely particulate solid.

fluid dynamics   the study of the properties and dynamics of fluids. Fluid dynamics includes rheology, acoustics, plasma physics, and quantum fluids.

heat capacity   the ratio of change in enthalpy to the change in temperature of a substance. The heat capacity is a measure of a substance's capacity to absorb heat.

heat exchanger   a process unit that warms one fluid and cools another, by transferring heat from the fluid at the higher temperature. The fluids remain isolated from each other; inlet compositions are the same as outlet compositions.

heavy ends   the components with higher boiling points, and thus lower volatility. The heavy ends come off the bottom of a distillation column. Also known as *heavies* or *bottoms*.

hold-up tank   a process unit that accumulates material and discharges material discontinuously. A hold-up tank might precede a batch reactor in a continuous process or it might be placed at the end of a process to fill shipping containers.

immiscible   two liquids that do not form a single phase when mixed. Oil and water are immiscible.

intensive property   a thermodynamic property that does not depend on the amount of a substance. Temperature and pressure are intensive parameters. If 1 kg of water at 20°C is added to 1 kg of water at 20°C, the temperature of the mixture is 20°C, not 40°C.

internal energy   an extensive property given by the sum of the electronic, vibrational, rotational, and translational energies of the molecules or atoms constituting the substance. The ratio of an incremental change in internal energy to an incremental change in temperature is the heat capacity at that temperature.

light ends   see *distillate.*

mass balance   an accounting for the mass entering, leaving, and accumulating in (or depleting from) a system.

mathematical model   an equation or set of equations that predicts the behavior of a phenomenon.

mole   the amount of a substance containing 6.022×10^{23} units, usually atoms or molecules. 1 mole of H_2O has a mass of 18 g.

monomer   from Greek, meaning one (*mono*) part (*meros*); a molecular building block that can be chemically linked to form polymers (thousands of units) or oligomers (tens to hundreds of units). Ethylene, $CH_2{=}CH_2$, is the monomer to the polymer polyethylene, $(-CH_2-CH_2-)_n$; tetrafluoroethylene, $CF_2{=}CF_2$, is the monomer to the polymer polytetrafluoroethylene (teflon), $(-CF_2-CF_2-)_n$.

oligomer   from Greek, meaning a few (*oligo*) parts (*meros*); a molecule with modest molecular weight (typically 100 to 1,000 amu) composed of several chemical units (monomers). See *polymer.*

open system   a system with flow of mass and/or energy across its borders.

operating cost   the expense of producing a commodity or providing a service, after equipment has been purchased and installed. The chief operating cost in a chemical process comprises reactants and utilities.

operating line   a straight line drawn on a equilibrium map of vapor composition vs. liquid composition, used in the graphical analysis of distillation columns and absorption columns. An operating line is usually devised from a point given by the compositions at the top or bottom of the column and a slope given by the ratio of flow rates moving up and down the column.

pi group   a Π group; a dimensionless product and ratio of parameters. A system may be characterized by the magnitude of its dimensionless group(s). For example, the magnitude of the Reynolds number indicates whether a fluid flow is laminar or turbulent.

plastic   a moldable polymeric material. A thermoplastic polymer, such as polyvinyl-chloride, PVC, becomes pliable when heated and can be reformed by reheating. Thermosetting polymers, such as epoxies or polyphenol-formaldehydes (bakelite), cannot be reformed by reheating.

plug flow reactor (PFR)   a vessel into which reactants are introduced at a constant rate and effluent is discharged at the same rate. However, the composition and temperature in a PFR vary along the length of the reactor, which is usually long and narrow like a pipe.

polymer   from Greek, meaning many (*poly*) parts (*meros*); a molecule with high molecular weight (typically 10^4 to 10^6 amu) composed of many chemical units (monomers). Common synthetic polymers are polyvinylchloride (PVC), $(-CH_2-CHCl-)_n$, polypropylene, $(-CH_2-CH(CH_3)-)_n$, and polystyrene, $(-CH_2-CH(C_6H_5)-)_n$. Common natural polymers are cellulose, RNA, and DNA. The two chief types of polymers are plastics and elastomers (rubbers).

pump   a process unit that increases the

pressure of a fluid. Common types are centrifugal and displacement.

rate equation   a mathematical expression for the rate of consumption of reactant in a chemical reaction. The rate equation usually includes reactant concentration(s), product concentration(s), and temperature.

reactor   a vessel in which contents are transformed by chemical reaction. Common reactor types are batch, continuous stirred-tank, and plug flow.

rectifying section   the trays (stages) in a distillation column above the feed tray, including the condenser.

recycle stream   a flow of material taken from the principal flow in a process and returned to an earlier point in the process. Material is usually recycled from reactors with low conversions and separators with poor separation.

reduced parameter   a physical parameter, such as length, multiplied and divided by other parameters to form a dimensionless group. For flow through a pipe of length l, the reduced length is l/d, where d is the pipe diameter.

reflux ratio   the ratio of the molar flow of liquid flowing down a distillation column to vapor flowing up; reflux ratio $= L/V$. The reflux ratio is less than one in the rectifying section and greater than one in the stripping section. Occasionally defined (elsewhere) as the ratio of liquid to distillate; reflux ratio $= L/D = L/(V - L)$.

residence time   the mean time a molecule spends in a given vessel. It is most useful in reactors, where the probability that a molecule reacts increases with time spent in the reactor. For a continuous reactor, the residence time will increase if the flow rate into (and out of) the reactor is decreased.

return on investment (ROI)   profit divided by capital cost.

rheology   broadly defined as the study of the deformation and flow of fluids. In common usage, rheology refers to the study of nonclassical and viscoelastic fluids.

rubber   an elastomer.

stage   a section of a distillation column or absorber in which two fluids contact, and components transfer between fluids to approach equilibrium compositions. In a distillation column, one fluid is liquid moving down the column and the other is vapor moving up the column. Also called a *tray* or *plate*.

static system   a system that may be at steady state or at equilibrium.

steady state   a system in which the conditions at any position do not change with time, although the conditions may vary with position. The height of the water in a river may be at steady state, although the water is higher upstream and lower downstream. A chemical process may be at steady state, although the chemical composition varies along the process; it varies from reactants at the inlet to products at the outlet.

Stokes's law   predicts the terminal velocity of a sphere moved through a fluid by gravity, given the sphere's diameter (D) and density (ρ), the viscosity of the fluid (μ), and the gravitational acceleration (g),

$$v = \frac{1}{18} \frac{g D^2 \rho}{\mu}.$$

stripper   a process unit that extracts (strips) a volatile component from a liquid mixture by contacting the liquid mixture with a gas stream. Bubbling air through a water/benzene mixture will extract (strip) the benzene.

stripping section   the trays (stages) in a distillation column below the feed tray, including the evaporator. In the lower stages, the vapor *strips* the light ends from the mixture.

sublimation   the transition from solid to vapor.

surge tank   a process unit with an unsteady flow rate in but a steady flow rate out. The quantity of material in a surge tank varies with time.

system   a subset of the universe demarcated by boundaries; the demarcation is usually made for the convenience of a mathematical model.

thermodynamics   the study of empirical relations between the various forms of energy. The term derives from the initial emphasis of the field: the study of heat engines and the interconversion of heat (thermo) and mechanical motion (dynamics).

tie line   a line on a phase diagram that crosses a two-phase region and connects coordinates on the two borders of the region, representing the two phases in equilibrium. On temperature–composition and pressure–composition phase diagrams, a tie line is always horizontal.

transient process   a process in which the conditions change with time.

tray   see *stage*.

turbine (or turboexpander)   a process unit that converts the pressure energy of a fluid into mechanical energy. Steam generated at high pressure in power plants is fed to a turbine to convert the pressure energy to mechanical energy, which is subsequently converted to electrical energy in generators.

universal law   a statement governing physical or chemical behavior valid at any conditions, at all times. The conservation of energy is a universal law.

viscoelastic fluid   a fluid whose resistance to motion depends on how much the fluid has been deformed. Silly putty® and molasses are viscoelastic fluids.

Subject Index

CPSIA information can be obtained at www.ICGtesting.com
Printed in the USA
BVOW060155240812

298719BV00002B/1/P